Knowing Life

Knowing Life examines the limits of dominant knowledge forms that contribute to current practices negatively affecting more-than-human beings, while also exploring alternative approaches to knowing that are capable of reducing harm and maximizing planetary thriving.

Specifically, this volume seeks multispecies answers to long-standing questions in Western philosophy: Who or what counts as a knower? What kinds of knowing are valid? Is knowledge a product of mind, body, or something else? Historically, these epistemic questions have been answered in ways that neutralize the knowing and knowledge contribution of and for more-than-human beings, as well as those on the margins of society considered less than "human." Consequently, these epistemic assumptions often support the destruction of ecological habitats, industrialization of food animals, widespread use of insect and plant toxins, water and air pollution, climate extinctions, ecological militarism, and the perpetual flow of living beings used for entertainment, research, clothing, companionship, and economic resources.

In this book, crosscultural and multidisciplinary contributors—including lesser-known global religious-philosophical accounts, philosophies of plant and insect life, race and disability studies, laboratory epistemology, embodied semiotics, and scholar-artists—challenge and expand these classical concepts through diverse modes of embodied engagement on multispecies knowing toward open futures of planetary co-flourishing.

Brianne Donaldson is Associate Professor of Religious Studies and Philosophy and Shri Parshvanath Presidential Chair in Jain Studies at the University of California, Irvine, USA. She is the author of *Creaturely Cosmologies: Why Metaphysics Matters for Animal and Planetary Liberation*, co-author of *Insistent Life: Principles for Bioethics in the Jain Tradition*, editor of *Beyond the Bifurcation of Nature: A Common World for Animals and the Environment*, and co-editor of *The Future of Meat Without Animals*, and *Feeling Animal Death: Being Host to Ghosts*.

Multispecies Encounters

Series editors:

Samantha Hurn is Associate Professor in Anthropology, Director of the Exeter Anthrozoology as Symbiotic Ethics (EASE) working group, and Programme Director for the MA and PhD programmes in Anthrozoology at the University of Exeter, UK.

Chris Wilbert is Senior Lecturer in Tourism and Geography at the Lord Ashcroft International Business School at Anglia Ruskin University, UK.

Multispecies Encounters provides an interdisciplinary forum for the discussion, development and dissemination of research focused on encounters between members of different species. Re-evaluating our human relationships with other-than-human beings through an interrogation of the 'myth of human exceptionalism', which has structured (and limited) social thought for so long, the series presents work including multispecies ethnography, animal geographies, and more-than-human approaches to research, in order not only better to understand the human condition, but also to situate us holistically, as human animals, within the global ecosystems we share with countless other living beings.

As such, the series expresses a commitment to the importance of giving balanced consideration to the experiences of all social actors involved in any given social interaction, with work advancing our theoretical knowledge and understanding of multispecies encounters and, where possible, exploring analytical frameworks which include ways or kinds of 'being' other than the human.

Published

Cat People: Human-Cat Interrelatedness in the Cat Fancy
Emily Stone

The Giant Squid in Transatlantic Culture: The Monsterization of Molluscs
Otto Latva

Human–Animal Relationships in Times of Pandemic and Climate Crises: Multispecies Sociology for the New Normal
Edited by Josephine Browne and Zoei Sutton

Knowing Life: The Ethics of Multispecies Epistemologies
Edited by Brianne Donaldson

The full list of titles for this series can be found here: www.routledge.com/Multispecies-Encounters/book-series/ASHSER1436

Knowing Life

The Ethics of Multispecies Epistemologies

Edited by Brianne Donaldson

Routledge
Taylor & Francis Group
LONDON AND NEW YORK

First published 2025
by Routledge
4 Park Square, Milton Park, Abingdon, Oxon OX14 4RN

and by Routledge
605 Third Avenue, New York, NY 10158

Routledge is an imprint of the Taylor & Francis Group, an informa business

British Library Cataloguing-in-Publication Data
A catalogue record for this book is available from the British Library

ISBN: 9781032659961 (hbk)
ISBN: 9781032659978 (pbk)
ISBN: 9781032659985 (ebk)

DOI: 10.4324/9781032659985

Typeset in Times New Roman
by Newgen Publishing UK

To

The rural Michigan land, its creaturely inhabitants, and indigenous ghosts who provided me years of refuge and presence when I needed it most. And to Tom and Sue who invited me there in the first place. My wish to know and be known by more-than-human others was awakened within those fields and woods. Those encounters travel with and through me always.

—Brianne Donaldson

Contents

Author Biographies

Elisa Aaltola is a Senior Researcher and adjunct professor in philosophy at the University of Turku, Finland. Her research has focused on animal philosophy and normative moral psychology. In addition to numerous papers, Aaltola has published 12 books on these topics, including *Varieties of Empathy: Moral Psychology and Animal Ethics* (2018) and *Animal Suffering: Philosophy and Culture* (2012). In 2021, she was awarded the Pro Animalia Prize for her life's work for animals, and in 2022, she was named the "Changemaker of the Year" in her native Finland.

Ralph R. Acampora is an Associate Professor of Philosophy at Hofstra University. He is the author of Corporal Compassion: Animal Ethics and Philosophy of Body (2006) and the editor of *Metamorphoses of the Zoo: Animal Encounter After Noah* (2010). He also serves as a review editor for the journal *Society and Animals*.

María del Rosario Acosta López is a Professor of Latin American Studies at the Department of Hispanic Studies at the University of California, Riverside. She teaches and conducts research on Aesthetics, Critical Theory, Political Philosophy, and, more recently, Decolonial studies, with emphasis on questions of memory and trauma in the Americas. Her most recent publications are devoted to Aesthetics of Resistance in Latin American Art, Decolonial perspectives on Memory and History, and Epistemic Injustice and Epistemic Violence. She is currently working on the manuscript of her next book, *Grammars of Listening: Philosophical Approaches to Memory after Trauma* (forthcoming in Spanish with Herder and in English with Fordham UP). https://profiles.ucr.edu/mariadelrosario.acostalopez

M.J. Barrett is an Adjunct Professor at the Graduate School of Environment and Sustainability, University of Saskatchewan, Canada. M.J's environmental scholarship combines intuition and intellect in teaching, research and human-nature relations. Her work takes seriously the agency and communicative capacity of the natural world; it also illustrates ways in which conventional thinking reinforces knowledge and species hierarchies. Publications focus on transformative sustainability learning and intuitive interspecies communication. M.J. holds degrees from Harvard, York, Queen's, and the University of Regina. Her goal is to nurture ways of knowing and being where the more-than-human world is respected as intentional, intelligent, and communicative. https://researchers.usask.ca/mj-barrett/index.php

Balbinder Singh Bhogal is Professor of Religion and the holder of the Sardarni Kuljit Kaur Bindra Chair in Sikh Studies at Hofstra University, NY. His research specializes in South Asian religions, particularly the Sikh tradition, from a decolonial perspective. His current book project—*Overcoming Colonial Modernity: Religion, Animal Sublime and Decolonial Sikhi(sm)*—builds upon past publications available here: https://hofstra.academia.edu/BalbinderSinghBhogal

Suzanne Bost is a Professor of English and an Affiliate in Women's Studies and Gender Studies at Loyola University Chicago; her work revolves around Latinx literature, feminist and queer theory, illness and disability, and archival practices. She is the author of three books—*Mulattas and Mestizas: Representing Mixed Identities in the Americas, 1850-2000* (2003), *Encarnación: Illness and Body Politics in Chicana Feminist Literature* (2009), and *Shared Selves: Latinx Memoir and Ethical Alternatives to Humanism* (2019)—and also co-edited, with Frances Aparicio, *The Routledge Companion to Latino/a Literature* (2012). Her current work focuses on decolonial feminisms and silence. See www.luc.edu/english/people/faculty/archive/suzannebost.shtml

Jacob Bull is a Senior Lecturer at the Centre for Gender Research, Uppsala University. A social and cultural geographer, Jacob's research is positioned at the intersections of Animal Studies, Gender Studies and Geography, with a focus on the role of animals in understandings of space, place and identity. He is lead applicant on the VR-funded project "Becoming human: gender theory and animals in a more-than-human world" and the FORMAS-funded project "Changing animal bodies: breeding responses to environmental, economic and social pressures."

Brianne Donaldson is an Associate Professor of religious studies and philosophy and Shri Parshvanath Presidential Chair in Jain Studies at the University of California, Irvine. Her books include *Creaturely Cosmologies: Why Metaphysics Matters for Animal and Planetary Liberation* (2015); *Insistent Life: Principles for Bioethics in the Jain Tradition* (2021, co-authored with Ana Bajželj) and the edited collections *Beyond the Bifurcation of Nature: A Common World for Animals and the Environment* (2014), *The Future of Meat Without Animals* (2016; co-edited with Christopher Carter), and *Feeling Animal Death: Being Host to Ghosts* (2019; co-edited with Ashley King). www.briannedonaldson.com

Che Gossett is the Associate Director of the Center for Research in Feminist, Queer and Transgender Studies at the University of Pennsylvania. Che co-edited a special issue of *Transgender Studies Quarterly* "Trans in a Time of HIV/AIDS" with Professor Eva Hayward in 2020, and their 2022 syllabus on trans and non-binary methods for art and art history co-authored with Professor David Getsy won the College Art Journal Award for Distinction. They have been a postdoctoral scholar at Columbia Law School, and a visiting fellow at Harvard Law School and the University of Cambridge.

Graham Harvey is a Professor of religious studies at The Open University, UK. His research largely concerns "the new animism," especially in the rituals and protocols through which Indigenous and other communities engage with the larger-than-human world. His recent teaching related work has involved a focus on foodways and on defining "religion" as sensual engagement with the world. His publications include *Food, Sex and Strangers: Understanding Religion as Everyday Life* (2013), and *Animism: Respecting the Living World* (2nd edition 2017). He is editor of the Routledge series "Vitality of Indigenous Religions" and the Equinox series "Religion and the Senses."

Kai Horsthemke is a visiting professor in the School of Education at the University of the Witwatersrand, South Africa, and a fellow at the Oxford Centre for Animal Ethics, United Kingdom. He is the author of five monographs, *The Moral Status and Rights of Animals* (2010), *Animals and African Ethics* (2015), *Animal Rights Education* (2018), *Indigenous Knowledge: Philosophical and Educational Considerations* (2021), and *The Meaning of Death* (2024). Together with Peggy Siyakwazi, Elizabeth Walton and Charl Wolhuter, he co-edited the first two editions of *Education Studies* (2013 and 2016, respectively). His recent edited book *Education, the Environment and Sustainability* was published by Routledge in 2024.

Wahida Khandker is a Senior Lecturer in Philosophy at Manchester Metropolitan University. She is the author of *Philosophy, Animality, and the Life Sciences* (2014) and *Process Metaphysics and Mutative Life: Sketches of Lived Time* (2020). www.mmu.ac.uk/hpp/staff/profile/index.php?id=86

Michael Marder is Ikerbasque Research Professor in the Department of Philosophy at the University of the Basque Country, Vitoria-Gasteiz, Spain. His writings span the fields of ecological theory, phenomenology, and political thought. He is the author of numerous scientific articles and eighteen monographs, including *Plant-Thinking* (2013); *Phenomena—Critique—Logos* (2014); *The Philosopher's Plant* (2014); *Dust* (2016), *Energy Dreams* (2017), *Heidegger* (2018), *Political Categories* (2019), *Pyropolitics* (2015, 2020); *Dump Philosophy* (2020); *Hegel's Energy* (2021); *Green Mass* (2021) and *Philosophy for Passengers* (2022), among others. www.michaelmarder.org.

Beatrice Marovich is an Associate Professor of theological studies, at Hanover College. She works at the intersection of theology, philosophy, and the environmental humanities. Her first book is *Sister Death: Political Theologies for Living and Dying* (2023).

Jo-Anne McArthur is an award-winning photojournalist, sought-after speaker, and the founder of We Animals, a non-profit organization whose mission is to advocate for animals through photojournalism. She has been documenting the plight of animals on all seven continents for over two decades. She is the author of three books, *We Animals* (2014), *Captive* (2017), and *HIDDEN: Animals in the Anthropocene* (2020), and was the subject of Canadian filmmaker Liz Marshall's acclaimed documentary, "The Ghosts in Our Machine." Jo-Anne is based in Toronto, Canada.

Juan Diego Pérez Moreno is a doctoral candidate in the Department of Spanish and Portuguese at Princeton University, working at the intersections between memory studies, ecopoetics, and decolonial studies, with a focus on modern and contemporary Latin American poetry and art. His dissertation traces a poetics of liberatory memory in the experimental poetry of Chilean poet and artist Cecilia Vicuña. Originally from Bogotá, Colombia, he currently teaches at Riverdale Country School in New York City. See https://princeton.academia.edu/JuanDiegoP%C3%A9rez

Angela Robinson is Assistant Professor of Gender Studies and Environmental Humanities at the University of Utah, where she also works with the Pacific Islands Studies Initiative. Her current book project examines climate change as a colonial regime and looks at more-than-human ontologies expressed through Indigenous cultural production as crucial forms of sociality that are vital to addressing climate change and its effect.

Jana S. Rošker is the first Slovene Sinologist, co-founder, and long-standing Head of the Department of Asian Studies at the University in Ljubljana (Slovenia). Her academic interests include Chinese epistemology, Chinese logic, and Modern Confucianism. In these research areas, she has published twenty books, and over one hundred and fifty articles and book chapters. She is chief editor of the journal *Asian Studies* and the founder and first president of the European Association of Chinese Philosophy (EACP). She has received several awards, such as the National Slovene Research Agency prize for extraordinary scientific achievements, the Golden Plague of the University of Ljubljana, the National Žiga Zois prize for research work of national significance, and the French-Taiwanese Cultural Foundation Award.

Musician and philosopher **David Rothenberg** wrote *Why Birds Sing*, *Bug Music*, *Survival of the Beautiful* and many other books, published in at least eleven languages. He has more than forty recordings out, including *One Dark Night I Left My Silent House* which came out on ECM, and most recently *Just Leave It All Behind* and *Homayoun*. He has performed or recorded with Pauline Oliveros, Peter Gabriel, Ray Phiri, Suzanne Vega, Scanner, Elliott Sharp, Umru, Iva Bittová, and the Karnataka College of Percussion. In 2024, he won a Grammy award as part of *For the Birds*. *Nightingales in Berlin* is his latest film, and his latest book is *Secret Sounds of Ponds*. Rothenberg is a Distinguished Professor at the New Jersey Institute of Technology.

Donovan Schaefer is an Associate Professor in the Department of Religious Studies at the University of Pennsylvania. He is the author of *Religious Affects: Animality, Evolution, and Power* (2015) and *Wild Experiment: Feeling Science and Secularism after Darwin* (2022), which won the Ludwik Fleck Prize from the Society for Social Studies of Science (4S) and the annual book prize from the International Society for Science and Religion. His research and teaching examine the role of affect and power in formations of secularism, religion, race, science, and material culture.

Gabriele M. Schwab is Distinguished Professor at the University of California, Irvine. She holds appointments in the departments of Comparative Literature, Anthropology, English, and European Languages and Studies. She received her Ph.D. in literary studies and critical theory at the University of Constance in 1976 and a Ph.D. in Psychoanalysis from the New Center for Psychoanalysis in Los Angeles in 2009. A recipient of a Guggenheim Fellowship and a Heisenberg Fellowship, her research interests range across critical theory, psychoanalysis, trauma studies, ecology, literature and anthropology, and 20th- and 21st-century comparative literatures. Monographs in English include *Haunting Legacies: Violent Histories and Transgenerational Trauma* (2010); *Imaginary Ethnographies: Literature, Subjectivity, Culture* (2012, Winner of the 2014 Choice Award for Best Academic Book); *Radioactive Ghosts* (2020); *Moments for Nothing: Samuel Beckett and the End Times* (2023). She is working on a new book project on Transspecies Imaginaries.

Bligh Somma is an Assistant Professor of Philosophy at Fordham University. His research focuses on animals, ethics, and psychology in philosophy of the Islamic world. His monograph *Models of Desire in Graeco-Arabic Philosophy: From Plotinus to Ibn Ṭufayl*, appeared in 2021 with Brill. His work appears in the *Journal of the History of Philosophy*, *Archiv für Geschichte der Philosophie*, and *Journal of Islamic Ethics*. fordham.academia.edu/BlighSomma

Daniel M. Stuart is an Associate Professor in the Department of Religious Studies at the University of South Carolina. He holds an MA in Sanskrit Literature and a PhD in Buddhist Studies from the University of California at Berkeley. His research focuses on the history of traditional Buddhist contemplative practices from their origins in premodern South Asia into the global present. He is particularly interested in how Asian thought systems, practice regimes, and cosmovisions converge in specific ways to form distinctive traditions of practice at particular moments in history. He is the author of five books: *Thinking about Cessation* (2013), *A Less Traveled Path* (2015), *The Stream of Deathless Nectar* (2017), *S. N. Goenka: Emissary of Insight* (2020), and *Insight in Perspective* (2024).

Laura Terrance is Akwesasne Mohawk from what is now known as Northern New York, Southern Ontario, and Southern Quebec. She holds a PhD in Gender Studies from the University of California, Los Angeles, and is currently an Assistant Professor in the Indigenous Studies Department at the

University at Buffalo. In 2012, she was awarded a master's degree in Ethnic Studies from the University of California, San Diego. Her areas of research interest include Indigenous studies, gender studies, critical race theory, social theory, knowledge production, Native literature, and Native film. Her manuscript, *Rotisken'ra:kéte: Violence and the Anti-Colonial*, considers the relationship between violence, subjectivity formation, and decolonization exploring representations of violence or threats of violence as retributive acts.

Acknowledgments

In a world of co-constitutive knowing and perceiving, an event like this text is indebted to many others. First, the contributions for this book were prepared and circulated for a virtual conference hosted through the Religious Studies Program at the University of California, Irvine (UCI) in November 2022. Thanks to all the contributors who joined to create a truly unique and meaningful gathering across miles, disciplines, and perspectives. Big appreciation also goes to Dwayne Pack and his IT crew—as well as administrators Marc Kanda, Michelle Spivey, and Cynthia Bee and graduate student Nikita Prokhorov—for making this international, multi-day event a success.

Second, several grants made this project possible: The UK-based Spalding Trust, UCI Humanities Center, University of California Humanities Research Institute, and the UCI Shri Parshvanath Presidential Chair in Jain Studies. The open-access version of this text is also generously supported by an additional UCI Humanities Center grant.

Third, I extend my personal appreciation to Fernando Villagómez, Linda Rabadi, Larry Farris, Jay Martin, Miriam Bender, and Susan Blakeley Klein for their support throughout my move to UCI, and encouraging me as I found a virtual foothold of belonging in this project.

The chapter contributed by Donovan Schaefer was first published as "Voracious Secularism: Emotional Habitus and the Desire for Knowledge in Animal Experimentation" in *Religion* 53:4 (2023), 700–723, doi: 10.1080/0048721X.2023.2258710. Thanks to Taylor and Francis for their permission to share an adjusted version here.

Introduction: From the Limits of Intelligibility to Perceptive Rearrangements

Brianne Donaldson

Limits of Intelligibility

A mentor suggested to me some years ago that privilege is not having to make oneself intelligible to anyone else. That conversation preceded (temporally) and exceeded (conceptually) the recent years of wokeness debates, cancel culture, and arguments over who does or doesn't have various forms of privilege, whether social, historical, gendered, geographical, racial, or otherwise. Perhaps it was the simplicity of the original claim that has helped it weather through time and trends in my memory, suggesting something of the insight's durability. And anyone who has struggled to make their concerns or existence intelligible to another who wields that simpler form of privilege is likely to experience something of its confusing futility. Perhaps there are times we ourselves have wielded that privilege as well, so that if something is not important to us, it is not important at all.

To be acknowledged as intelligible is not merely a theoretical stamp of approval; it is a profound form of cosmic belonging with serious moral implications. In the absence of being intelligible, one can be diagnosed as terminally unknowable, invalid, unrecognizable, incoherent, or even nonexistent. That which is unintelligible can be excluded from circuits of communication and cast outside of grievability altogether. As Judith Butler writes of human communities during war, "An ungrievable life is one that cannot be mourned because it has never lived, that is, it has never counted as a life at all" (2010, 38).

How much more so, then, do the limits of intelligibility contribute to current levels of violence impacting more-than-human entities and multispecies networks? While epistemology—or the study of knowledge, mainly gained through "human" observation and reasoning—is a standard aspect of Euro-American philosophical accounts, it is rarely examined within a multispecies context. On the contrary, anthropocentric knowing is baked into the Western philosophical inheritance of the Scientific Revolution, especially the metaphysical dualism of Descartes which bifurcated the world into thinking, rational subjects (*res cogitans*) on one hand, and non-thinking, material objects (*res extensa*) on the other. The epistemologies resulting from this persistent dualism hinder genuine ecological perceptivity by dividing the world into hierarchies of knowers-known, active-passive, rational-irrational, and subject-object. Such binary structures not only neutralize modes of

DOI: 10.4324/9781032659985-1

knowing in a multispecies universe packed full of dynamic life forms ranging from energetic pulses and quarks to endangered gray whales and communicating families of redwood trees, but they also neutralize world visions and lifeways that don't share the same dualist foundations.

The twentieth-century English mathematician and philosopher Alfred North Whitehead (1861–1947) aimed to combat the Cartesian split of rational human and valueless matter by cultivating his own Process philosophy of organism to reconceive the true relation of existent entities to one another. By formulating every aspect of existence as a perceptive processual *event* rather than a fixed substantial *thing*, Whitehead asserted in his "First Harvard Lecture" that "Reality is not static: it is a process of becoming" (1924, 16). In a reality of becoming rather than fixed being, the cognitive "I think" subject is no longer most ontologically basic. Rather, subjective feeling—Whitehead calls it "prehension" (from the Latin: "to grasp," which he modified from the term comprehension)—grounds the subjective experience of *every* becoming event (Henning 2023, 11). As a subject, each event prehends, relates to, and is affected by other datum; but as an object, each event is datum for other subjective future occasions which together constitute the "creative advance" (Whitehead 1978/1929, 228). In his magnum opus *Process and Reality* (1929), Whitehead calls this the "reformed subjectivist principle" (160), meaning that all events are both prehending subject of others *and* prehended object by others. Whitehead is unambiguous in his claim, that "apart from the experience of subjects there is nothing, nothing, nothing, bare nothingness" (167).

The epistemic effects of such a thoroughgoing universal subjective perception irrevocably overturns the knower-known bifurcation. Though my discussion of Whitehead's metaphysics is severely abbreviated here, it suffices to say that to know—or prehend—in the Process philosophy of organism describes a *constitutive* relation rather than a detached cognitive observation in which a subject *here* knows an object *out there*. If every event is characterized by the ontologically-basic fact of subjective prehension, knowing is not something a subject *has*; rather, knowing defines what a subject is. As Whitehead puts it, "*how* an actual entity becomes constitutes what the actual entity *is*" (1978/1929, 166).

Subjectivity, in this view, is neither best understood in cognitive terms nor is it limited to rational human beings. Certainly, Whitehead is not alone in this view. There are diverse accounts and unscripted experiences of knowing and being known by life—many present in this text—which circumvent the latent dualism in contemporary philosophical, political, religious, and economic arrangements. The accounts in this collection rarely assume a knower-known binary, but offer perspectives that grapple with constitutive multispecies relations. "Every organism requires an environment of friends," writes Whitehead in *Science and the Modern World* (1967/1925), "partly to shield it from violent changes, and partly to supply it with its wants" (206). Multiple robust accounts are needed to conceptualize this give-and-take universe that are not only one-sided.

This brief look at one alternative foundation for knowing life in Whitehead's Process philosophy of organism invites us to investigate other unexamined assumptions, such as dualism and its resultant anthropocentric accounts of knowing or knowledge-production, that contribute to the ongoing destruction of ecological habitats, mass killing of industrial-raised animals for food, widespread use of petroleum-based insect and plant toxins, water and air pollution, climate extinctions, and socially-sanctioned use of living beings for entertainment, scientific progress, companionship, economic resources, or many other forms of normalized violent interactions. This alternative foundation also invites us to wonder how other approaches to epistemology might transform ethical responsiveness within the multispecies co-constitution of our experience and entangled planetary existence.

Historically, epistemic questions—such as who or what counts as a knower? What kinds of knowing are valid? Is knowledge a product of mind, body, or something else?—have been answered in ways that subordinate the knowing and knowledge contribution of, for, and with more-than-human beings, as well as those on the margins of society considered less than "human." From this vantage point, one can understand the refrain of new materialists seeking to "make matter *matter*"—in all its various pulsations and embodiments—as a call for wider modes of perception beyond traditional notions of epistemology that remain, as Nietzsche might suggest, "human, all too human" (2009/1878).

That the "human" itself is a contested term and site is not a new assertion. From antiquity to the present, many *Homo sapiens* have not been recognized as intelligible (enough) to be deemed fully "human"—as true in Aristotle's "barbaric mistake" as it persists in the modern prison-industrial complex or at nation-state boundaries where "aliens" dwell as unfamiliar, dangerous, and unwanted. Yet, perpetual hostilities of privileged in-groups toward outsiders seems a fairly trivial drama when contrasted to the evolutionary tendency of dominant societies to render the entire universe unintelligible. The very energetic multiplicity which birthed our possible existence and remains intractably part of "our" becoming—"all the living things that man does not recognize as his fellows, his neighbors, or his brothers," says Jacques Derrida (2008, 34)—is bizarrely excised from prevailing epistemic accounts such that knowing life in its epochal immensity and infinite planetary vitalities is simply "not important at all if it is not important to me "

Two visions appear from my recent experience which illuminate the practical costs of such unintelligibility. The first, when I was shadowing a physician directing the Disorders of Consciousness (DoC) program at the Texas Medical Center (2017–18). On rounds one morning, we stopped at the door of a female patient who, after being deprived of oxygen due to a post-labor blood clot, suffered an anoxic brain injury that left her seemingly nonresponsive. It was not until the DoC team experimented with various sitting positions, including supporting the new mother's arm that she was able to reliably produce a finger motion for yes/no, one of the many ways by which responsive awareness can be affirmed. Current research shows a misdiagnosis rate upwards of forty percent for vegetative states (also called unresponsive wakefulness syndrome), meaning that thousands of patients have covert responsiveness that is not intelligible to current medical modes of knowing, resulting in their being housed in facilities, removed from life support, and missing critical windows for rehabilitation. The process of detecting responsiveness in such patients reveals how current medical sense-making does not adequately anticipate covert communication in patients with profound cognitive and physical impairments that press them beyond the edge of "human" attribution. Such intelligibility quests require personal resistance to the status quo, professional persistence, and resources of time and space to experiment beyond current epistemic modes.

The second vision was awakened in my memory from news stories released in early 2023 of undercover video footage of "humane" carbon dioxide chambers to stun pigs in the U.S., U.K., and Australia. These videos reveal the thrashing terror of pigs gasping for breath and struggling to escape while being suffocated in a CO_2 "gondola" for the seemingly universal desire for bacon. In 2017, I had been inside a large pork processing plant in rural Illinois that also uses CO_2 stunning. This visit was part of my wider research into global and multispecies entanglements of industrial agriculture. I entered the slaughterhouse with a colleague's environmental sustainability class where we toured the facility in reverse. We started with the packaging section, where familiar parcels fell off a conveyor belt into boxes as "chops," "links," "bacon"—shrouding bodies, as Carol J. Adams puts it, "with gastronomic language, so we do not conjure dead, butchered animals, but cuisine" (2015, 21). We walked through the deboning section, the loin section, the shoulder section—each area named for what immigrant hands dismembered. We watched the spine removal, the waxy refrigerated halves and headless carcasses suspended in the smokehouse, the head removal, the suction gutting, hair fired off in a furnace—all an incredible feat of mechanization—until we finally got to the live animals at the end. A sea of faces, bodies, and squeals—nearly 12,000 pigs are trucked into that slaughterhouse every day. We were not permitted to see the pigs offloaded from the truck, as this is when they vocalize the most; although, from a nearby strip mall where I filmed on numerous occasions, you could certainly hear them and feel their rumble of resistance.

Concurrently, several recent ethology, or animal behavior, studies—many completed since 2010—describe the cognitive complexity of pigs, including memory, spatialized and symbolic learning, discerning other beings and objects, perspective taking, self-awareness, time perception, emotion, personality, and curious play, among other features (Marino and Colvin 2016). In a series of *New York Times* letters to the editor grouped together under the title "Thinking About Pigs Before They're Food," contributors repeated many of the now-popular phrasings that pigs are as "intelligent as a 3-year-old child" and "smarter than the dogs we share our homes with," and "even play video games" ("Thinking About" 2012)! Each of these empirical accounts appears to validate what E.B. White showed in his 1952 children's book *Charlotte's Web* that pigs are not unthinking objects on the wrong end of Descartes' dualism. Rather, Wilbur the pig, as well as Charlotte the spider and the farmer's daughter Fern who both try to save the pig from slaughter, are entities capable of perceiving and knowing in distinct ways that defy the passive roles set out for them in the book's social narrative. In a note to his publisher Ursula Nordstrom a few weeks before the book went to press, White described that his primary reason for writing *Charlotte's Web* was his desire to save a pig's life after a lifetime of farming:

> A farm is a peculiar problem for a man who likes animals, because the fate of most livestock is that they are murdered by their benefactors. The creatures may live serenely but they end violently, and the odor of doom hangs about them always. I have kept several pigs, starting them in spring as weanlings and carrying trays to them all through summer and fall. The relationship bothered me. Day by day I became better acquainted with my pig, and he with me, and the fact that the whole adventure pointed toward an eventual piece of double-dealing on my part lent an eerie quality to the thing. I do not like to betray a person or a creature, and I tend to agree with Mr. E.M. Forster that in these times the duty of a man, above all else, is to be reliable. It used to be clear to me, slopping a pig, that as far as the pig was concerned I could not be counted on, and this, as I say, troubled me. Anyway, the theme of "Charlotte's Web" is that a pig shall be saved, and I have an idea that somewhere deep inside me there was a wish to that effect.
>
> (Popova n.d.)

White's effort, like the ethology studies mentioned, illustrates what the CO_2 chamber attempts to destroy: the insistence of life-seeking entities which is otherwise all too clear with any modest amount of attention. The erasure of multispecies knowing is not merely a passive forgetting, although critical, embodied, and even fictional recollections offer a needed booster shot of memory. As Alice Walker recounts in her short essay "Am I Blue?" written around the edges of a horse paddock containing Blue, "I was shocked that I had forgotten that human animals and nonhuman animals can communicate quite well; if we are brought up around animals as children we take this for granted. By the time we are adults, we no longer remember" (1989, 5).

The limits of intelligibility, whether at the bedside of a patient experiencing traumatic brain injury or in the technologies of destruction at the slaughterhouse, can be attributed to oversight or forgetfulness. And they can also be deliberate eradications and concealments—forms that Boaventura de Sousa Santos called "epistemicide," or the destruction of knowers or knowledge systems (2007, 149), through edifices erected to obscure, often paired with political, institutional, or habitual justifications that camouflage the optional as necessity.

Perceptive Rearrangements

The book includes eighteen chapters and four "Praxis" segments that proceed in four sections: (I) Living Concepts, (II) Growing Perceptions, (III) Freeing Subjectivities, and (IV) Experimental Responses. In each section, authors demonstrate how diverse modes of multispecies knowing dismantle certain axioms and proliferate new terms, fresh combinations, and unlikely crossings through which other knowledges can be accessed, understood, or responded to. The four sections emphasize, in overlapping manners, seven goals of multispecies epistemologies:

1. overcoming rationalism's separation of thought and matter (Descartes) to explore how living entities have their own experiences, beyond reason alone, that produce knowledges and capacities, variously defined and (often) inadequately perceived by others;
2. confronting ontological mind/matter dualisms presuming knower-known, Self-Other, subject-object, internal-external, nature-culture, with entangled multispecies ontologies that destabilize dualities and/or transform them into mutually requiring relations;
3. by contrasting the subjective turn in Western philosophy suggesting that mind produces the world (Kant), multispecies epistemologies explore how the world produces the mind, and how subjectivity presupposes, obscures, or even denies relationality;
4. by expanding classical empiricist claims that we gain knowledge solely through the senses (Locke and Hume), multispecies epistemologies account for modes of perception beyond five senses—especially the dominance of vision and its role in linguistic representation—as well as modes of perception that do not presume a normative "human" body, universal sense, or univocal notion of consciousness;
5. by illuminating taxonomies of "species"—including a proper "human"—that consolidate power, justify violence, naturalize subordination, and perpetuate colonial tendencies, multispecies epistemologies explore dynamic explanatory modes of careful attention to the particularity of living beings, including those strange, unfamiliar, and even unrecognizable to the senses;
6. by resisting knowledge as merely a task of logical or linguistic analysis in search of true propositions, multispecies epistemologies require vulnerable exposure, feeling with, and learning from the final facts of existent entities, possibly resulting in what Val Plumwood calls a "shocking reduction" of one's fixed sense of self, meaning, language, and ethical aims (2002);
7. by acknowledging the contributions and constraints of Western epistemologies toward moral thinking about more-than-human beings—often limited to rights- or ownership-paradigms—multispecies epistemologies stretch toward greater cross-cultural, religious, transdisciplinary, activist, and artistic contributions alongside diverse forms of embodied and mystical engagement that eschew a centralized geography, legal framework, language paradigm, cultural norm, or religious-philosophical context.

Living Concepts

Section I is prefaced with a practical zine exploring "Threshold Concepts for Multispecies Education" summarizing M.J. Barrett's five years of research developing teaching strategies for sustainability education in which more-than-human beings are acknowledged as research collaborators and co-instructors in collective learning. Readers are encouraged to implement these strategies into their own practice and teaching.

Ralph Acampora (Ch 1) argues that somatic vulnerability and semiotic spontaneity, rather than consciousness, grounds multispecies ethics. What characterizes "life" itself is not the *logos*-demands of autobiography, but an embodied solidarity rooted in semiotic signification and interpretation. Ethical action in this semiotic state is not a matter of logical argument but a fundamental body-mind experiential process that arises from semiotic altruism, whether through intercorporeal sense-making or its imaginative extension to beings or systems of signification beyond direct contact.

Beatrice Marovich (Ch 2) redefines life as interwoven with death, such that awareness of death is no longer a lever separating proper "man" from "animal" as laid out by Heidegger or Catholic scholasticism. Rather, various orientations toward death, such as an opossum "playing dead," shows how several species have a "minimal concept of death"—per Marovich's original translations of Spanish philosopher Susan Monśo—that can be utilized to craft shared rituals and attentive practices of protection in our time of manmade mass death.

In his exhumation of "voracious science," Donovan Schaefer (Ch 3) confronts a dichotomy between secular-as-rational, and religion-as-emotive, revealing the affective role of desire-to-know at the heart of scientific research on animals. Drawing upon outsider ethnographic studies of laboratory workers and insider accounts of cultivating a research attitude that justifies violence, Schaefer illuminates how the thrill and pleasure of knowing produces secular-scientific dispositions immune to animal suffering and resistant to challenges toward entrenched practices of harm.

Wahida Khandker (Ch 4) revisits the concept of "nature" as a synchronization of living and non-living processes. Exploring a lesser-known relation between biologist-mountaineer Julian Huxley and process-philosophers Henri Bergson and Charles Hartshorne, Khandker first places readers on a hiking path, encountering and acclimating to trekking pitfalls that exemplify modes of multispecies synchronizations characterizing what otherwise appear as separate creatures, climate forces, fungi, or habitats. The concluding investigation into how birdsong continuously recreates a physical environment suggests attuned practices of harmonic adaptation over disharmonic disruptions as essential social-geographic synchronizations amidst climate-change perils.

Kai Horstemke (Ch 5) argues that African expressions of sustainability-through-social knowledge—such as *I am because we are*—are easily reduced to idealized jargon meant to safeguard anthropocentric "indigenous knowledges" or "epistemic diversity" at the expense of biodiversity and individual more-than-human lives. If knowledge is one integrated multiplicity that includes the perspectival livingness of all existent entities, there can be no "local" or "indigenous" knowledges, but rather socially-located beliefs, skills, and practical know-how that can be more or less adequate in apprehending knowledge's living totality. Horsthemke argues that African relational thought and practice need to be more radical—and radically egalitarian—across species.

Growing Perceptions

Section II begins with a practical toolbox for reworking land acknowledgments. Angela Robinson and Laura Terrance clarify problems with performative land acknowledgments and provide reflection questions to help readers reanimate land and think spatially. These suggestions encourage individuals and institutions to move beyond a stewardship model of land-as-passive object to land—with its plant-animal partnerships—as creative caregiver for people as well. Moreover, being present to the spatial realities and reciprocal formations of place draws attention from the historical focus on temporal progress that continues to justify settler-colonial violence toward the land and its inhabitants.

Exploring the limits of memory and sense-making María del Rosario Acosta López and Juan Diego Pérez Moreno (Ch 6) trace how survivors' unprecedented experiences of violence are not intelligible through available grammars. To receive "unheard of" testimonies, such as from the the *jepirachi* trade wind of the indigenous Wayuu survivors in Colombia whose silencing motivates the authors' inquiry, demands an aesthetic reconfiguration of the colonial sensorium and epistemological regimes that have heretofore excluded testimonies of communities and lifeforms silenced by the nature-culture divide.

Plant philosopher Michael Marder (Ch 7) turns his vegetative analyses toward the Biblical Tree of Life in the *Book of Zohar* to argue that the divine prohibition of consumption was not about eating fruit from the tree of knowledge of good and evil, but with eating (and knowing) that disrespects the unity of existence. By elucidating the kabbalistic emanation of *šeḵīnah* that can either receive the manifold flows of life or deaden them, Marder illuminates an overlooked vegetal wisdom—a move from ecological exile to replanting the world—at the heart of a Western Judaic tradition.

Brianne Donaldson (Ch 8) demonstrates how the ancient Jain tradition, through its account of universal sentiency, prefigures Bentham's utilitarian calculations of pain/pleasure, enabling knowledge of other beings through analogy with oneself. Moreover, through its account of *un*knowing one's birth state through rebirth, the tradition prefigures John Rawls' notion of human-only justice through the "veil of ignorance," enabling knowledge of reciprocal suffering of and *as* others. These early accounts of Jain multispecies knowledge as co-arising with right view and right conduct overlap contemporary feminist insights into "onto-ethics-epistemological" world-making.

Animism scholar Graham Harvey (Ch 9) reminds readers that animism is not a theory that everything is alive, but rather describes "etiquettes" of response-ability toward multispecies others. Harvey exchanges hierarchical categorizations of life for assessments of distance, intimacy, and expectations. Exploring animist interactions with Anishinaabe stones, children collecting maple syrup, Māori sweet potato harvest, and other narratives of "knowing life," Harvey advocates better questions in anthropological studies to transform researchers' epistemic-orientation from classification to flows of mutual communication and collaborative learning.

Freeing Subjectivities

Section III opens into the camera lens of renowned photographer Jo-Anne McArthur, inviting readers toward "Knowing Animals in the Anthropocene through Animal Photojournalism." The evocative images—in turn heartrending or fortifying—reveal hidden subjectivities usually concealed from perception and demonstrate the revelatory power of animal photojournalism to ignite a public quest for disclosure and change.

Jacob Bull (Ch 10) then shows how prevalent notions of gender impact and often misdirect animal welfare practices. Through case studies, Bull explores the gendered justifications given to support killing a young male giraffe in a Copenhagen zoo, fencing guidelines for his own heritage breed pigs that overlook female desires, and original research on bee reproduction methods in which sex cannot be considered an act of singular entities but a colony-wide process.

Sinologist Jana S. Rošker (Ch 11) follows with a richer understanding of intersubjective knowing in Chinese epistemologies rooted in a moral and social correspondence between an organically openly structured external world and similarly structured internal conceptualization. This learning-based account of knowing as a "fusion of aesthetic realms (*jingie*s)" is not dependent upon objectivity and representational naming; rather, it is conditioned by understanding and experiencing the distinct aesthetic realms in which the knower and known are embedded—and perpetually dissolving into one another—which Rošker exemplifies in two stories from the Daoist work *Zhuangzi* about a seabird and happy fish.

Che Gossett (Ch 12) examines Marxist capitalism as an animalizing violence that dispossesses the worker. By routing this insight through Black radical thought and cinema, Gossett uses the films of Charles Burnett—*The Horse* and *Killer of Sheep*, as well as a documentary on the activist (or terrorist?) group Earth Liberation Front—to identify relations between blackness, animality, the underclass, and their ongoing legacy in carceral democracy, to advocate new narratives crafted by multispecies storytellers.

Balbinder Singh Bhogal (Ch 13) invites readers to radically decenter the "I think" of colonial and European epistemologies by which thought creates a thinkable "outside" in the first place. In the "taskscape" of Sikh practice, one reforms ego-thinking through connecting to animal perception and mystic intuition that, when re-integrated, produce an ever-transforming liminal human here and now—enlivened, rather than merely enlightened. While individuation of ego allows knowing *about*, Bhogal pushes multispecies knowing toward mystic classifications of knowledge *through acquaintance* and knowledge *by merger*, where cosmic knowing of One-as-radically-interdependent-reality is perceivable only through *un*selfing ego-representation and *un*thinking the "I think."

Bligh Somma (Ch 14) supplies an under-explored Islamic account of agency and responsibility within Faḫr al-Dīn al-Rāzī's 12th-century multispecies epistemology. Founded in wider Qur'ānic accounts of animal actions and capacities that exceed narrow Aristotelian versions of the same, al-Rāzī justifies animal intelligence while retaining nuanced distinctions between universals, meaning, and self-awareness. Using examples from original translations—as well as a vignette with his own dog companion, Somma concludes with al-Rāzī's compelling argument that animals' unique knowledge of God offers a shared—if not identical—sphere of animal and human knowing from which to reconsider multispecies ethical obligations.

Experimental Responses

Part IV opens by transporting readers to the underwater worlds of pond sounds with musician David Rothenberg. Prevented from performing during the Covid-19 lockdown, Rothenberg—known for his co-compositions with whales and birds—turns to "Making Music with Mysterious Species in Ponds" in and around upstate New York. Using hydrophones to explore the watery "undersound," Rothenberg provides visual sonograms, lively descriptions, and QR codes to aurally access these collaborative multispecies recordings.

Daniel M. Stuart (Ch 15) launches readers into the Buddhist cosmovision, challenging persistent accounts depicting Buddhist action as mostly mental or internal. Through original translations, Stuart explores an early text-complex detailing a theory of karma as it intersects a variety of relations that practitioners are encouraged to develop with nonhuman beings. These diverse practices of knowing and embodiment bind together internal and external activities through multispecies relations understood to impact the health and welfare of the entire cosmos.

Alisa Aaltola (Ch 16) suggests attention and love as an epistemic alternative to the rational analysis often undergirding animal ethics. Identifying strengths and limits of Platonic love, Aaltola draws upon the insights of Simone Weil and Iris Murdoch to reframe love-as-attention rooted in an epistemic state of *un*selfing (with meaningful parallels to Bhogal's and Donaldson's chapters) in which love is more than self-satisfaction. Universal love, in conjunction with ethical reflection, demands an epistemic and moral method of attention to the *real*-versus-idealized experience of any individual, the interconnections shared between beings, and the particularity of any being's complexities, including ugliness or destructiveness, beyond a generic "good."

Gabriele Schwab (Ch 17) issues a call toward transspecies care and equitable biopolitics requiring fluid epistemologies that perceive organisms on their own terms and understand the "human" through its cellular entanglements with other species. Schwab turns toward the tiny—including analyses of molecular/genetic transformations, cohabiting with viruses, and cultivating non-anthropocentric grammars with fungi—for strategies to shape new ecologies of mind and bio-logics capable of resisting destructive fusions of empire and capitalist commodification.

The book culminates with a spirited methodological imperative to conduct research in more living and lively ways. With the Gloria Anzaldúa archive as backdrop, Suzanne Bost (Ch 18) welcomes the outdoor squirrel, buzzing overhead lights, and torn-page doodles as contributing colleagues to collective study. Swapping pragmatism for more ambitious aims of impossibility, Bost follows multiple knowledge pathways, traces speculative imaginings for alternative futures, and pursues a pluralization of theories in service of shapeshifting epistemologies that multiply verbs of knowing.

The excesses of existence—the insistent "more than"—that defies capture in dominant epistemic regimes can explode into existential wonder, emergent affects, and new perceptual skills that thought alone cannot deliver. Knowing life is not only about perceiving things *out there* currently excluded, but also about cultivating new capacities *in here* to know and be known in paradoxical, unregulated, and unsanctioned modes. Ultimately, any ethics of multispecies epistemologies does not merely require new ways of *thinking about* the world, which retains the separative illusion of subject/object, matter/spirit, or inside/outside, but requires experiments of *participating within* and *belonging to* the world through persistent attention, free-ranging feelings, untethered memories of what could have been otherwise, cross-cultural receptivity, resistant imaginings, risky border crossings, intuitive communications, and undoings of Self and thought toward a multispecies merging adequate to life's dynamic destructions, creations, and adaptive revisions. In light of ongoing destruction of more-than-human beings and ecosystems—which is synonymous with self destruction—we do not need fewer practices for knowing life, but many, many more.

Brianne Donaldson
Claremont, California
July 2024

Bibliography

Adams, Carol. 2015. *The Sexual Politics of Meat, 25th Anniversary Edition: A Feminist-Vegetarian Critical Theory*. New York: Bloomsbury Revelations.

Butler, Judith. 2010. *Frames of War: When is Life Grievable?* New York: Verso.

Derrida, Jacques. 2008. *The Animal that Therefore I Am*. Edited by Marie-Louise Mallet. Translated by David Wills. New York: Fordham University Press.

Henning, Brian G. 2023. *Value, Beauty and Nature: The Philosophy of Organism and the Metaphysical Foundations of Environmental Ethics*. Albany: SUNY Press.

Hickey-Moody, Anna and Tara Page. 2015. "Introduction: Making Matter Matter." In *Arts, Pedagogy and Cultural Resistance: New Materialisms*, edited by Anna Hickey-Moody and Tara Page, 1–20. London: Rowman and Littlefield.

Marino, Lori and Christina M. Colvin. 2016. "Thinking Pigs: A Comparative Review of Cognition, Emotion, and Personality in Sus domesticus." *International Journal of Comparative Psychology* 28. https://doi.org/10.46867/ijcp.2015.28.00.04.

Nietzsche, Friedrich. 2009/1878. *Human, All Too Human: A Book for Free Spirits*. Translated by Helen Zimmern. Amherst: Prometheus Publishing.

Plumwood, Val. 2002. "Prey to a Crocodile." *Aisling Magazine* 30. Accessed August 12, 2022 at www.aislingmagazine.com/aislingmagazine/articles/TAM30/ValPlumwood.html.

Popova, Maria. n. d. "E. B. White on Why He Wrote *Charlotte's Web*, Plus His Rare Illustrated Manuscripts." *The Marginalian*. www.themarginalian.org/2013/10/15/e-b-white-on-charlottes-web/.

Santos, Boaventura de Sousa. 2007. "Beyond Abyssal Thinking: From Global Lines to Ecologies of Knowledges." *Review (Fernand Braudel Center)* 30, no. 1: 45–89.

"Thinking About Pigs Before They're Food". 2012. *New York Times*, February 26, 2012. www.nytimes.com/2012/02/27/opinion/thinking-about-pigs-before-theyre-food.html
Walker, Alice. 1989. *Selected Writings 1973–1987*. San Diego: Harvest/HBJ Publishers.
Whitehead, Alfred North. 1924. "First Harvard Lecture," September 25, 1924. *Whitehead Research Library*. Accessed July 10, 2024 at https://wrl.whiteheadresearch.org/items/browse?advanced%5B0%5D%5Belement_id%5D=50&advanced%5B0%5D%5Btype%5D=is+exactly&advanced%5B0%5D%5Bterms%5D=First+lecture.
———. 1967/1925. *Science and the Modern World*. New York: The Free Press.
———. 1978/1929. *Process and Reality: Corrected Edition*, edited by David Ray Griffin and Donald W. Sherburne. New York: The Free Press.

Section I

Living Concepts

PRAXIS A

THRESHOLD CONCEPTS FOR MULTISPECIES COMMUNICATION ZINE

M. J. Barrett

This zine explores a set of threshold concepts identified as central for transforming human-nature relations and countering dominant narratives of separation. Threshold concepts are troublesome, transformative, and often irreversible concepts that help define a field of study. Key concepts to be explored include: (1) there are different ways of knowing; (2) we can communicate with nonhuman nature and nonhuman nature can communicate with us; (3) transrational intuition and embodied knowing are valid ways of knowing; and (4) worldview is the lens through which we experience reality. The zine reflects on the place of these concepts for cognitive justice, and most significantly, for their impact on relations with other species. This analysis has significant implications for ways in which we conceive of, connect with, and relate to those who are more-than-human—as well as what is lost when we do not engage with and model for others, these threshold concepts-in-action.

DOI: 10.4324/9781032659985-2

Threshold Concepts for Multispecies Communication

M.J. Barrett

The threshold concepts in this zine are based on five years of research funded by the Social Sciences and Humanities Research Council of Canada. We developed teaching approaches for education in which more-than-human beings are acknowledged as having agency as research collaborators and co-instructors on our collective learning journey. For the full manuscript, see Barrett et al. 2017.

"Threshold concepts" are troublesome, transformative ideas that often challenge or affirm identities, help integrate complex understandings, and once grasped, can be irreversible (Meyer & Land 2006) .

THRESHOLD CONCEPTS

Threshold concepts can be ***Aha! Moments*** that shift paradigms

In a world where everything seems to be falling apart, and the calls for new paradigms increase, **threshold concepts** provide important direction and focal points for our energy and imaginations. They identify places where we, or our students, might get stuck in the liminal space between old and new ways of thinking as we work towards multispecies knowing and more ethical relations with other-than-human kin.

1 WE CAN COMMUNICATE WITH NONHUMAN NATURE & NONHUMAN NATURE CAN COMMUNICATE WITH US

Perhaps the most troublesome one of the lot, this first threshold concept frames everything to come.

2 THERE ARE DIFFERENT WAYS OF KNOWING

Coming to see how we continuously rely on and accept dominant forms of knowledge can help us recognize how many ways of knowing there are. For instance, researching with more-than-human kin may require one to be attentive to intuition and embodied knowing, while letting go of socially-inscribed assumptions about what ways of knowing have been given more status: rational thinking, written texts, and so forth.

Academics and those in the sciences can struggle with different ways of knowing if they have internalized a view that only humans create and preserve knowledge. Imagine how the field of biology would transform if direct communications with the plants, animals, stones and whole ecosystems being studied were fully embraced in the natural sciences, and we trained students to engage with them ethically, and skillfully.

TRY THIS

List all the ways of knowing that you use in your personal life. Then ask yourself: how many of these do I use in my academic/professional life? Is there a difference? Why?

3 WORLDVIEW IS THE LENS THROUGH WHICH WE EXPERIENCE REALITY

A key goal of transformative learning is "seeing our worldview rather than seeing with our worldview" (Sterling 2010, 23). By recognizing the assumptions of one's own paradigm, it becomes easier to be more open to things that are outside of it. Worldviews and the many discourses that reinforce them, actively create and maintain our notions of legitimate knowledge and ways of being in the world. Those who have travelled or been immersed in cultures different from their own, have often already grasped this concept.

Author's reflection

I used to talk about the importance of epistemological pluralism—the need for multiple ways of knowing to address the complex social and environmental problems in which we humans are entangled . . . .

But I no longer believe that is sufficient, as it misses the power dynamics associated with the "add and stir" mentality often associated with pluralistic approaches—like the idea that adding a woman or person of color to a committee is the resolution to achieving gender equity or combatting racism. While sometimes helpful, it is far from enough. Simply adding a place at the table for more-than-human voices is insufficient.

4 TRANSRATIONAL (TRANSPERSONAL) INTUITION AND EMBODIED KNOWING ARE VALID WAYS OF KNOWING

Transrational knowing goes beyond rationality but it is not counter to it (Astin 2002). Valuing and using knowledge gained through gut feelings, dreams, visuals in the "mind's eye" or "mind's ear" and other forms of intuition is an important threshold to cross for multispecies knowing.

Similarly, recognizing that one's body is a valid source of knowing can be a troubling and transformative concept, essential for communicating with those who are 'more-than-human'. In academia, approaching bodily knowledge as a topic of study and source of knowledge is a strong—and important—challenge to dominant epistemologies.

TRY THIS

Listening to plants: Sit quietly with a plant, either indoors or outdoors, and ask if it is thirsty, needs more sunlight, nutrients, is happy with where it is, etc. Pay attention to the sensations in your body, including an inclination to go and get water, to move it to another location, or to simply smile (perhaps it is perfectly happy with its current situation).

5

All knowledge is situated (Haraway 1988), constituted not only in relation to each person's individual history and epistemological assumptions but also co-constructed through relationships with other humans, ideas, institutions, and the other-than-human beings with whom they share their lives.

KNOWING IS RELATIONAL

The recognition of knowledge-as-relational can often be irreversible; once grasped, one can no longer see oneself or one's knowing in isolation. This concept can also provide meaningful linkages to the other concepts and vice versa. For example, accepting that there are many different ways of knowing and understanding whether one's worldview is oriented around relationships—or not—enhances understanding of knowledge-as-relational.

6

DOMINANT BELIEFS AND DISCOURSES CAN UNDERMINE OTHER WAYS OF KNOWING AS VALID

Discourses are powerful and reproductive habits of thought that reproduce social norms, rewriting particular beliefs in and on our bodies and minds. They create the dominant mini-narratives and expansive meta-narratives in which we are marinated because they are embedded in our daily speech, actions, and physical spaces.

A helpful way of thinking about discourse is as a well-worn bicycle trail. As you try to move out of the deep rut of a trail to explore new territory, the pull of gravity—which is the power of discourse—keeps you in the trail's groove until you encounter, and move through a portal to escape gravity's pull.

A Cautionary Note:
THE MAP IS NOT THE TERRITORY

The explanations we provide and the theories we develop are not the same as experiencing ethical engagement with more-than-human kin. In fact, sometimes the very act of explaining, naming, and theory-building risks erasing the potentials of actively and authentically being with and hearing their voices.

M.J. Barrett

BIBLIOGRAPHY

Astin, Alexander W. 2002. "Is Spirituality a Legitimate Concern in Higher Education?" Keynote Address Presented at the Spirituality and Learning Conference, 18 April 18, 2002, San Francisco, CA.

Barrett, M. J., Matthew Harmin, Bryan Maracle, Molly Patterson, Christina Thomson, Michelle Flowers, and Kirk Bors. 2017. "Shifting relations with the More-than-human: Six Threshold Concepts for Transformative Sustainability Learning." Environmental Education Research 23, no. 1: 131–43.

Haraway, Donna. 1988. "Situated Knowledges: The Science Question in Feminism and the Privilege of Partial Perspective." *Feminist Studies* 14, no. 3.: 575.

Meyer, Jan and Ray Land. 2006. "Threshold Concepts and Troublesome Knowledge: An Introduction." In *Overcoming. Barriers to Student Understanding*, edited by Jan Meyer and Ray Land, 3–32. New York: Routledge.

Sterling, Stephen. 2010. "Transformative Learning and Sustainability: Sketching the Conceptual Ground." *Learning and Teaching in Higher Education* 5, no. 11: 17–33.

1 Precursor to Knowledge

Semiosis as Ground of Multispecies Morality

Ralph R. Acampora

In this chapter[1] I will explore and assay the theoretical and experiential relevance of knowledge's precursor, semiosis, to animal ethics and to interspecies morality. We must first have meaning before we can assert or assess knowledge. Zoosemiotics is here taken as a theory or inquiry about the meaning-structures found in the animal kingdom. From the vantage of disciplinary history, it has been associated with cognitive science and ethology as well as philosophy of language and philosophy of mind. Lately, as in the work of Jesper Hoffmeyer, the broader conception of biosemiotics has been developed with affinities for somatic studies and philosophy of body. Animal ethics—at least in its first-wave instantiation among analytic/Anglophone philosophers—has usually been understood to offer (or contest) criteria that widen the scope of moral standing to include (some/many/most/all) animals other than humans. Recently, a number of alternative (continental European, feminist, and postmodern) approaches to the field have tried to lay bare the axiological and/or ontological presuppositions for the intelligibility and practicality of relating to other animals as moral beings, as patients mostly, but sometimes as agents too.

What I would like to propose—in an experimental vein, on a trial basis as it were—is that the fact or activity of semiosis may be taken normatively as a crucial element in the definitional framework of ethical considerability and/or that it may also be taken descriptively as a significant part of the proto-ethical ontology out of which interspecies morality arises. One way to conceive of these gambits is to think of them as attempts to straddle an inherent ambiguity in the very concept of meaning, which can carry at least two senses, at least as conveyed in the English language: (1) that of linguistic (or otherwise semantic) significance or equivalence—e.g., the word "sister" typically *means* "female sibling," and (2) that of practical importance or evaluational esteem—e.g., as in locutions such as "she *means* a lot to x" (where x refers to some person, process, or system). In general, the phenomenon that could serve as a bridge between these two senses of meaning is that of *salience*, inasmuch as both signs and values literally or figuratively "stand out" in or from an otherwise undifferentiated field of function or experience.

Likewise, in both cases there is always an effectively interpreting entity to or for whom any salient phenomenon appears. And (only) when a saliency makes a

DOI: 10.4324/9781032659985-3

felt difference to or by a fleshly interpreter, the emergent meaning matters morally, primarily because it is experienced in or through a corporeal condition of finitude (which is essential to constituting the vital needs and exposure to harm/demise that ground the very intelligibility of ethical standing). In a nutshell, what I here venture is that all beings, and only those beings, who are both somatically vulnerable and semiotically generative ought to be considered moral patients within the sphere of ethical concern, and that those live bodies who consciously control their interpretive activities can be thought of as candidates for membership in the community of moral agency, that is, if they show signs of volitional self-guidance and interest in fellowship.

Somatic Vulnerability and Semiotic Spontaneity, Rather Than Consciousness, as Grounding Animal Ethics

Elsewhere, against the psychocentric grain of most animal ethics, which emphasizes animal consciousness or subjectivity (as in standard utilitarian and deontological accounts), I have argued that (some state or property of) mental consciousness as such does *not* ground the concept of moral standing; rather, the presence of a live body (*Leib*) is an essential feature of any ethically considerable being (Acampora 2006, Ch 4). Yet I admit this feature, though necessary, is not by itself enough to explain moral standing. Another aspect should enter the picture, namely some structure of (spontaneous) semiosis. The notion of semiosis I have in mind here is basically Jesper Hoffmeyer's account: "According to the biosemiotic perspective, living nature is understood as essentially driven by, or actually consisting of, semiosis, that is to say, processes of sign relations and the signification—or function—in the biological processes of life" (2008, 4).

To better understand the genesis of the ethical claims that animate bodies present to moral agents, we need to consider how meaning (difference that makes a difference) orients living existence. Because the orientation of lifestyle reflects the valence(s) of an organism's teleology (or aim toward living), we would be able to track the presence of (welfare-)interests in various entities and thereby get a handle on the distribution of ethical considerability at large. My tentative claim, then, would be that existential constitution, or existence as such, as a somatic/semiosic (the latter term denoting "meaning-generative") complex is necessary and sufficient for a being to have good of its own, on the basis of which it becomes intelligible and appropriate to speak of the being's weal or woe, benefit or harm.

Which kinds of entity would this picture of moral standing rule in, and which would it rule out? The class of moral patients, on this account, would turn out to be fairly biocentric, including almost all human beings, most nonhuman animals, and perhaps some other forms of life as well. Kinds of entity that would not be recognized as having non-derivative moral standing would include mountains and minerals (at least at the spatio-temporal scale on which moral agency is exercised, terrains and chemicals appear not as somatic (*leiblich*) and semiosic but rather as corporeal (*körperlich*) and causal respectively); typical machines (including even artificially intelligent computers, so long as they are operating as non-sentient assemblages); laboratory tissue and cell-cultures generated and controlled directly by means of first-order (non-recursive/a-cybernetic) technological contrivance since such semiosis as exists therein is derivative not spontaneous, nor is there present organismic sentience; and angels and gods (if there are any transcendent entities, they would rate as non-finite "super-sentient" beings above and beyond rather than beneath or below the bounds of moral sense; in other words, because of their peculiar metaphysical status, immortal/invulnerable spirits on this model would not call for moral recognition because they have no needs). The listing of illustrations and series of corresponding explanations just given are not exhaustive, but should suffice to provide the gist of applying the relevantly exclusionary criteria at stake.

Plants, Viruses, and Other Gray Areas of Semiosis

Examples of entities in a gray area of quasi- or semi-ethical status could include (somewhat controversially, and possibly misdiagnosed) totally stultified humans; likewise, those caught in unresponsive wakefulness syndrome (UWS) might not be considered genuinely semiosic beings although it is important here to flag the substantial level of UWS's misdiagnosis (Vanhaudenhuyse et al. 2018). Artificial animals (such as xenobots made from frog stem cells and A.I.) or other bionic life forms also present difficulty for ethical classification; relevant questions to ask on a case-by-case basis would include: how sensitive is this creature? Does it interpret stimuli by and for itself?

Perhaps most intriguingly, plants in general pose challenges to moral ontology, as pointed out by Michael Marder's chapter in this book. One wonders whether botanical "irritability" bespeaks full-fledged semiosis or if it is only proto-semiotic? On this point Jakob von Uexküll's (1982/1940) ambivalence is noteworthy: he seems to indicate a pan-biotic scope for the phenomenon of meaning and yet makes a distinction between the animal's living environment (*Umwelt*) and the plant's residential envelope or dwelling-case (*Wohnhülle*) that would appear to introduce gradations of semiosis whereby the latter form of habitat involves only liminal or borderline sign-exchange. Likewise yet more recently, Kalevi Kull has implied that wherever we find a biologically functional circle there too is an *Umwelt*—yet I would hesitate to go so far if a given *Funktion-Kreis* were (shown) to be attained via algorithmic mechanism alone, and indeed Kull himself displays a similar reluctance when he allows that semiosis (taken to be coextensive with *Umwelten*) "could be defined as the appearance of a connection between things, which…do not interact or convert each other through direct physical or chemical processes" (1998, 299). As Michael Marder puts it, "the plant seems to muddle conceptual distinctions and to defy all established indexes for discerning different classes of beings" (2011, 89). On the one hand, Kull has argued persuasively on the basis of vegetative needs that phytosemiosis is a reality at the botanical level (2000, 326); on the other hand, few, if any, botanists would be prepared to recognize plants' *felt* needs.

Similarly, with respect to fungi and prokaryotic/eukaryotic life forms, Kull alludes to "microsemiosis" at the cellular level of enzyme and membrane management (2000, 326) while Günther Witzany charts multicellular/interorganismic sorts of communication based on allegedly sign-mediated formation of microbial colonies (2005, 70). And yet the experiential dimension of living bodiment, which would make these examples clearer candidates for inclusion as moral patients, seems absent to most observers. Lastly, viruses have notoriously troubled ontological categorization for biologists and others. While viruses are expertly efficacious in the molecular de- and re-codings necessary to function as (ersatz?) parasites and proliferators of disease, they nonetheless do not ordinarily reproduce on their own; so are they alive or not, somatic or corporeal?

Semiosis Versus Mainstream Animal Ethics Criteria

How might the view I have sketched compare to some of the standard accounts already available in the animal ethics canon? First, let's look at Peter Singer's utilitarian position on animal liberation. The view I have adumbrated above would be congenial with, yet give a different spin to, his focus on sentience as the operative criterion for moral standing. Flat-footedly, it may seem that I have simply re-described sentience as somatic vulnerability and then added another criterion in the form of semiotic spontaneity. However, what I should like to say instead is that sentience typically has another face than the familiar one of passive sensation. When an entity suffers or feels pain, it doesn't merely undergo an experience or simply register a given feeling; rather it must *interpret* certain stimuli *as* negative, and so sentience itself can be seen as shot through with semiosis. This semiotic aspect can figure on different levels—perceptual, emotional, imaginative—in different kinds of beings. From this perspective, then, it might be best to speak of *sensibility*, a term that already carries at once desired connotations both of somatic receptivity and semiotic construal. Of course, these connotations could fade when we have occasion in colloquial parlance to use the associated adjective "sensible" (often designating practicality), and so in ethical discourse, we would have to take care to understand "sensible beings" as fleshed makers of meaning (whether or not they are reasonable by disposition or capacity).

Tom Regan's deontological position on animal rights is another prevalent view that may be worthwhile to compare with my account of semiosis. His approach is based on a criterion of subjectivity, according to which rights-bearers have moral standing just in virtue of being "subjects-of-a-life." Who are subjects in this sense? When speaking informally, Regan was sometimes heard to say that subjects-of-a-life are not merely biological creatures but *biographical* ones as well.[2] This characterization sounds as if it may overlap a zoosemiotic portrait—after all, it might be argued, any animal who is a generator of meaning must *ipso facto* have a perspective from which an intelligible story could be told of its life. More formally, however, Regan's stance occupies a position intermediate between the views of zoosemiotics (as I have rendered that above) and of what might be called narratological anthropocentrism, according to which the horizon of moral standing encircles only *auto*biographical beings, whose life-stories can be told not merely *about* them but also *by* them*selves*. By technical definition, subjects-of-a-life for Regan:

> have beliefs and desires; perception, memory, and a sense of the future, including their own future; an emotional life together with feelings of pleasure and pain; preference- and welfare-interests; the ability to initiate action in pursuit of their desires and goals; a psychological identity over time; and an individual welfare in the sense that the experiential life fare well or ill for them, logically independently of their utility for others, and logically independently of their being the object of anyone else's interests.
>
> (1983, 243)

In this more detailed account, a sense of the psychic richness and fairly robust sapience intended by Regan comes through. It becomes understandable why it is that he applies the criterion unambiguously only to mammals, though he believes birds probably belong also, and reptiles/amphibians should be given a benefit of doubt in virtue of their uncertain status. From the vantage I am scouting out here, Regan's standard puts an undue emphasis on mentation, or other cognitive criteria, as such and hence expresses a sufficient yet unnecessary condition for ethical considerability.

Semiotics Versus Analytic Ethics

Whether or not my hypothesis could be validated in the context of recent or contemporary analytic ethics, however, I would also like to venture the proposal that the phenomenon of semiosis set in somatic experience is at least a very helpful hint at (if not a definitive hallmark of) value-generation—and so it can be seen as a primary aspect of the phenomenological grammar peculiar to the evolution of ethics (as regards interspecific morality in particular). I turn in this direction because I have misgivings about the whole objectivistic enterprise of fixing a single, specific property of individual entities that would putatively require—from the perspective of abstract rationality—certain ethical norms in respect of them. If we think of moral significance as itself a kind of semiotic valence, then the holy grail of objectivist normativity would be figured as the notion that reality has some built-in ethical code according to which natural signs of ethical import could be read off by any being with sufficient sapience to engage the requisite activities of interpretation.

The problem for me with this vision of an inherently morally meaningful universe is that, by my lights, no rendition of it has withstood trenchant critique. In the continental traditions of European thought, the project of universalist morality was cut loose from its axiological anchors by Nietzschean skepticism; in the methodology of Anglophone philosophy, normative ethics was capsized, if not sunk, by J.L. Mackie's ethical nihilism. The closest anyone I'm aware of has come to establishing an objective foundation for morality is David Gauthier, who argues that conative intentionality *ipso facto* constitutes evaluative activity that logically must respect similar activities conducted by others and that a full-blown contract theory of ethics can be built on this basis (1986, Ch 6). I used to think Gauthier's position could be refuted by appealing to the contingency of goal-directed behavior, most radically that there is no necessity of engaging such whatsoever (i.e., Camus' point that opting out of life altogether is always an option for autonomous beings). Yet I have come to recognize that even inaction entails some sort of conative exercise of intentionality, for instance, enacting the will to veto animal instincts of hunger and self-protection. Still, I do not see that Gauthier's argument succeeds. Although he may have demonstrated that any rational agent must enter into at least a minimal framework of morality, he leaves the supposition of rationality itself as an undischarged assumption. In other words, the decision to continue operating in accord with the reasoning conventions to which one has been acculturated is a non- or a-rational choice not entailed by the application of those conventions to the fund of publicly available knowledge.

Nor can it be held self-evident that we must be rational because otherwise the world would make no sense and our very survival would thus be forsaken—since, to mix Shakespearean and Kantian allusions, it may well be the case that life is indeed sound and fury signifying nothing and that Hamlet's suicidal query cannot be settled within the bounds of pure reason alone. Ultimately, the postulates of the world's intelligibility and life's desirability are surds (non-rational properties) that cannot be objectively justified. To exist and orient themselves, autonomous beings possessed of self-consciousness must make meaning and set values (and so exercise conative intentionality)—but their most originary gestures herewith are radically underdetermined by (because prior to) inferences from facts.

Against the background of these reflections, if not precisely by consequence of them, my hunch is that morality has more to do with a felt awareness of responsibilities that emerge from actual and potential relationships; it is primarily a somatopsychic phenomenon or experiential process, not a metaphysical proposition in the first place and only later a matter of logical argument. So now, let us look at the existential context of ethical valuation from phenomenological and hermeneutic perspectives.

A Semiosis of Altruism and Intercorporeal Sense-Making

In the lifeworld of beings who are concerned and discourse about such things, then, the concept of morality centrally refers to a positive or benign regard for others—what we can name loosely *altruism*. Granting this conception as a definitional given, we may then ask how does such a regard come to be? Using his biosemiotics lens, Hoffmeyer speculates that it arises from recognizing a life's irreplaceability—and that this recognition is itself rooted in the intercorporeal dialogics of semiotic individuation (2010). To put it succinctly, I constitute myself in somatic transaction with the other, who in the process becomes somebody who must be acknowledged as sharing a physical exposure that calls for care similar to my own (Hoffmeyer 2008, 325). To flesh out the workings of this sympathetic transactivity, it will be helpful for us to consider the nature of animacy as such—and for this the work of another leading light in biosemiotics can serve as a launching pad.

In theorizing what he refers to as "semiotic embodiment" (2007, 467), Claus Emmeche takes the position that meaning-laden automotion is at the heart of animacy:

> A living being that is not simply an organism but is also an animal, is emphatically "animated", that is, it realizes a specific form of movement which is not reducible to pure physical change or basic biologic processes (like metabolism). Movement, understood as living change and as semiotic, when realized by self-moving organisms (i.e., animals), crucially depends on mediating all the elements of quality, physical force, and interpretation…
>
> (2007, 465)

To spotlight the relevance of motility (the ability to move spontaneously and actively) for understanding live bodiment continues a concern in Continental thought that goes back at least to Merleau-Ponty (1962) and Hans Jonas (1966). What these thinkers wished to underscore was that being self-mobile holds greater significance for an organism's mode of existence than superficial attention to mere locomotion might indicate: beyond obvious changes in location (conceived as quantitative alterations of spatial coordinates), auto-mobility affects qualitative sense of place and effects a generative locus of semiosis. For that latter upshot especially, we will now take our cues from Maxine Sheets-Johnstone, a contemporary hermeneut of body whose independent work deserves more recognition than the usual run of academicians has been wont to grant. As we shall see, her insights are consonant with the embodied cognition models of human psychology and animal ontology lately rising from the neurosciences, linguistics, and eco-phenomenology.

Two Husserlian phenomenologists, Javier San Martin and Maria Luz Pintos-Penaranda, observe that animality entails an "originally material interrelation with the world"; they continue, "Being a body [*Körper-Leib*] implies, for every animal subject, that her or his own experience of reality means being open to something that is beyond herself, open to something that is not herself, to the experience of the other-than-me" (2001, 357). Sheets-Johnstone goes a step further when she affirms the fundamental place of *motile* bodiment in the lives of social animals such as human beings—in resolving philosophy's so-called problem of other minds, she notices that "moving bodies are the foundational dynamic channel by which we forge understandings of others, coming to know their beliefs, judgments, and values through their movement" (2008, 213). Moreover, she extends this insight in an ethical direction when she links animacy to fellow-feeling:

> [O]ur spontaneous capacity to move ourselves and to be both kinesthetically aware of the dynamics of our own movement and kinetically aware of the dynamics of the movement of others is the foundation of our interanimate sense-makings. That capacity generates the possibility of affect attunement,

> and in turn, the possibility of empathy. Animation thus indeed holds the key to understandings of intersubjectivity.
>
> (212)

In this way, following Sheets-Johnstone, we can appreciate the importance of a prelinguistic "bodily logos" that builds on the evolutionary basis of species-typical patterns of motion, gesture, sound, and posture that function as mutually recognizable "consigns" (2008, 295ff.). Because it departs from the heretofore dominant models of language-based identity and sociality, this account constitutes a sea change in self-understanding and semiology of community. Neo-pragmatist philosopher Mark Johnson frames the shift as moving from an objectivist theory of meaning to an embodied/experiential view. In the former, meaning is borne by arbitrarily conventional symbols (words and sentences), which are grounded by propositions representing literalistic concepts that mentally mirror states of affairs in the external world (Johnson 2007, 272). According to the latter scheme, meaning "arises through embodied organism-environment interactions in which significant patterns are marked within the flow of experience…as we engage the pervasive qualities of situations and note distinctions that make sense of our experience and carry it forward" (273). The process of semiosis involves not only formalistic operations of discursive logic but also imaginative and felt dimensions of experience—and it remains body-based throughout, says Johnson, "because the semantic and inferential structure of [even] abstract concepts is drawn from our sensorimotor interactions, typically by cross-domain mappings (conceptual metaphors)" (273).

What is most interesting for the purposes of our present discussion is that a lifeworld mediated by the affect attunement embedded in such patterns of semiosis is already pregnant with socio-ecologic growth into moral community—insofar as "[t]hese beginning empathic associations that develop among like creatures, these *intercorporeal sense-makings* that arise spontaneously…describe an intersubjectivity that is in actuality a thoroughgoing *intercorporeality*, a way of living together in harmonious understandings" (Sheets-Johnstone 2008, 294). I have written a book on this matter myself (Acampora 2006), but here I will defer to Sheets-Johnstone's compact explanation that "what in fact grounds morality and makes it fully real, is foundationally one's own body, one's own *physical* body which is vulnerable and…affectively experienced as exposed, assailable, open to pain, disease, suffering, and death"; this experience, coupled with an acknowledgment that other live bodies (with whom I am semiotically acquainted via similarity of physiognomy and comportment) are also physically exposed and existentially vulnerable, can carry self-regard into concern for others (2008, 299).

To be clear, Sheets-Johnstone insists that a temporally expansive consciousness of bodily vulnerability is uniquely human, other animals' sensitivities in this respect being limited to here-now contexts (just as nonhumans may seek in the moment to escape potentially lethal circumstances but yet have a mentally anguished, dreadful prospect of their entire lives being subject to temporal finitude or mortality) (298). However, I would say that such a delimitation is relevant only to attribution of moral agency, not to defining the boundary moral patients as a class of its own—which may well extend considerably past the set of our conspecifics; and Sheets-Johnstone herself would appear to admit as much when she alludes to animate sympathy developing into a sensibility of "common creaturehood" beyond humanity as such (299).

That semiotics could have ethical import is not an unprecedented proposition. Consider, for example, Jonathan Beever's dissertation (in philosophy at Purdue University in the United States), which drew out implications of the Peircean approach for animal ethics (2013). Dario Martinelli, in his overview of the field, claims that zoosemiotics has a latent moral agenda on behalf of animal welfare or protectionism (2010, Ch 5). Susan Petrilli and Augusto Ponzio inflate semioethical regard to the scope of biosphere and beyond: "We believe that semioethics is the widest perspective possible that human beings may now reach in their role as semiotic animals and, therefore, as *cosmically* responsible agents" (2008, 137). The reason for this cosmic responsibility is rooted in human nature, according to Petrilli, "The capacity to 'answer for,' to 'account for' structural to human beings implies responsibility for life on earth in its globality" (2011). Yet somewhat curiously, though Petrilli thinks that semioethics is influenced by biosemiotics and bioethics, she nonetheless yokes the semioethical project to a hyperbolic humanism wherein "the whole planet's destiny, in the last analysis, is implied in the choices and behavior of human beings" (2011). It is as if Petrilli (and Martinelli as well) are so impressed by the anthropogenesis of semioethics that they risk sliding back into anthropocentrism—hence their predilections toward a kind of stewardship ethos. My approach diverges from this thrust, inasmuch as I wish to highlight the ethico-political aspect of solidarity with other creatures—we are not so much their caretakers as their partners in the terrestrial drama of semiotic evolution.

A Proto-Ethical Ontology of Multispecies Solidarity

It remains to be seen how the horizon of common creatureliness grows beyond the rather limited set of just those beings with whom I've *actually* had semio-somatic congress. Here, there is a role for imaginative cognition and reasoning: having had the experiences discussed above, I can then think through a process of argumentation by which I realize that there are indefinitely many similarly constituted entities with whom I and others *could* come into like relations. This activity itself should not be allowed to devolve into fanciful abstraction, and so it is best conducted with a grounding in the kind of participatory observation practiced by multispecies researchers in the line of Jakob von Uexküll, a method which attempts "reconstruction of the *Umwelt* of another living being [via] the ratification of the decoding processes in its behavior after the enumeration of the signs the living being is capable of receiving, as well as the code by means of which it interprets th[os]e signs" (T. Uexküll 1982, 4).

It is important to understand that neither *Umwelt* analysis, nor the ethical reasoning based upon it, ever stray far from bodily experience. This is because sensorimotor sensations are linked to abstract cogitation via image schemas that allow for cross-modal perceptions to function as conduits for intellection. Mark Johnson defines image schemas as "dynamic, recurring patterns[s]of organism-environment interactions" that synthesize "the most fundamental structures of perception, object manipulation, and bodily movement, given that human bodies share several quite specific sensorimotor capacities that are keyed to the size and constitution of our bodies and to the common characteristics of the various environments we inhabit" (2007, 136–37).

As I see it, at least two such schemata are salient to the development of morality: the *container* schema, which is based on our handling and management of tangible holders; and the *center-periphery* schema, which is "tied to the fact that our perceptual [chiefly visual] fields have focal areas that fade off into a vague horizon of possible experiences that are neither currently in focus nor at the center of our conscious awareness" (Johnson 2007, 137). The former is extrapolated when we think of ethical significance or considerability—that is, we engage moral semiotics—in terms of inclusion/exclusion to or from practical accounting appraisals of relevant interests; the latter is utilized when we discriminate and evaluate different gradations of moral worth or status.

Usage of schemata such as these is, according to Johnson (by way of insights from linguist George Lakoff), facilitated by employment of ideational metaphor, such that "what we call abstract concepts are defined by systematic mappings from body-based, sensorimotor source domains onto abstract target domains" (2007, 177). For instance, the notion of ethical significance comes to language via one of the principal conceptual metaphors in general currency—*categories are containers*: the source domain is the bodily experience of holders, the target domain is the set of considerable entities in moral deliberation, and the relevant mapping instantiates the law of excluded middle via the metaphorical logic of quasi-containment, whereby "an entity either falls within a given category or falls outside of it, but not both at once" (Johnson 2007, 180). In addition, this conceptual metaphor (category = containment) makes it possible to envision and express moral progress in terms of (re)defining ever-larger sets of moral patients within a more and more voluminous scope of categoreal containment.

The preceding analysis also brings into view another aspect of the moral ontology I have been charting in the course of this chapter, namely that semiosis itself—and even more so semiotics (as the reflective study of semiosis)—puts one into commerce with systems of signification (and meta-systems of interpretation) evolved and operating before and beyond the individual. Especially for human animals, communication is mediated by a rich nexus of symbolic/linguistic machinery, the lifeblood of which is transpersonal interactivity. What this means is that the semiosic agent cannot function solo, must become involved in, and adapt to a larger domain of sociality—and what this means, in turn, is that the wheels of ethics are greased for pro-social/altruistic dispositions and tendencies (capable of interspecies extension). Hence, we can see that the expansion of moral community under discussion in this chapter is not a contrived or *ad hoc* extension but is rather prefigured in the very ontology with which we have been dealing—as one commentator on Uexküllian biosemiotics has put it, "an organism is never just one. Instead, each organism has a context, and *Umwelt* in which it lives, and, in being so [contextualized], the organism is always already more than itself" (Buchanan 2008, 29).

The two-tiered bridge between ontology and morality is built of epistemology and axiology, and it spans a sea of skepticism about animal representations: can we really portray animal beings and their interests with enough cognitive confidence to validate the ethical and political positions we may take up in respect of them? It seems to me that a responsible answer to this kind of query must avoid both the complacency of naively realistic scientism and cynicism of hyper-ironic constructivism. I have addressed this issue at length elsewhere, and so I will repeat only my provisional conclusion in the present context:

> Obviously and unavoidably, our representations of other organisms are socially constructed—yet, recognition of that does not commit us automatically to believing animals themselves (or any natural entities) are created by us. One cannot, of course, express anything about the bare world of nature without dressing it up in linguistic or artistic clothing; nonetheless, the postulate that another reality—besides myself or us—is a subject and party to (not merely an object in or construction of) my/our discourse and deeds is more plausible than the idea that I/we make the world entirely out of words and/or actions. . . . "[T]ranslucent" mediation of cognition [can] represent nature sufficiently well for us to arrive at value-laden yet non-arbitrary views of animals as they are—not "in themselves" but in relation to their human, organic, and geographic environments.
>
> (Acampora 2001, 302)

Indeed, the rigorously representational contributions found in this very book may be read as exemplary efforts or object lessons in just that kind of enterprise.

Finally, and to end on an ethical note explicitly, yet unabashedly, normative, the point should be emphasized that the intellectual process by which the class of moral patients is enlarged does not have to be mutual to secure its status; this is very important to recognize, because insisting otherwise would leave us, in the words of another moral philosopher, with "a focus on moral agents that is at best unwarranted, as it blurs the distinction between the possibility of morality and the object of morality, and at worst sinister, as it sanctions, via the idea of reciprocity and contract, lesser protection for the weak" (Cavalieri 2009, 34)—which is to say, that a semio-generative creature is still somebody who counts ethically even if not possessed of intelligence sufficient to enact moral agency. There is a crucial respect in which any-body who enacts signification is alive, because the dialogic play of semiosis is a (perhaps even *the*) structural dynamic that animates an otherwise inert object; consequently, it makes sense—literally and figuratively—to converse with suchlike bodies as fellow "personalities" *who* matter morally.

Notes

1 Most of this chapter was previously published, under different editing, as "The (Proto-) Ethical Significance of Semiosis: When and How Does One Become Somebody Who Matters?" In *The Semiotics of Animal Representations*, edited by Kadri Tüür and Morten Tønnessen, 343–62. Leiden: Brill, 2014.
2 Personal conversation with author, no date recalled.

Bibliography

Acampora, Ralph R. 2001. "Representation Cubed: Reviewing Reflections on Animal Imagery." *Society and Animals* 9, no. 3: 299–307.
———. 2006. *Corporal Compassion: Animal Ethics and Philosophy of Body.* Pittsburgh: University of Pittsburgh Press.
———. 2008. "Animal Constructs and Natural Reality." *Humana Mente* 7. Accessed December 1, 2023. www.humanamente.edu/Issues/Issue7.html.
Beever, Jonathan. 2014. "Zoosemiotics and Bridging the Value Gap." *Semiotica* 198: 121–35.
Buchanan, Brett. 2008. *Onto-Ethologies: The Animal Environments of Uexküll, Heidegger, Merleau-Ponty, and Deleuze.* Albany: SUNY Press.
Cavalieri, Paula. 2009. *The Death of the Animal: A Dialogue.* New York: Columbia University Press.
Chandler, Daniel. 1994 *Semiotics for Beginners.* New York: Routledge. www.visual-memory.co.uk/daniel/Documents/S4B/.
DeGrazia, David. 1996. *Taking Animals Seriously: Mental Life and Moral Status.* Cambridge: Cambridge University Press.
Emmeche, Claus. 2007. "A Biosemiotic Note on Organisms, Animals, Machines, Cyborgs, and the Quasi-Autonomy of Robots." *Pragmatics and Cognition* 15, no. 3: 455–83.
Gauthier, David. 1986. *Morals by Agreement.* Oxford: Clarendon Press.
Hoffmeyer, Jesper. 2008. *Biosemiotics: An Examination into the Signs of Life and the Life of Signs*. Translated by Jesper Hoffmeyer and Donald Favareau. Edited by Donald Favareau. Scranton: University of Scranton Press.
———. 2010. Lecture, *International Seminar on Social and Ecological Ontology,* Pieve Tesino, Italy, July 9.
Johnson, Lawrence. 1991. *A Morally Deep World: An Essay on Moral Significance and Environmental Ethics.* Cambridge: Cambridge University Press.
Johnson, Mark. 2007. *The Meaning of the Body: Aesthetics of Human Understanding.* Chicago: University of Chicago Press.
Jonas, Hans. 1966. *The Phenomenon of Life: Toward a Philosophical Biology.* New York: Harper and Row.
Kull, Kalevi. 1998. "On Semiosis, *Umwelt*, and Semiosphere." *Semiotica* 120, no. 3/4: 299–310.
———. 2000. "An Introduction to Phytosemiotics: Semiotic Botany and Vegetative Sign Systems." *Sign Systems Studies* 28: 326–50.
Kull, Kalevi and Peeter Torop. 2003. "Biotranslation: Translation Between *Umwelten.*" In *Translation Translation*, edited by Susan Petrilli, 313–28. Amsterdam: Rodopi.
Llewelyn, John. 1991. *The Middle Voice of Ecological Conscience.* London: Palgrave-Macmillan.
Mackie, John Leslie. 1977. *Ethics: Inventing Right and Wrong.* New York: Viking Press.

Marder, Michael. 2011. "Plant-Soul: The Elusive Meanings of Vegetative Life." *Environmental Philosophy* 8, no. 1: 83–99.
Martinelli, Dario. 2010. *A Critical Companion to Zoosemiotics: People, Paths, Ideas* (Biosemiotics 5). Dordrecht: Springer.
Merleau-Ponty, Maurice. 1962. *Phenomenology of Perception.* Translated by C. Smith. London: Routledge and Kegan Paul.
———. 1968. *Resumes de cours, College de France, 1952–1960.* Paris: Gallimard.
Nietzsche, Friedrich. 1989. *Beyond Good and Evil: Prelude to a Philosophy of the Future.* Translated by Walter Kaufmann. New York: Vintage.
Petrilli, Susan. 2011. "The Responsibility of Power and the Power of Responsibility: From the 'Semiotic' to the 'Semioethic' Animal." *SusanPetrilli.com.* Accessed November 12, 2022. https://susanpetrilli.com/reading-online.html.
Regan, Tom. 1983. *The Case for Animal Rights.* Berkeley: University of California Press.
Rothfels, Nigel, ed. 2002. *Representing Animals.* Bloomington: Indiana University Press.
San Martin, Javier and Maria Luz Pintos-Penaranda. 2001. "Animal Life and Phenomenology." In *The Reach of Reflection*, edited by Steven Crowell, Lester Embree, and Samuel J. Julian, 342–63. Boca Raton: Center for Advanced Research in Phenomenology.
Sheets-Johnstone, Maxine. 2008. *The Roots of Morality.* University Park: Penn State University Press.
Singer, Peter. 2011. *The Expanding Circle: Ethics, Evolution, and Moral Progress.* Princeton: Princeton University Press.
Steiner, Gary. 2008. *Animals and the Moral Community: Mental Life, Moral Status, and Kinship.* New York: Columbia University Press.
Taborksy, Edwina. 2022. "Terminology." *SEE: Semiosis, Evolution, Energy.*
Tompkins, Peter and Christopher Bird. 1979. *The Secret Life of Plants.* New York: Harper and Row.
Uexküll, Jakob von. 1982/1940. "The Theory of Meaning." *Semiotica* 42, no. 1: 25–82.
Uexkull, Thure von. 1982. "Meaning and Science in Jakob von Uexküll's Concept of Biology." Introduction to special issue of *Semiotica* 42, no. 1: 1–24.
Vanhaudenhuyse, Audrey, Vanessa Charland-Verville, Aurore Thibaut, Camille Chatelle, Jean-Flory L. Tshibanda, Audrey Maudoux, Marie-Elisabeth Faymonville, Steven Laureys, and Olivia Gosseries. 2018. "Conscious While Being Considered in an Unresponsive Wakefulness Syndrome for 20 Years." *Frontiers in Neurology* 9, no. 671: 1–9.
Witzany, Günther. 2005. "From Biosphere to Semiosphere to Social Lifeworlds." *Triple C: Cognition, Communication, Co-operation* 1, no. 1: 51–70.
Wolfe, Cary. 2010. *What Is Post-Humanism?* Minneapolis: University of Minnesota Press.

2 Weaving Life and Death Back Together

On Multispecies Mortality

Beatrice Marovich

In the mid-1980s, contemplating Hiroshima and Auschwitz, the philosopher Edith Wyschogrod described the contemporary world as one of manmade mass death—a world of death on an unthinkable scale, technologically produced and executed. Anticipating what would eventually become a biopolitical and necropolitical narrative mode, Wyschogrod (1985) argued that the mass annihilation of "untold numbers of persons" creates not only new social forms but also "a death-world made up of mini-cosmoi of the living dead embedded in larger societies" (2). In these mini-cosmoi, "vast numbers of persons are simply marked for annihilation as part of an impersonal process of destruction" (35). This situation—the condition of manmade mass death—obviously changes how we think about our politics, our collective lives. But it also changes how we think about death and our own mortality, Wyschogrod argued. "What is the meaning of death in the twentieth century, when millions of lives have been extinguished and the possibility of annihilating human life altogether remains open? Is there an art of dying, which remains useful in this time and circumstance?" (1).

Her question was, it seems to me, a bit rhetorical. In her reference to "an art of dying" she was invoking an old, Western, philosophical ideal of "the good death" as one that is free of pain and suffering, and can serve as a measure of a life well-lived. In the death of a philosopher like Socrates, we see the model of a man who faces his own individual mortality with courage and authenticity. This courage and authenticity bestow his death with meaning, and even goodness. Wyschogrod took issue with idealizations of this take on finitude, arguing that it was fundamentally anachronistic in a world of manmade mass death. Death in Western thought, she argued, "despite varied articulations, has been dominated by a single pattern which depends upon interpreting the self as a cognition monad and the process of dying as requiring the behavior appropriate to a rational subject" (Wyschogrod 1985, x). This sort of self-oriented, rational contemplation of death becomes meaningless as mass death "lights up" the "horizon from which death can be interpreted," Wyschogrod argues. In this new light, the death of an individual due to accident, old age, or natural hazards can barely register. Individual death becomes insignificant (13). What is an individual death in the face of collective annihilations? A serene, self-oriented, contemplative, and rational view of mortality only makes sense when

DOI: 10.4324/9781032659985-4

one can presume the relative stability of the world—the social collective—that grounds one's own existence. It is a privilege. And the privilege itself comes apart at the seams—exposing itself as privilege—in conditions of manmade mass death.

One might argue that, in her contemplation of manmade mass death as a twentieth-century phenomenon, Wyschogrod was quite late to the scene. Manmade mass death had been unfolding already for centuries in the wake of the Atlantic slave trade and settler colonialism. Countless people had already faced the reality of manmade mass death, prior to the twentieth century. One might also argue that Wyschogrod's scope was limited, in her contemplation of manmade mass death as a phenomenon relevant only to human finitude. By the 1980s it was becoming clear that mass species extinction was a terrifying reality, poised to intensify. In this sense, perhaps, Wyschogrod herself was still too intellectually constrained by the Western tradition she critiqued. After all, this Western philosophical tradition had long maintained that personal mortality was something that only mattered to men, and not to animals. If it is the case that Wyschogrod was conceptually constrained by the intellectual limits of this tradition, she was certainly not alone. This way of thinking—this Western philosophical contemplation of death—has left deep traces on the hearts and minds of many people who have inherited its thought patterns. But perhaps the time for disinheritance is now.

In what follows, I want to take this condition of manmade mass death as a kind of backdrop. And I want to echo Wyschogrod's critique of the limits of a view of mortality that makes of it a rational, self-oriented contemplation of my own individual death. In a world of manmade mass death, living in the wake of horrific twentieth-century conflicts as well as ongoing legacies of the Atlantic slave trade, settler colonialism, and mass species extinction, death is not merely a matter of individual or personal concern. Mortality today is something more and more of us are forced to contemplate on the level of collectives—including multispecies collectives. To think about mortality as *mine*, rather than *ours*, in this condition, is restrictive. Our collective mortality can matter to us very much, personally and individually. So, this is not to suggest that personal feelings or individual deaths are of no consequence. Of course they are! Instead, this is a critique of contemplative forms, or contemplative patterns. To die a good death like the men of Western philosophy—in the grip of one's own reason, with cognitive and rational faculties intact—was to die in the form of a contemplative ideal. And it is this individualized and self-oriented contemplative model that has become a relic of a bygone era.

I argue that thinking about death from the perspective of a multispecies epistemology—a multispecies account of the knowledge of death—can unsettle this old contemplative model and amplify the resonance between radically different approaches to mortality. If it will become possible to break out of this other, more ancient, paradigm for reflecting on mortality in order to think about death as a collective phenomenon, there are important lessons to be gleaned from a multispecies epistemology of death. Taking the dissonances and resonances of a multispecies perspective into account, when thinking about mortality, can serve as a reminder that death is something complex and uncontainable. Death is not something to be tamed. It is not a phenomenon that rational contemplation can simply manage or maneuver. Rather, mortality is a lived and embodied phenomenon that is understood and coped with in radically different ways in different communities, in different species, and in different bodies.

If there is any chance of being able to cope with our mortality on a collective level, then those of us who have inherited these older mortal (Western) reflections that scholars like Wyschogrod critique must become better able to embrace the radical plurality of approaches to death, and understandings of death. We may, that is, have to find ways of coping with its uncontainability. Despite what Western philosophy may have taught, there is not one art of dying. There is not one way to cope with death. Rather, there are many models—perhaps infinite models—of response to learn from and attune to. To embrace this view does not solve the problems that we are confronting; it does not make these conditions of manmade mass death disappear. But, as Wyschogrod argues, an old tradition of thought—an ancient Western tradition of thought—is losing its potency, its vitality, its applicability, or its ability to illuminate something real and true about life, death, and our experience of each of them. In the wake of this loss, or this failure, at least we can be reminded of the fact that there are so many other forms of wisdom to turn to, and to lift up, both human and more-than-human.

Man, and the Whatever the Animal Lacks

Western philosophy has traditionally offered what feels like an endless series of assertions that Man is something different from the Animal. I use the singular, gendered, capitalized forms of both Man and Animal here to underscore the fact that for centuries, this philosophical tradition employed these figures actively and with a straight face. For those of us working in the wake of this tradition, thinking is still impacted by these acts of figuration. The conceptual figure of Man really did, and was designed to, subordinate sexual differences as they exist within human cultures, and to speak more specifically about the life (or mind) of Man. Likewise, the conceptual figure of the Animal really did, and was designed to, elide biological or species difference within more-than-human life worlds. The point, for most of the history of Western philosophy, was not to create an understanding of humans that accounted for our many differences (gender, sexual, racial, religious, ethnic, national, geographic, economic, etc.). Nor was it to create a complex conceptual language with which to map out our many relations with beings and entities in more-than-human life worlds. The point was, instead, to figure out what Man was by articulating what he was not: the Animal. This, as I have already mentioned, generated a long series of assertions about where this line—between Man and Animal—should be drawn.

The crucial distinction between Man and Animal tends, more often than not, to hinge on a kind of *lack* in Animal life. The Animal is different from Man because the Animal *lacks* something crucial that Man has. This "thing" that Man possesses (and the Animal lacks) is only rarely physical. One exception, of course, is the assertion that Man has the capacity to walk on two feet, which the Animal allegedly lacks. But physical capacities, such as bipedalism, can be easily witnessed in the life of birds, apes, kangaroos—even cockroaches! So philosophers have tended to prefer distinctions that are more psychic than physical, more spiritual than material. These are not so easily refuted with physical evidence.

Language, for instance, is a capacity that many Western philosophers have claimed is lacking in Animal life, being proper instead to the life of Man. Martin Heidegger, in his quest to prove that the human is "essentially something other than an animal organism," argues that animals lack language (1946, 247). Heidegger is also quite clear that, by language, he means something other than *communication*. Language "is not the utterance of an organism," he writes; "Nor is [language] the expression of a living thing" (248). It is not symbolism, nor can it be reduced to signification. Instead, for Heidegger, language is more extremely abstract. It is the vague and mysterious "clearing-concealing advent of being itself" (248). Language is bound up with being. If language were simply communicative, we might easily witness language at work in more-than-human worlds. So, for Heidegger, language is the very thing that gives humans access to (allows us to tap into, to see, to understand, to describe) the extreme abstraction of being itself. Language opens the door to the world, for Man This lack of language, or this lack of access to being, gives the Animal a kind of intellectual poverty, in Heidegger's view. Or, to put it another way, Heidegger philosophically impoverishes animal lives.

The point of this impoverishment, suggested Jacques Derrida, was simply to conceptually conscript the Animal to a condition of privation: a state, or existential condition of lack, absence, destitution, or need. This "poverty" of Animal life does not, Derrida notes, serve to place the Animal in a hierarchy, or a system of rank where it sits always just below Man (2008, 155). Rather, Heidegger seeks to explore the ways in which the Animal is "deprived," so that he can articulate something about Man's own capacities. What is Man? Simply that ethereal thing the Animal is deprived of. This conceptual move, to deprive the Animal of an abstract capacity that is then used to constitute Man, locks this figure (existentially) into a state or condition of privation. What makes Man something other than the Animal is the fact that Man has a cunning strategy for evading, or lifting himself out of, privation. This state of privation then serves as a kind of black mirror against which the figure of Man can witness (admire and describe) the faint glimmer of his own almost indiscernible reflection.

Heidegger is often described as a thinker who was critical of the metaphysics of Western philosophy, and the Christian ideas that had, for so much of its history, animated this form of thought. But his "destruction" of the Western history of ontology (its study of being itself) was less an attempt to reject this metaphysics than it was a quest to "stake out the positive possibilities of that tradition," as he put it (Heidegger 2008/1962, 44). With this gesture of conscripting the Animal to a state of lack or privation, Heidegger is repeating a more fundamental gesture within the history of Western metaphysics. He is describing what privation is, and staking out a path by which Man might raise himself above it. Man can make something (very abstract, barely even perceptible) of himself, as long as he can hover just above that lack.

Few thinkers have theorized privation as exhaustively as the scholastic theologian Thomas Aquinas (1225–74 CE). The figure of privation helped Aquinas articulate a range of ideas about sin, and evil. Aquinas affirmed the fundamental *goodness* of God's creation by arguing that the good is real while evil is simply the privation of the good—a lack of goodness (2003, 60). In this sense, there is nothing in creation that can be classified as essentially evil because evil does not have a reality. But all things in creation, to the extent that they begin to fall short of the good, can be rendered evil, almost as if the good seeps out of them like air from the hole in a tire. This also meant that, for Aquinas, evil was potentially everywhere, especially in nature. He classified any act of nature that was in some sense incapacitating (pain, sickness, disability, injury, decay, and death) as evil. He also, to be sure, distinguished between what he called *malum poenae* ("evil of penalty") and *malum culpae* ("evil of fault").[1] So natural evils were simply *things that happened*—they were not the fault of any individual action but of original sin itself, committed by the original Man and passed inexorably down. Natural evils were forces beyond individual sin.

Individual human beings could do nothing to stop these evils from their natural course of progression. There is no stopping death and decay. But this evil dimension within nature, and nature's tendency to endlessly punish Man with incapacitation, did have the consequence of solidifying a kind of bond between nature and the nihil that annihilates. The privation of nature (its death, decay, and dissolution) could not be stopped by Man. "Death and dissolution are natural to human beings by reason of a necessity of matter," wrote Aquinas (2003, 250). Still, he assured Man that it was possible to hover just above that privation, even if it wasn't *in this life*. "Death and dissolution are contrary to our nature insofar as immortality is natural for us," Aquinas reassured (250). Because Man has the capacity to know God, to understand God's eternal law, Man could live in accordance with, or attune himself to, this eternal law. Provided that Man was in tune with this law, he was assured something special (non-privative) in this eternal life with God. Moreover, through contemplation of God, he could get in touch with the non-privative in the here and now. The Animal was, for Aquinas, a thing of nature—not essentially evil, but simply destined (like all of nature) for the death, decay, and dissolution that was a part of natural evil. The Animal was not capable of contemplating God's eternal law. Man, while made up of natural stuff, had an immortal soul—that abstract *something* that enabled contemplation and lifted him up just enough to keep him from being swallowed up by privation.

The immortal soul of Man also makes death into a crucial boundary line between Man and the Animal. While both Man and the Animal have bodies that die, Man has the capacity to rise above the privation that includes (and makes use of) death. Death, for the Animal, is the final end. The Catholic theologian Karl Rahner described death—in classical Christian thought—as a form of "dematerialization" (1973, 46). That is to say, death was the moment when theologians believed that "the spiritual principle in man" was finally "considered to be freed from the confines of matter and corporeal existence" so that he could be "transformed to the state of free spirit" (45). For Man, the death of his body allows the most abstract and non-material thing within him to continue to rise up above privation. Following this line of logic, Rahner writes, death is not the end of existence for Man "nor is it a mere passage from one form of existence to another"; rather, "death is the beginning of eternity" (28).

Heidegger does not appear, on the surface, to be offering a similar formulation. His philosophy has shed many of the terms and categories of this Christian theological viewpoint. Heidegger is not interested in theorizing Man's relationship with God, nor does he offer any promises about Man's access to eternity. But Heidegger is interested in what will give Man access to something just beyond privation. When Heidegger speaks of Man's existential being as a being "toward death," he is not suggesting that Man collides with death and stays with it, or unifies himself with it. Rather, he is suggesting that Man (with his access to language) is getting something *out of death* or making something of the nothing that death is. For Heidegger both Man and Animal are finite mortals. But, as Derrida notes, for Heidegger only Man "*has* this finitude" (2008/1962, 150; emphasis added). Only Man can *grasp* it in some way, objectify it, understand it so as to become other than it. Man takes a kind of nosedive into death, grabs something from it, and flies beyond it. Man is the one who knows death well enough to evade it. Through his language, Man can grasp the fact that he will die. So it is that Man can *really live* a life that matters, because he knows that he will die. The Animal, who does not know death, cannot even be murdered, because it has never *really lived* in the first place.

The view of the human, of the more-than-human, of sexuality, and of race that inevitably emerges from this patterning built up around Man and Animal offers a way of thinking about life and mortality that is founded upon an existential state or condition of privation. And this state of privation relies upon forms of violence to take its particular shape. It leads not only to a misrecognition of how other earthlings live, and die, but also forms of violence against other humans: a stripping away of some capacity that seeks to relegate them to a condition of fundamental lack. Man relies upon the idea that *he* knows death, that *he* understands his death in some essential way. Because of this knowledge, this understanding—this *epistemology of death*—Man believes himself to be a form of life that really matters. Man knows death in ways that what is other to him—such as the Animal—allegedly does not. But Man does not know death as he thinks he does.

A Minimal Concept of Death

Spanish philosopher Susana Monsó argues that Western philosophical views on life and death have given rise to an over-intellectualization of the concept of death. This over-intellectualization, she charges, has supported philosophical arguments that animals (unlike humans) lack a concept of death. Concepts, as Monsó defines them, are "the basic components of many of our thoughts" (2021, 28; trans. Marovich). So to suggest that *only* humans have a concept of death is, in essence, to suggest that animals lack a basic building block that would allow them to comprehend the fact of death. Animals, from this point of view, do not understand what death is. So, death is of no consequence to animals. Such logic often leads to the moral or ethical argument that animal lives are of less value. Monsó develops a minimal concept of death that, I think, offers seeds for a multispecies framework for thinking about our mortality.

To begin with, Monsó notes, Western philosophical thought has tended to present concepts of death as binary: Either you understand what death is, and can reflect on the fact that you will die, or you don't. But this sort of view does not allow for any kind of gradation when it comes to modes of understanding death. When we consider how human children learn about death, however, it's quite clear that such gradation exists. "Babies are not born with a concept of death," Monsó writes, "and neither do children develop this overnight" (2021, 77). Indeed, children must be exposed to both a range of social teachings as well as a host of personal experiences before death begins to register as meaningful or even comprehensible. Neither would it be honest to argue that all human adults possess a comprehensive understanding of death. Indeed, death remains one of the greatest enigmas in most human communities. For this reason, Monsó argues that it is better to think about concepts of death as existing on a kind of spectrum, with the most abstract and complex understandings at one end, and the most concrete and minimal understandings on the other (78). Many humans living on Earth today, especially those in wealthy communities in the United States and Europe with access to reliable medical care, encounter the death of other human beings only infrequently. But for many other people in the world, as well as other animals, death is not something abstract to be encountered only via contemplation or theory, but something concrete that they are experientially exposed to on a regular basis (138–39). Therefore, the level of abstraction that death takes on can vary greatly, even in human communities.

Rather than begin discussions of, and reflections on, death at the most abstract, complex, and theoretical end of the spectrum, Monsó argues for beginning with the most simple, basic, minimal concept of death. This minimal concept of death, as Monsó describes it, can be attributed to an animal when it appears to be capable of classifying some individuals as dead, and of understanding "dead" as a property that applies to those beings who should otherwise be expected to exhibit traits associated with "living." Moreover, any animal who possesses a concept of death should be able to understand that these traits and characteristics associated with "dead" are irreversible (Monsó 2021, 88). Monsó also argues that there is a "Holy Trinity" of factors that must be present for an animal to be capable of developing a concept of death: cognition, experience, and emotion (130). The latter two factors are significant insofar as Monsó highlights that affect and embodiment (not just our intellectual capacities) play a key role in the development of concepts of death. Notably, one thing that Monsó leaves *off* of this trinitarian shortlist is sociability. An animal does not need to be a social animal, with strong social bonds to other animals of its species, to understand and respond to death, she suggests (136). Therefore, the emotions that an animal might display as it reacts to the death that it cognitively observes and corporeally experiences do not need to be grief, mourning, or loss. It might, instead, be something like disgust or fear.

Monsó holds up the opossum as an example in order to argue that minimal concepts of death are widespread in animal worlds. Opossums are among a set of animals who "play dead," or engage in acts of thanatosis, in order to evade their predators. When an opossum plays dead, its entire body goes rigid and its mouth falls open. At times it may begin to drool and its eyes can even glaze over, as if lifeless. Opossums also release a liquid, from a gland near their tails, that gives off the odor of rotting flesh. This act of thanatosis appears to be largely involuntary on the part of an opossum, a kind of instinctual response to the feeling of being in mortal danger. Monsó is not arguing that this act of playing dead gives insight into the concept of death in the life of an opossum. She is not, for instance, arguing that thanatosis proves that an opossum clearly has a sense of personal mortality (which would be evidence of a complex concept of death). Rather, what Monsó argues is that this act of playing dead indicates just how prevalent a minimal concept of death must be among predators. This is a strategy that has evolved in the life of opossums because it appears to work. Thanatosis is a kind of game, or performance, that creates a cognitive and emotional reaction in their predators, who (based on their prior experiences) identify this animal as one who is dead (Monsó 2021, 212). Why would this act of playing dead matter, that is, why would it be part of the evolutionary heritage of opossums unless it could serve as a kind of semantic signal that creates the convincing illusion of death? And why would opossums possess this ability if it were not the case that an understanding of death was widespread among their community of predators?

There are, I think, a number of interesting elements to extrapolate from Monsó's analysis. First, of course, is Monsó's own motivation for staging this argument in the first place. She suggests that death has served philosophically as a dividing line between humans and other animals for so long that it has come to feel natural. But Monsó argues that this alleged distinction between humans and other animals, in which humans are presented as the only species who understand the fact of death, must go the way of other walls that were built to cordon off the human. "All other walls have fallen," Monsó argues, "and this one should too" (233). For Monsó the purpose of toppling this wall is not simply to develop a more capacious understanding of other animals, although that is one meaningful benefit. Rather, Monsó believes that there is something humans can gain from accepting the fact that animals understand death as well. She suggests that in unsettling the boundary between human and animal, often drawn along the line of death, we might also unsettle *our own* understanding of death. In reconciling ourselves with the fact that we are not unlike other animals in our knowledge of death, perhaps we can also "reconcile ourselves with our animal condition," she suggests; and perhaps even "reconcile ourselves with our mortality" (233).

I am less optimistic than Monsó about the potential to reconcile oneself with mortality. I tend to view affective dispositions toward the fact of death to be generally unstable. The meaning that one makes of death can vary greatly from individual to individual depending upon many factors such as age, gender, race, economic status, and more. And the way that we make meaning out of death can vary from moment to moment. Humans are incredibly fluid and changeable. I also think that there can be socially protective factors at play, in our refusal to be entirely reconciled to the fact of death. There is a deep tenderness toward one another that we expose when we refuse to accept that death is an outcome of a given situation. We are powerful when we protect one another against death, when we resist it. Instead, what I find interesting in Monsó's analysis is that her proposal that we begin philosophical contemplations of death with a more minimal concept of death opens up the sense of what death is, and how to respond to it. Beginning with a minimal concept of death does not erase any of the complexity or abstract mysteriousness within it. Rather, it encourages us humans, especially those of us who have inherited those old Western views of death, to widen the scope of our vision. It encourages us to be open to more, to see more, to learn more. Why would this matter? Because in a world of manmade mass death, more is needed. Resources are drying up, dying out, expiring. Old traditions are going bankrupt. Yet, still, more can be done, and more is being done. The radical differences emerging from the simplest and most minimal concept of death can be a reminder of this.

Weaving Life and Death Back Together

For billions of years on the planet, life and death have been working together collaboratively to create processes of creation, destruction, emergence, and change. Multispecies anthropologist Deborah Bird Rose describes life and death working together to sustain "a family of life on Earth, a family that is always changing, always finding connections, generating fit, seeking an always shifting balance in an Earth system that is itself far from equilibrium" (2011, 20). Our many relations within human and more-than-human worlds are sustained not only by life (which is obvious) but also by death. As Earth-born creatures, Rose writes, "we are fragile creatures, dependent on resilient, but endlessly shifting, Earth systems" (59). Nothing can change the fact that we are all "inexorably embedded in and sustained by webs of life and death" (59); which is to say that "death is completely central to life for us and many other beings" (61). Death, in Earth history, is something that brings us together and something that collaborates with life to make change.

While there is a time and a place for resisting death, the conditions of life itself are such that it is not something to resist absolutely; death is not a force to demonize. Rather, Rose suggests, in our human and more-than-human communities we are called to reaffirm life when death tears someone from us. One of the ways that we tend to do this, in so many human communities, is to affirm the importance of generational time—the collective temporality that exceeds the temporality of an individual's embodied life. In the wake of death—in rituals like a wake—what is affirmed is this alternative dimension of temporality that can contain the lives of those we love even in the face of death. In this generational temporality, we can continue to find one another across life and death again and again. Generational temporality is a fabric that is made up of a set of weaving patterns; it exists because, for so many thousands of years, human rituals have been weaving life and death back together again in the face of dissolution. Generational temporality affirms the presence of a multispecies collective, woven of the living and the dead, in the wake of any individual death.

In this world of manmade mass death, the generational dimension of temporality is precisely what we are losing sight of, or losing touch with. Rose calls this loss, meaning the death of both individuals as well as generationality itself, "double death." This double death, writes Rose, "contrasts with the ecological and evolutionary contexts in which death is immanent in and necessary to life. Double death breaks up the partnership between life and death" (2012, 128); it amplifies death, so that it cannot connect itself to life again (127–40). When the partnership between life and death is fractured, the multispecies collective that is comprised when life and death are woven back together again is also frayed. The erosion of generational temporality is the erosion of bonds between the living and the dead. This is the erosion of a multispecies collective that can hold us in that space beyond life and death.

This is, in many ways, a hopeless situation. I write about it not because I labor under the illusion that I have any sort of concrete solution. Like Rose and others, I can see this trouble clearly, and I know that I have no remedy to fix it. This trouble now is bigger than us and bigger than so many imagined fixes. But perhaps, as the theorist of more-than-human frameworks of thought Donna Haraway suggests, what is called for in urgent, troubling, times like this is not an act of fixing the future, not an act of "clearing away the present and the past in order to make futures for coming generations" (2016, 4). Perhaps, instead what is called for is simply, "staying with the trouble"; that is to say, making unexpected collaborations, turning neither to despair nor hope, but to our senses (1). To remain attached to ongoing pasts, here within our thick present, and to move toward possible futures. To tell one another stories, so that we can sense something beyond hope and despair (130).

When we remember that mortality is not simply *mine* (that it is not a matter of my own individual fate or fortune), but *ours*, it can throw us back into a terrifying awareness of this trouble. This awareness highlights and underscores the shadows of double death around us and reminds us of the threats to the dimensions of generationality that connect us, not simply with our family trees in human life, but through family resemblances we sense and witness between ourselves and other earthlings. Yet our awareness of this trouble can also remind us that this potently charged past—these billions of years of evolutionary heritages—are still pressing up against us. Those who have died are gone, in one sense. But in other ways, they remain with us. We can still sense these connections at work. We smell these connections at work in the processes of decay and germination that unfold around us. We can feel these connections at work in the tricky survival strategies of opossums.

When we sense these ever-present connections to death, we are also reminded of the fact that death is not simply *mine*, in the form of an individual subjective struggle, but *ours*—something that all of us earthlings do and struggle with and against together. Mortality is collective. This fact is terrifying at the same time as it is relieving. And none of this makes death—let alone the double death that surrounds us as individuals and as collectives—dissipate or disappear. But we can lean into this multispecies generationality in order to tell ourselves stories in order not to despair, as we float into and toward so many possible futures. Enduring in this way is also to resist the isolations of double death.

Note

1 For a discussion of this, see Brian Davies' introductory commentary in Aquinas (2003, 21).

Bibliography

Aquinas, Thomas. 2003. *On Evil*. Translated by Richard Regan and Brian Davies. Oxford University Press.

Derrida, Jacques. 2008. *The Animal That Therefore I Am*. Translated by Marie-Louise Mallet. New York: Fordham University Press.

Haraway, Donna. 2016. *Staying With the Trouble: Making Kin in the Chthulucene*. Durham: Duke University Press.

Heidegger, Martin. 1946. "Letter on 'Humanism.'" Translated by Frank A. Capuzzi. Accessed August 18, 2022. https://archive.org/details/heidegger-letter-on-humanism-capuzzi

———. 2008/1962. *Being and Time*. Translated by John Macquarrie and Edward Robinson. New York: Harper Perennial.

Monsó, Susana. 2021. *La Zarigüeya de Schrödiner: Como Viven y Entienden la Muerte Los Animales*. Translated by Beatrice Marovich. Madrid: Plaza y Valdés.

Rahner, Karl. 1973. *On the Theology of Death*. New York: Seabury Press.

Rose, Deborah Bird. 2011. *Wild Dog Dreaming: Love and Extinction*. Charlottesville: University of Virginia Press.

———. 2012. "Multispecies Knots of Ethical Time." *Environmental Philosophy* 9, no.1: 127–40.

Wyschogrod, Edith. 1985. *Spirit in Ashes: Hegel, Heidegger, and Man-Made Mass Death*. New Haven: Yale University Press.

3 Voracious Secularism

Emotional Habitus and the Desire for Knowledge in Animal Experimentation[1]

Donovan Schaefer

In the agony of death a dog has been known to caress his master, and every one has heard of the dog suffering under vivisection, who licked the hand of the operator; this man, unless the operation was fully justified by an increase of our knowledge, or unless he had a heart of stone, must have felt remorse to the last hour of his life.

(Darwin, *The Descent of Man and Selection in Relation to Sex* (1871))

Fun

A bronze statue erected as a memorial in a public square becomes a focal point of controversy. Young people march against it. Eventually, the powers that be side with the statue's opponents, whisking it away in the dead of night.

This scenario is a familiar one. But every detail is strange. The statue in question was erected over a century ago, in 1907, and removed just three years later. The location was London, not the American streets where demonstrations against the long shadow of racism have recently become increasingly visible. Moreover, the young protestors in 1907 were all from a particular stratum of society—students studying medicine—at a time when higher education was a prerogative tightly held by the upper classes.

The most significant difference of all, though, was that the bronze statue was not of a politician or national hero. It depicted only a dog.

The dog had no name. He was a captive of William Bayliss and Ernest Starling, scientists at University College London (UCL). In early 1903, he was vivisected—surgically cut open while still alive—in front of a classroom that included two women, Lizzy Lind af Hageby and Leisa Schartau, suffragettes and animal rights activists. The women published an exposé of the procedure in a book released later that year, *The Shambles of Science* ("shambles" being a slaughterhouse). They contended that the animal was not anesthetized and was suffering as he was sliced into. The scientists claimed the dog was sedated according to law. Not in dispute was that the dog had already been vivisected twice in earlier procedures and kept alive in a mutilated state for months. A highly public dialogue played out around the case in the ensuing years. In 1907, a coalition of socialists, feminists, and

DOI: 10.4324/9781032659985-5

anti-vivisectionists helped raise a statue to the dog in the working-class neighborhood of Battersea. A few years later, the rioting London med students compelled the authorities to take it down.[2]

The first editions of *Shambles* dedicated an entire chapter to describing the procedure at UCL. The title of this chapter was a single word: "Fun." The chapter's premise was that the operating theater was not, as one might expect, a scene of staid academic discourse—or even a somber reckoning of the scientific necessity of a gruesome procedure. It was, instead, a kind of carnival. The authors describe an atmosphere of "humor" and "excitement," the attending students frequently breaking out into "roars of laughter and uproarious stamping of feet" (Lind af Hageby and Schartau 1903, 23). While the dog lay bleeding to death on the table, the lecturer entertained the students with a digression on another group of animal subjects whose esophagi were cut open, with the result that:

> the food taken fell down on the floor instead of passing into the stomach. The dogs ate and ate and ate—they were frightfully hungry—and were much surprised to see the food fall out; they tried again with the same result. They could go on like that for hours.
>
> (Lind af Hageby and Schartau 1903, 25)

To which the authors added, sardonically: "How comical! How clever of the physiologist who tried this! Aren't animals stupid?" (Lind af Hageby and Schartau 1903, 25).

What does the festive, pornographic atmosphere of the exhibition room at UCL in 1903 tell us? What light does it shed on the make-up of the "smooth, cool man of science" (Kean 1995, 21)? How does it relate to the student reaction to the statue of the murdered dog later erected in Battersea? Psychologist and former animal researcher John Gluck has written that "in my experience, animal researchers' disdain for the concept of animal rights exceeded the level of derision evolutionary biologists expressed at proponents of intelligent design" (2016, 262). What all of these responses on the part of animal researchers point to is a continuum of structures of feeling surrounding the work of science. These tendencies form the core of a kind of secular body: a set of dispositions, affects, and formations of emotional habitus that consolidate the sense that the scientific body is an *ultra-human* body—a body fundamentally separate from animal life. This chapter investigates the desire of science—a desire that drives important scientific achievements, but also horrors. This desire, what Aristotle called the human "desire to know," is directly linked to a history of extraordinary racialized and gendered forms of violence throughout the history of science (Aristotle 1933, 2). Today, the most salient arena of scientific violence is a sprawling infrastructure of experimental research using nonhuman animals.

This chapter considers how science, in disavowing the emotional machinery at the heart of its project, ends up producing regimes of spectacular violence that are largely kept out of sight. And it connects this desire at the heart of science to the literature emerging from the subfield of secularism studies on the "secular body." The suggestion advanced here is that animal experimentation produces a set of affects that consolidate one version of the secular body. In a version of what Pierre Bourdieu would call "habitus"—and specifically the "emotional habitus" theorized by scholars like Deborah Gould and Monique Scheer—affective practices shape a formation of subjectivity that can be understood as a distinct secular body, a kind of more-human-than-human (and definitely more-than-animal) self that understands itself as immune to feeling even as it is actually profoundly determined by it.

This ultra-human secular body is fascinated by the knowledge that can be produced through animal experimentation, and this fascination produces an immune response that makes it difficult for animal experimenters to meaningfully absorb ethical challenges to their actions. Charles Darwin, quoted in this chapter's epigraph, meditates on the moral torment of the animal experimenter who receives a plea for mercy from a suffering creature, claiming that only a stony conviction that the knowledge to be derived from the procedure is just will exempt the scientist from guilt (1882, 70). But as we've already seen, animal experimentation is not feelingless: it is often done in a mood of excitement and exhilaration, not resignation. The irony, of course, is that the fact that scientific knowledge-production is embodied—a device of feeling, not the unfurling of detached reason—is also what makes us (not just scientists but all of us as we think and feel our way along) animal.

My goal here is to avoid a full-bore conversation about the ethics of animal testing and instead focus on the disciplinary practices and corresponding affective formations—both instantiations of the secular body—that make animal experimentation not only possible, but a source of uproarious joy, from the theater at UCL to today. My focus is not on the presumed right to test on animals, but on the pleasure that cements that sense of right and the rage that rises to defend it. In an age when the moral voice of science as a tool to combat misinformation around climate change, pandemics, vaccines, and the vestiges of scientific racism is more urgently needed than ever, this chapter proposes a radical reckoning with how science's violent engagement with animal bodies becomes part of the broader landscape of the secular.[3]

The chapter begins with a reflection on how the field of secularism studies has pivoted to an interest in the formation of "secular bodies"—disciplined subjects that express the secular priorities of rational sovereignty and emotional neutrality. But this professed neutrality is, I go on to show, a sham, based on an untenable split between thinking and feeling, which is disrupted by attention to the affective dimensions of scientific knowledge-production itself. The secular body, then, is actually a mode of playing up one set of emotional priorities (the fascination with scientific knowledge-production) at the expense of others (such as compassionate concern for science's victims). After briefly reviewing the history of animal testing over the past half-century, I turn to two perspectives on the disciplinary practices that make animal experimenters: one a view from the outside (Lesley Sharp's anthropological study of workers in animal laboratories) and one from the inside (John Gluck's memoir of his own career in animal science).

Secular Bodies

The subfield of secularism studies (sometimes "critical secularism studies") has roots in anthropology, sociology, literary theory, and philosophy.[4] But it has also been well-received in religious studies, where it has dovetailed with a long-standing attention to the discourse and practice of the putative separation of the religious from the secular.[5] In brief, these overlapping fields can be collectively summed up as a series of reflections on how "the secular" is procedurally defined—in culture, in history, in law, and in politics—in contradistinction to "the religious," but with an eye to how the bounding line between them is itself contingent and locally determined. So to make a "formation of the secular" (to use Talal Asad's phrase) is also, in complementary fashion, to make a formation of religion. The figure we zoom in on can be either a face or a vase, but the squiggly line between them is what does the work.

The formula of what we might call the "disenchantment" of the secular is articulated by Charles Taylor, who characterizes the secular subject as a "buffered self" (2007, 27). By this he means that as "modernity" advances, subjects are increasingly oriented to the intellectual factor of life and progressively more detached from bodies and emotions. Taylor describes it like this:

> A modern is feeling depressed, melancholy. He is told it's just his body chemistry; he's hungry, or there is a hormone malfunction, or whatever. Straightaway, he feels relieved. He can take a distance from this feeling, which is declared ipso facto unjustified…Here is the contrast between the modern, bounded self—I want to say "buffered" self—and the 'porous' self of the earlier, enchanted world.
>
> (2011, 41)

Although the common sense of Enlightenment rationalism might see this as a trajectory of improvement, Taylor wants to portray the rise of the buffered self as a situation of loss. But in either case, the basic coordinates of a separation between thinking moderns and feeling nonmoderns are preserved.

This framework has been challenged by postcolonial scholars like Talal Asad. Asad's work, like Taylor's, sees the secular not as neutral, but as a new situation that is produced by evolving dynamics of modernity—"neither continuous with the religious that supposedly preceded it (that is, it is not the latest phase of a sacred origin) nor a simple break from it (that is, it is not the opposite, an essence that excludes the sacred)" (Asad 2003, 25). But crucially, for Asad, the secular is not just a new intellectual condition. "I take the secular," he writes, "to be a concept that brings together certain behaviors, knowledges, and sensibilities in modern life" (2003, 25). Rather than a detachment from embodiment, secularity brings with it its own templates of action and feeling—ways of being embodied. Asad calls for a study of the secular plural, what he calls "formations of the secular," that will be studied with a new "anthropology of secularism" (2003, 17). This has prompted some scholars to ask if we can even speak of a kind of *secular body*, "a particular configuration of the human sensorium—of sensibilities, affects, embodied dispositions—specific to secular subjects" (Hirschkind 2011, 633).

The conventional formula for dividing the religious from the secular takes religion to be emotional, embodied, and contingent while the secular is rational, intellectual, and universal. However, secularism studies literature directly challenges this orientation. What I would call a *material secularism* perspective observes that secularism comes along not just with a set of discourses, paradigms, beliefs and disbeliefs, but with its own ways of being embodied. Side-by-side with a discursive analysis, then, this orientation notes that the line between secular and religious is blurry, and that we should expect the secular to be determined by material culture, embodied practice, and emotion just as much as we would find on the side of what gets called religion. Religious studies scholar Ann Pellegrini refers to this as "feeling secular" (2009); anthropologist Saba Mahmood has called these "secular affects" (2009).

We can see here how the scope of interest of secularism studies coalesces with the literature on "habitus" in the sociological lineage of Pierre Bourdieu. Bourdieu defines habitus as "systems of durable, transposable dispositions…principles which generate and organize practices and representations that can be objectively adapted to their outcomes without presupposing a conscious aiming at ends or an express mastery of the operations necessary in order to attain them" (1990, 53). In other words, habitus is a collection of socially engrained habits that incline bodies toward certain actions, feelings, and strategies of interpretation. This converges with what Michel Foucault calls a "technology of the self"—a set of practices designed to fashion subjectivity by working on the body (1988). Following Deborah Gould and Monique Scheer, we might extend this concept to identify the cultivation of *emotional habitus*—trained ways of feeling and experiencing the world (Gould 2009, 32; Scheer 2012).

Secular habitus, then—no less than religious habitus—has (is) a body. And some secularism studies scholars have stressed the importance of secular feelings as part of this picture. The secular body is not a body rinsed of emotions.[6] It is, instead, a body defined by a particular affective orientation to the world. For scholars working in this vein, the emphasis is always on secular formations as a plurality. There is no one way to be secular; but there are traits of the secular body that crop up as patterns and recurring motifs. In her article "Religious Reason and Secular Affect" (2009), Saba Mahmood concludes by exploring the notion that formations of the secular are often defined by their attachment to intellectual inquiry itself. Although "critique" is often represented as defined precisely by its distance from emotional attachment (just as "secular" is distant from "religion"), Mahmood suggests that we need to think of critique as mobilizing its own attachments, including an emotional investment in intellectual activity itself (2009, 861). What Immanuel Kant calls the "motto of the Enlightenment"—*sapere aude*, "Dare to Know"—is not just an abstract imperative (Kant n.d.). It's an emotional manifesto, an affective blueprint for the excitement to be derived from knowledge production itself.

Feeling Scientific: Caviar for the Men in the Laboratories

Taylor's concept of the buffered self is best understood not so much as a historical type as a myth—and a myth that is particularly effective at denying its own self-determination. There is only ever a rebalancing of feelings, never feeling's elimination. Crucial to understanding this reorientation when we think about secular affects is the recognition that the secular is uniquely concerned with knowledge production, especially as this authorizes the secular as the repository of pure reason. This dovetails with research in philosophy of science and science and technology studies that has suggested that science itself has an emotional engine at the heart of it—an excitement in the exploration of concepts and information that drives scientific inquiry forward.[7]

This claim flies in the face of our conventional wisdom about the nature of scientific inquiry, which is often defined as maximally distant from emotion. The secular body of the scientist is assumed to be composed by a set of habits that programmatically detach it from emotion—while also configuring it along gendered and raced lines (Ferreira da Silva 2007; Lloyd 1993; Wynter 2003). As Sylvia Wynter writes, a single idea of the human—"Man"—"overrepresents itself as if it were the human itself," exclusively responsible for "securing the well-being, and therefore the full cognitive and behavioral autonomy of the human species itself/ourselves" (2003, 260). Jewish studies and animal studies scholar Aaron Gross writes that the "isolation of the human in the human sciences is predicated upon various arguments for human uniqueness that are not intelligible without animal others" (2014, 8). But we could say the same about the natural sciences. To be a scientist, qua secular body, is to be consummately male, consummately white, consummately human.

This confidence in the detachment of scientific rationality from emotion is even given support from some scientific quarters. In the mid-twentieth century, psychologist Paul MacLean posited his influential theory of the "triune brain," a neuroanatomical model dividing the brain into three distinct layers: the "reptilian" brainstem, the "limbic system"—a set of structures in the medial region (the *limbus*, "rim") between the neocortex and the brainstem that were the seat of emotion—and the neocortex, the domain of reason itself (Pogliano 2017, 343; see also Holden 1979, 1066). Not only, then, was thinking foundationally different from feeling, humans were foundationally different from animals by virtue of our exclusive property of a fully developed neocortex. This portrait of different brain structures for different faculties neatly aligned with the modern myth that placed thinking and feeling, mind and body, human and animal on separate planes. Rather than an innovation, it was just part of the infrastructure of the conventional wisdom.

However, MacLean's model has become a *bête noire* of contemporary psychology, firmly clung to by the public in spite of direct contradiction by decades of scientific research. Psychologist Lisa Feldman Barrett calls it a "zombie idea"—a relentless, shambling force kept alive by popular opinion, irrespective of the fact that scientific evidence buried it long ago (Barrett 2019). MacLean's triune brain model can't explain, for instance, how birds, whose brains lack the mammalian neocortex, exhibit "remarkable cognitive abilities that far exceed the more basic, species-specific behaviors involved in reproduction of the species and in maintaining homeostasis" (Butler 2009, 1185). More pressingly, the bulk of neuroanatomical research of recent decades has focused on brain connectivity—networks—rather than discrete regions. Simply put, the brain is too radically interconnected for an anatomical thinking/feeling split to make sense (Duncan and Barrett 2007; Pessoa 2008; Phelps 2006). A. B. Butler concluded that the triune brain hypothesis is really a way of dressing up a *scala naturae*—"a linear ranking of animal taxa, with fish near the bottom, reptiles not much higher, nonhuman mammals higher and ranked in sequence…with humans at the top" (2009, 1190)—in a pseudoscientific costume.

In a chapter of Charles Darwin's *Descent of Man* (1871) titled "Comparison of the Mental Powers of Man and the Lower Animals," Darwin points out that "it is often difficult to distinguish between the power of reason and that of instinct," citing early experiments on animal cognition (1882, 75). For instance, monkeys in one study, when given a series of folded pieces of paper—some of which contained lumps of sugar and the others containing wasps—learned to hold the parcels to their ears to assess the possibility of threat (Darwin 1882, 78). In other words, they used a synthesis of learning and feeling to better navigate their situation. Darwin also asserts that "All animals feel *Wonder*, and many exhibit *Curiosity*," suggesting continuity between the intellectual emotions of humans and those of animals (1882, 71; emphasis original).

American psychologist William James moves in a similar direction. He goes further, though, in specifying not only that thinking and feeling are in a dynamic relation with each other, but that the act of knowledge production yields emotional excitement. "Certain sequences of ideas charm us as much as others tire us," he writes, adding, "It is a real intellectual delight to get a problem solved, and a real intellectual torment to have to leave it unfinished" (James 1884, 189). This leads James into a complex reflection on the development of our intellectual habits. He speculates that we discipline children to squeeze their physical emotions into mental conduits, producing hyper-sophisticated, specialized palates of taste that are particularly advanced among artists and scientists (199). As he explains, "the 'marvels' of Science, about which so much edifying popular literature is written, are apt to be 'caviare' [sic] to the men in the laboratories" (203). The cultivation of our intellectual aptitudes is also a cultivation of our emotional habitus. The secular scientific body is marked by a "taste" for scientific work, with the advance of knowledge itself becoming a source of excitement.

A century later, neuroscientist Antonio Damasio began his career by, in part, mapping the flaws in MacLean's topography of brain structures and their correlating functions (Damasio and van Hoesen 1983). Damasio's subsequent research took square aim at the thinking/feeling divide both in science and popular understanding. In his 1994 pop science book *Descartes' Error*, he insisted that "nature appears to have built the apparatus of rationality not just on top of the apparatus of biological regulation, but also *from* it and *with* it. The mechanisms for behavior beyond drives and instincts use," he added, "both the upstairs and the downstairs: the neocortex becomes engaged *along with* the older brain core, and rationality results from their concerted activity" (Damasio 1994, 128; emphasis original). Not only that, Damasio, like James, describes how science as a bodily act is something we can cultivate a taste for, what he calls a "passion for reason," a "drive that originates in the brain core, permeates other levels of the nervous system, and emerges as either feelings or nonconscious biases to guide decision-making" (1994, 245).

Even MacLean's canonical triune brain image is more fluid than it looks at first glance—maybe more so than MacLean realizes. The eighteenth-century philosopher David Hume wrote that our sense of whether a piece of information floating in our mind's eye is true is derived from feeling. "If I see a billiard-ball moving towards another," Hume argued, "I can easily conceive it to stop upon contact. This conception implies no contradiction; but still it *feels* very different from that conception, by which I represent to myself the impulse, and the communication of motion from one ball to another" (2007, 35; emphasis added). In a 1979 interview, MacLean admitted that "as far as we have been able to tell to this day—and this is probably the most fundamental thing we have learned or ever will learn about the nervous system—the only part of the brain that can tell us what we perceive to be real things is the limbic brain" (Holden 1979, 1068). He later wrote that "without a co-functioning limbic system, the neocortex lacks not only the requisite neural substrate for a sense of self, of reality, and the memory of ongoing experience, but also a feeling of conviction as to what is true or false" (MacLean 1990, 578).

On the other side of the equation, he also readily observes human cognition shot through with affective elements: "The young people flocking to Woodstock reminded MacLean of pigeons whose homing instinct had gone awry. And committee meetings…are hardly different to him from the struttings and preenings and status displays that go on in a monkey colony" (Holden 1979, 1067). As science writer Constance Holden coyly noted, "it is tempting to look for the reptile embedded in human behavior, and MacLean is always ready to yield to temptation" (1979, 1067). MacLean's own devout reiteration of the modernist consensus of a thinking/feeling binary is itself an affectively determined belief. Even the classic divided-brain framework, it seems, ultimately lapses into an understanding of knowledge as fundamentally emotional.

It's worth noting here that nonhuman animals, too, take pleasure in learning about the world. The philosopher of science Michael Polanyi, like James, affirms that science is defined by our "intellectual passions" (1962, 141). "The affirmation of a great scientific theory," he insists, "is in part an expression of delight" (141). But the pleasure of scientific knowledge-making in humans is paralleled by the fascination animals have with learning about and exploring their worlds. "These intellectual tastes of the animal prefigure," he writes, "the joys of discovery which our articulate powers can attain for man" (141). Although Polanyi sees the intellectual passions that define human scientific activity—the caviar of the laboratory—as more advanced than these animal engagements with the world, he nonetheless models a form of multispecies knowing, a shared modality of deriving joy from the world through learning about it. The common property that defines human and animal experience (in contradistinction to other domains of life) is this unique foundation in feeling.

Laboratory Violence and the Fiction of "Shared Suffering"

Sipping wine at a conference reception, a former animal researcher told me she had walked away from her field because she could no longer perform depression experiments on laboratory rats. In the original formulation of this experiment, a depression-like state is induced in animals by dropping them into a beaker of water and forcing them to paddle to stay afloat. They are removed just before drowning, but then the process is repeated. The animal is left physically intact but emotionally destroyed. The original inventors of the method wrote that it evoked "a state of despair in which the rat has learned that escape is impossible and resigns itself to the experimental conditions" (Porsolt, Le Pichon and Jalfre 1977, 730–32). It's a razor-sharp mechanism of psychological torture, expressly designed to manufacture a laboratory specimen of depression for use as a research tool. The former researcher told me she couldn't do the work anymore. It kept her up at night. (Sleeplessness, it seems, is an occupational hazard of contact with animal experimentation[8]). Then she added something strange. Her eyes widened and her face tensed up. "But sometimes I wonder," she said, "what would have happened if I'd been able to finish my research. *I still want to know*."

One of the most influential terms circulated in recent conversations about animal testing in the humanities is Donna Haraway's notion of "sharing suffering" as a form of multispecies knowing (Haraway 2008, 69). Haraway's focus is on pushing back on the simplistic framing that sees animals who live and die in laboratories as inert objects that are purely passive as the labor of science is done to them. Building on frameworks from science and technology studies that emphasize the agency (or actancy) of both humans and nonhumans, she proposes a dynamic relationship between lab animals and human lab workers. "Lab staff could not function as researchers," she writes, "if they did not function as caregivers Dogs could not work as models if they did not work as patients" (59). As an ethical prescription, she calls for a new focus, on the part of animal researchers, to the suffering endured by animals. "Human beings' learning to share other animals' pain nonmimetically is," she proposes, "an ethical obligation, a practical problem, and an ontological opening" (84).

Haraway's call for a new paradigm of "sharing suffering" when it comes to animal pain is on the one hand, welcome. But it's also facile. What makes it facile is that it ultimately rests on a carefully crafted fantasy that gently hushes murmurs of pain. As Sharp writes, Haraway's "celebratory assertions at times overshadow the serious realities that pervade lab animals' lives, whose emotions remain elusive, for whom 'sacrifice' or killing is an inescapable endpoint, and whose needs are eclipsed regularly by those of researchers" (2019, 40). Scientists are not sufferers. Nor do they approach their research from a position of affective neutrality, from which empathetic engagement with immiserated laboratory animals would be an easy reach. Instead, scientists begin in a state of excitement about the work they do. Haraway's contribution to contemporary theory is unmatched; but her work is also marked by a powerful tendency to push feeling into the background, muting its significance for any understanding of subjectivity (Schaefer 2015, 35). It's this tendency in her work that makes it possible to evoke a fantasy of scientists as "suffering" rather than as fascinated, excited, and enchanted as they chew their caviar. Haraway herself authorizes this detachment from ethical concern: "Curiosity, not just functional benefit," she insists, "may warrant the risk of 'wicked action'" (Haraway 2008, 70). In Haraway's view, the intellectual pleasure scientists derive from their work is itself a justification for animal pain.

Haraway's framework, I would suggest, is its own form of secular embodiment: a cultivated emotional habitus that disguises its own emotional investment behind a grid of perpetual self-affirmation. In his book *Voracious Science and Vulnerable Animals*, ex-animal researcher John Gluck observes that animal experimenters are made, not found. In order to become a successful animal experimenter, one must, he argues, go through a process of "erosion" (Gluck 2016, 13). "[T]hose attracted to a career involving research on animals," he writes,

> must undergo an emotional and ethical retraining process every bit as important as their scientific training. They must learn to put aside identification with animal pain and suffering and replace it with a passion for advancing scientific knowledge. They must come to believe that they are joining the ranks of a special corps of truth-seekers and improvers of the human condition who may assume their work to be justified on its face, no matter the cost in animal lives.
> (Gluck 2016, 13–14)

In other words, scientists must cultivate a certain kind of secular body—a voracious one, in which some affective priorities (knowledge-making) are systematically elevated just as others (compassionate concern for nonhuman animals) are eradicated.

I intend to survey two approaches to these practices of self-cultivation—what we could call secular spiritual exercises—one etic and one emic. The etic approach is Lesley A. Sharp's ethnographic study of moral reasoning among animal experimenters; the emic approach is Gluck's own memoir of his exit from a career in lab animal research. Before I get to this in the next section, though, I want to talk about animal experimentation over the last half-century. Lind af Hageby and Schartau, the feminists who shone a light on the festival atmosphere of the vivisection theater, called animal experimentation a "shambles"—an abattoir (Lind af Hageby and Schartau 1903). This situation of excruciating abuse of animals in experimental contexts continued throughout much of the twentieth century. As late as the 1970s, Gershon Berkson's experiments on wild rhesus macaques involved a procedure of capturing and surgically blinding infant macaques. Their reactions were then monitored, as well as the reactions of other (non-mutilated) macaques toward them. The young macaques, "groping for roots as they went, had great difficulty finding their way through the mangrove," though Berkson's own research also showed that "Their mothers stopped often and waited for them, and the group as a whole was extra vigilant and alarmed if the blind infants were approached by human observers" (de Waal 1997, 226–27, fn 12; see also Berkson 1973).

In the same decade, John Kalaska and Bruce Pomeranz at the University of Toronto began a series of experiments on kittens to test neurological development after their forepaws had been denervated. In the methodology section of their 1979 article on paw denervation, they describe how:

> Each nerve was exposed through a small incision, doubly crushed about 2 mm apart with watch-makers forceps, and tightly ligatured at each crush with 6-O silk. The nerves were transected 2 mm distal to the ligatures and a length of the distal stump was excised. The proximal stumps were each deflected into a tear made in the belly of a nearby muscle. Incisions are sutured, and the wounds healed without scarring.
>
> (Kalaska and Pomeranz 1979, 619)

In essence, the kittens were tranquilized, then, on waking, found they had permanently lost the use of their forelimbs. I first learned about this experiment as an undergraduate in an introductory biopsychology lecture; our professor laughed as he described the bewilderment of the cats waking up to discover they had been crippled.

Meanwhile, MacLean was defining the perimeter of the brain's neuroanatomical structures through lesion experiments on a variety of animal species like squirrel monkeys, *Anolis* lizards, and hamsters. His primary technique was to saw open their skulls and destroy parts of their brains with tools, though in some instances, vacuums were used to pull chunks of the brains of mice out through their ears, shattering their eardrums in the process (Holden 1979, 1068). The mutilated creatures were then reintroduced to their cages and their warped, tottering behaviors studied to see what had changed. By cutting open the skulls and damaging the temporal lobes of monkeys, he boasted, he could get them to "eat nuts and bolts and feces indiscriminately because the part of the brain that says 'this is it' (food) is out of order" (Holden 1979, 1068). This casual degradation and harm of animals—tricking them into eating pieces of metal and shit—was enthusiastically offered as evidence of the ontological supremacy of human beings.

Animal experimenters claim that the excesses of pre-1980s science have been pruned back, that animal testing now is carried out sparsely and with far less gratuitous violence. The problem with this claim is that it rests on a careful curation of what kinds of experiments are discussed: experiments that trigger gut revulsion are downplayed, while animal misery continues to be mass-produced in subtler forms. The Forced Swim Test mentioned above, for instance, is still in wide use in both teaching and research contexts. In its original formula, this experiment was used to create a state of severe depression in animals, with an eye to using them as models for testing drugs that could help develop a pharmaceutical solution for clinical depression in humans. In other words, the horizon of sheer hopelessness experienced by depressed people is foisted on laboratory animals by systematically terrorizing them to the extent that they lose all hope of feeling safety or peace or self-determination.[9]

These animals don't suffer in an obviously gruesome way. But the only way they could be useful for human depression research is if their emotional parameters were *basically like ours*, which means they must feel the suffocating despair of depression just as we do. This careful selection of *invisible* suffering is analogous to how solitary confinement has emerged as an "acceptable" torture technique in contemporary U.S. prisons (Guenther 2013; Schaefer 2015, 92–93). Forced Swim Tests don't trigger a twinge of outrage on the part of outside observers. They are an unseeable torture—but a torture all the same. Experiments like this are even replicated in undergraduate teaching contexts, throwing into question the premise that the findings of these experiments are only used to advance essential research.[10]

In another twisted irony, violent animal experiments often reveal exactly these tendencies of our fellow mammals to exhibit care and concern for one another. Berkson's finding in his experiment with blinding baby monkeys, for instance, revealed that "the blind babies were never left completely alone," as other monkeys—including unrelated adults—remained with them at all times to assist them (de Waal 1997, 227, fn 12). A more recent experiment took newborn macaque monkeys away from their mothers and replaced the infants with various toys. The yield of this casually horrific procedure? Experimental confirmation that the mothers would form preferential "attachment" to soft toys rather than hard toys (Livingstone 2022). As with solitary confinement, these animals may well have their physical needs met, but the *invisible* violence of isolation or the removal of one's children—a pain that is only made possible by our nature as mammals, the status as intrinsically social beings we share with many other creatures—is an endlessly deep wound (Ferdowsian, et al. 2022).

The particulars of individual studies aside, one of the most brutal—and under-attended—features of contemporary animal experimentation is the "procedural harm" inherent in the mass production of laboratory animals requires extensive breeding, transportation, and housing programs (Halteman 2011, 123). At every link of this chain, animals are caged, isolated, and separated from their relations and companions. Solitude and confinement are psychically crushing for all mammals (and many non-mammalian species, too). Yet these conditions are blithely reproduced on a monumental scale throughout the animal experimentation industrial complex. It's another source of invisible torment; no reflex revulsion is triggered, but the damage is no less real.

And beyond all of these troubling examples of "less-violent" animal studies, there is still a constant carousel of extremely violent, invasive animal experiments taking place today behind closed doors. It's just that animal experimenters who work on, for instance, brain lesioning experiments, have learned not to circulate images and videos of their research subjects running around with their skulls half-sawed off and pieces of metal sticking out of their exposed brains, all the better to minimize the possible twinge of sympathy that might shake public confidence in the benignity (or necessity) of animal testing. Elon Musk's Neuralink corporation is believed to have killed 1,500 animals between 2018 and 2022, with a high-pressure culture of research results leading to reports of gruesomely botched experiments. Findings from a complaint against Neuralink stemming from the corporation's collaboration with the University of California, Davis, documented these incidents:

> One rhesus macaque monkey's nausea was "so severe that the animal vomited and had open sores in her esophagus before she was finally killed."...Surgeons used an unapproved adhesive to fill open spaces in an animal's skull, created from implanting the Neuralink device, "which then caused the animal to suffer greatly due to brain hemorrhaging."...[There were also] instances of animals suffering from chronic infections, like staph infections where the implant was in their head. There were animals pulling out their hair and self-mutilating, which are signs of really poor psychological health in laboratory animals and are very common in rhesus macaques' and other primates.
>
> (Torrella 2022)

Images of what goes on behind closed doors at Neuralink have never been released. Musk, a master of curating the visual ecology around his enterprises in order to build media interest, has kept images of the animal test subjects in his facilities tightly under wraps, only releasing pictures of (physically) intact animals in carefully choreographed media events.

Fashioning the Secular Selves of Animal Experimenters

My interest here is in *how* animal experimentation produces a set of affective disciplines that create a kind of secular body—a body formed by a masked desire for knowledge and an indifference to animal pain. Just as we can inculcate depression in rats, a certain genre of scientific practice nurses a bullish confidence in one's own intellectual mission that brooks no challenge to its prerogative to commit acts of violence in the service of the exhilarating quest for scientific knowledge. Rather than thinking of this through the tenuous framework of "shared suffering"—as Haraway would have it—it is much more effective to imagine it as a kind of scientific habitus or technology of the self. It is an emotional habitus that is designed to format how scientific bodies feel (or not) about the work they do with animals. In all these instances, what's interesting is how the process of producing secular bodies may represent itself as an evacuation of "emotion" from the process of reasoning. But it is, I would argue, actually about playing up some emotional priorities—like the exhilaration that comes from scientific discovery—while dialing down others—like compassionate response to suffering creatures.

With what remains of this chapter, I want to explore two perspectives on animal experimentation—one etic and one emic—that draw exactly the links between affect, scientific practice, and the fashioning of secular selves that I'm interested in here (Gluck 2016, 14; Sharp 2019, 5). As Sharp writes, "the making of the moral self in science is often inescapably 'sentimental'" (Sharp 2019, 143).

Sharp's overriding interest in her book *Animal Ethos* is the complexity of the moral practices surrounding interactions with laboratory animals.[11] She's fascinated by the disjunct between what she calls "ethics" (a top-down package of formalized procedures conveyed through mandatory codes of practice) and "morality" (a bottom-up process of reflection and response that diverges from the prescribed ethical code, often leading to the impromptu implementation of higher standards of care for captive animals) (Sharp 2019, 8). My interest, however, lies in one of the less explicit dimensions of her project, specifically, her subtle diagram of how practices of ethical self-formation produce secular bodies able to render themselves immune to the moral hazard they find themselves immersed in. And especially how we can understand these training protocols as cultivating emotional habitus.

Sharp corroborates the understanding of science's emotional aspect that I have explored here. Animal experimenters, she observes, find their intellectual work deeply enjoyable and exciting. For instance, she reports that when she asks researchers to recall a "favorite animal," they almost invariably gush not so much about close bonds they formed with captive animals, but rather about the quality of scholarship they were able to produce on their backs; they "nearly always offer sentimental narratives of creatures whom they regard as exceptional 'workers' who 'produce reliable' or 'strong' data" (Sharp 2019, 138).

The pleasure of knowledge-production provides a powerful ballast that makes it possible—sometimes even easy—to blot out the moral injury of lab work. Sharp never lays out an inventory of the disciplinary practices animal researchers use to shape their responses to their circumstances, but they're there. And many of these seem explicitly designed to minimize the possibility of moral encounters with animals—zeroing out the possibility of forming emotional bonds with them—while dialing up the excitement of making knowledge from them. "At the onset of their careers," she writes, "lab researchers must master a two-pronged approach to animal experiments: as they master specialized skills, they must also learn to do so with emotional detachment. Not unlike novice clinicians, strong sympathetic responses of queasiness, sorrow, guilt, or fear signal that one lacks the emotional backbone to work productively in a lab environment" (Sharp 2019, 56).

For example, Sharp notes that "coding systems" in labs "are designed not only to assist in the quantification of animals as data, but as a deliberate strategy to depersonalize human-animal interaction to prevent emotional attachment and favoritism" (2019, 173). And by the same token, "many labs maintain labor hierarchies that keep contact hours specifically between researchers and animals to a minimum" (173). This is why animal caretakers are forbidden from taking pictures of the captive animals they interact with every day (181). This also seems to be the heart of the long-simmering controversy about giving captive animals names, which is formally disdained (but informally permitted in some contexts) (175): "sentimental naming practices are discouraged in large part to assist researchers in shielding themselves from the emotional responses that death's inevitability can generate" (126).

Many of the self-fashioning practices Sharp describes are what I would call discursive practices; that is, they discipline language to shape the self's emotional habitus. Sharp's term of art for this is a "sanctioned lexicon"—an enforced way of talking about something (2019, 12). Interestingly, Sharp herself is skeptical of the claim that "semantic policing" can "alter thinking" (236), but I think the situation is more complex than that. My suggestion would be that regulating what words are used—how other creatures or one's own experiences are described, for instance—will shape one's pattern of future responses. This shaping is not total (no project of self-fashioning is, as we'll see when we turn to Gluck, below). There is always a moral remainder that can't quite be scrubbed away. As Sharp writes, "the seemingly unemotional, rational tone accompanying such demonstrations and descriptions [of the killing of laboratory animals] may belie deeper, private ambivalence, questioning, or conflict" (2019, 107). But the influence of words on self-formation shouldn't be understated, either.

One of the most prominent ways animal researchers train themselves to be a certain kind of compassionless secular subject is by tightly regulating the use of the word "kill" with respect to animals in their care. "One never speaks of 'killing' animals," Sharp notes. "Instead, they are 'culled,' 'sacrificed,' 'euthanized,' or 'terminated,'" often with a context-specific connotation: "one 'culls' a 'batch' of newborns, 'sacrifices' or 'euthanizes' a research subject, and 'terminates' an undervalued animal" (2019, 12). "Sacrifice," she observes, has risen to the top as the preferred euphemism among lab animal researchers (131). As she points out, this rephrasing attempts to implant a deep sense of value in the act of killing laboratory animals (132). But even this word ends up supplanted by the even more neutral "sack." "Sack," as one of Sharp's interviewees explains, almost totally erases the fact that an animal's death is involved, musing how "In football you 'sack a player,'—you get rid of it and knock it out of the game" (58). Even after they are killed, animal researchers will speak of the dead bodies of animals as "samples" or "data" rather than the veterinarian's parlance of "corpses" or "remains" (127). The intense emotional tincture of words like "kill" and even "sacrifice" is carefully swept under the rug, I would suggest, as a device to form the animal experimenter's emotional habitus.

Sharp notes that animal experimenters like to tell her (and themselves) that animals in labs live "better" lives than animals in the wild (2019, 141). This repetition is, it seems to me, also a ritual of cultivation of a secular emotional habitus. The constant reaffirmation of the refrain helps to grind down the sharp edges of the horrific moral decision-making that is the quotidian reality of laboratory life. But this claim is directly undermined by another observation she makes: the contrasting relationships animal experimenters have with pets. She notes that most animal experimenters have an inventory of species they will *never* work with (64). These *non grata* species almost always correspond to species they have close relationships with outside their lab. One researcher tells a conference, while showing off a picture of his dogs, "I could never have dogs in my lab because I love these two mutts as much as my own children," and in a private interview another tells Sharp about his Belgian Shepherd, noting he would "never work with dogs because it's too close to home" (65). "Rarely," she notes, "did I find someone who would voluntarily work in the lab with a species they had at home" (66). These contradictions—alongside the pervasive epidemic of compassion fatigue—seem to undercut the constant self-reassurance that animals in labs are living well.

Sometimes, the procedure for self-fashioning secular bodies is not to extinguish the moral response, but to build an easy path to circumvent it. One of Sharp's findings in the labs is a complex set of strategies designed to blunt the horror of regular killings. The main strategy here is distancing: when an experimenter or technician has worked closely with an animal for long enough to have formed a bond with the creature, they are absented from the work of killing them. This is sometimes about a simple division of labor: someone else—someone personally unfamiliar with the animal about to die—comes in and carries out the kill (Sharp 2019, 138). But it often goes even further than this. Those who have had close contact with certain animals are often granted paid leave to stay home on days when their closest companions are to be killed (186).

New institutional efforts to manage "compassion fatigue" among animal workers run along a similar track of mitigating the damage of morally toxic conditions (and thereby making it easier to work longer in them) without changing the underlying situation. For instance, one lab "places heart-shaped stickers or notes on the enclosures of animals slated to be euthanized" (Grimm 2023). The speedbump of moral injury isn't so much smoothed over in these practices as it is, as far as possible, skirted, allowing the overall violence of lab animal experimentation to remain intact. One of Sharp's recommendations in the book's conclusion is for a more robust ethical training for animal experimenters, focused not just on academic technique but on administration of care—in other words, for a rearranging of the foundational emotional habitus of this kind of secular body (Sharp 2019, 232).

A subplot of Sharp's book is the pronounced stratification between researchers and technicians. At the level of background, disposition, and priorities, researchers (usually academics who design and manage experiments then communicate the results to others) and animal technicians (often coming from non-scientific backgrounds, whose responsibilities entail constant interaction with captive animals) have divergent backgrounds and outlooks. One of the most striking differences she finds is that the former group—researchers—almost invariably had very little friendly contact with animals as children. "Roughly eighty percent of the researchers I interviewed," she estimates, "did not have pets when they were growing up, so their first close-up encounters with animals (most notably frogs, rabbits, cats, and fetal pigs) were during dissection assignments in biology labs in secondary school and, later, with live animal subjects in college and beyond" (Sharp 2019, 65).

The animal technicians, by contrast, were far more likely to have had pets growing up, and it is these workers who often do extensive research and lobbying to advocate for better welfare and enrichment standards for the animals in their care (Sharp 2019, 170). They also have much higher rates of burnout, often stemming from "compassion fatigue," which Sharp characterizes as "the emotional trauma entailed in regularly euthanizing animals (and not necessarily those one has come to know well), working in an environment where one is undervalued, and assisting those who regard animals merely as 'resources' or 'data points'" (2019, 158). Sharp doesn't say it, but it seems like a reasonable inference that what we're looking at here are very different patterns of self-formation dictating who lands in what station in the animal experimentation machine. A general indifference—or at least a non-socialization—to animals seems to be a *de facto* prerequisite for a certain comfort level with animal experimentation on the part of academic overseers. And not only that, the vast discrepancy in the everyday encounters of researchers and animal technicians leads to widely divergent experiences of the work itself. Compassion fatigue afflicts the on-the-ground workers—and is actually exacerbated by the aloofness and indifference of those who only skim the surface of what actually goes on in the lab (Sharp 2019, 170–72).

It is no diminution of Sharp's achievement to say that her perspective on her subjects is highly filtered. In her acknowledgments, she notes that she believes the labs that gave her access to their interiors—both physical and psychic—were probably already operating at high standards of welfare and enrichment (Sharp 2019, xii). She also documents a kind of *omertà*—a code of silence—among animal experimenters, effectively amounting to a complex system of barriers that prevent outsiders from seeing or knowing what they really do (101, 162). This extends to Sharp's suspicion, at times, that a space has been rearranged for her benefit to present a certain image of an animal experimenter's disposition (170). This raises the concern that what Sharp is seeing and reporting is accurate from her eyes, but also represents a highly selective view, the tip of the iceberg of what aspects of the self-cultivation required to function in animal laboratory contexts are in use.

This is why it's important to supplement Sharp's perspective with an emic account. Gluck's *Voracious Science and Vulnerable Animals* (2016) offers a very different perspective—that of an insider with an unfiltered and unvarnished view into the world of animal experimentation, albeit one a few decades out of date. In it, he reviews, with measured horror, his own career, first as a graduate student of primatologist Harry Harlow, then an animal researcher in his own right in the 1970s, then a clinical psychologist advocating for sweeping reforms in animal experiment standards in the 1980s. He talks about the training he went through as a future animal researcher, the product of a subculture awash in toxic masculinity, designed to grind down compassionate responses. He was trained to kill rats by hurling them at brick walls, for instance (Gluck 2016, 61), and graduate students were instructed to press iron rods into the ear canals of rats until their eardrums blew (34). He calls it a process of moral "erosion," but we can just as well think of it as a systematic inculcation of habits that, in Michel Foucault's words, allow budding animal experimenters to "transform themselves in order to attain a certain state of happiness, purity, wisdom, perfection, or immortality" (Foucault 1988, 18). "By the time I had finished my undergraduate education and started graduate school," Gluck writes, "my professors—and the overall research context into which I threw myself—had exorcised my sentimental concern for animals' welfare and constructed for me a new belief system in which there was really no such thing as the animals' perspective" (2016, 14).

Gluck's career is marked by profoundly disturbing encounters with the horrors of laboratory life. But the accumulation of these incidents will eventually stir his rebellion against the ethical formation he himself has undergone. He finds a rhesus monkey who has been abandoned in a cage in the basement of a laboratory building for years, kept alive by the vet techs but unable to move, his arms a mass of sores and scars from self-mutilation. Rather than leaving him to suffer as his colleagues have done, he euthanizes him (Gluck 2016, 112). He hears that a monkey kept chained up outside his local hardware store in New Mexico had been put into a fight with a dog and had their throat ripped out (137). One of Gluck's undergraduates resigns from his lab, telling him "I can't shock these monkeys anymore" (144). A vet tech goes rogue, shutting down an experiment using monkeys outfitted with poorly manufactured and painful self-injection devices. Gluck meets a dog who has been blinded in an experiment gone wrong. Even under the best of conditions, he realizes, laboratory animals "led stunted, unfulfilled lives of boredom punctuated by episodes of fear and pain" (xii). Finally, he exits animal research altogether, retraining as a clinician. As part of the process, he is forced to confront the emotional trajectory of his life—the debts he has incurred from a career in animal science. And so he writes his book.

Gluck notes that Augustine observed in humans a drive towards knowledge—the *libido sciendi*—that has "an essential carnal quality, generating a strong desire that can escape the bounds of reason" (2016, 4). This is what Gluck means by *voracious science*. It's a desire to know that *wipes out* other cares—for family, for health, for other people, for nonhuman animals.[12] The secular body is built by carefully rearranging the animal experimenter's emotional habitus. Certain feelings are elevated while others are steadily stilled. The desire for knowledge—the exhilaration of doing science—is cultivated to such an extent that it eclipses moral responsiveness. Seen through this prism, the experimental protocols devised by animal experimenters are not just individuated actions; they are also modes by which the indifference toward animal pain is cultivated and consolidated. It's a closed circuit of steadily advancing moral erosion.

Why, then, as Gluck claims, are critics of animal testing hated even more than creationists by scientists who rely on animal experiments for their careers (2016, 262)? The obvious answer is that their livelihood is at risk; the collapse of the animal testing economy would put them out of a job. Others would emphasize the pursuit of medical benefits, the lives undoubtedly saved and extended through the visitation of violence on nonhuman animals. But I think these answers capture, at most, a fraction of the story. What we're seeing is not just a counterattack to a challenge to one's *actions*. It's an immune response, stemming from an emotional habitus—a deeply engrained feature of a particular kind of secular body—a deep bitterness toward the very prospect of having the thrilling, ever-expanding horizons of knowledge cut short. The deep, vicious resentment is about the removal not just of a paycheck—or even of the prospect of better medicine—but of a source of pleasure—of deep joy and excitement in the adventure of scientific discovery, an adventure carved out in blood.

Conclusion: How Much Pain Is Too Much Pain?

I once worked with a student who had experienced significant health challenges in her first year of university. She spent a long time in hospital. When I first met her, she was in her third year—still struggling, but slowly and surely putting her health back together.

As her fourth year dawned, she came to my office for a start-of-term check-in. It was autumn. She was wearing several layers of clothing—scarves, a hoodie, an open light jacket. Up to this moment when this student had spun into my office she was often on the brink of panic. Today was different. She had a tranquility about her—a focus and determination I hadn't seen before. She sat down across from me. Before she said anything, a tiny brown face peeked out from the fold of her hood, followed by a smooth body that quietly scampered across her shoulder and slipped into her scarf. A second little face looked out from her pocket and sniffed the room. She said they'd be frightened if I got too close, but the creature in her hand seemed delighted to have her gently brush him with her finger. She scooped one of the creatures up and held him out to me in her cupped hands. Her rats. Her friends.

Rat is a word of revulsion. It's a twisted insult, the word we pin to those we most loathe. A rat is a traitor, a dirty thing. Rats don't live; they infest. Vectors of disease, blazons of grime and dire poverty. To be *ratty* is to be grimy, tattered, revolting. When we say *rat*, we curl our lip. It's so easy to despise rats. They're focal points of contempt. They could never be loved.

The rats in my student's pockets were healers. Even well-cared-for domestic rats don't live long—a year or two at most. As her rats passed on, she replaced them with new ones, always in a pair. When I think about the rats and mice left to paddle for their lives in smooth jars, manufacturing depression through the repeated inculcation of hopelessness, or massacred by the millions after single-use experiments, furry bodies flung in garbage bags and then incinerated, so easy to forget, I think about these creatures. They were creatures with their own horizons of joy and pain. They were companions and friends.

What does it take to wash out that horizon of joy and pain? How do we make it impossible to love rats? I would argue that this is one of the forms of the secular body. Thinking of science as pleasurable—as a thing we do with our animal bodies, rather than with transcendent reason—prompts us to keep a particular set of concerns in front of us: concerns about how the pleasure of knowledge-making—in both the sciences and the humanities—can become so alluring that it overrides all care, sometimes translating into violence. Gluck notes that animal testers hate animal rights activists—who reject a single domain of experimental practice—even more than they hate creationists—who reject the whole apparatus of the modern life sciences. Why is it that the scope of disagreement fails to correlate to the degree of animosity expressed? It is, I argue, because these reactions are not isolated responses to particular situations. They coalesce into a durable set of embodied habits and dispositions—an emotional habitus—that reflexively confronts any challenge to its constitution as an existential threat capable of draining joy from the world.

Joseph LeDoux, whose work has also been built on brain lesioning experiments with rats, calls the brain a "hot dog in a bun," and invites us to imagine pulling it apart in our hands (1996, 87). It's a startling image, both profoundly playful and extravagantly violent, a giddy fantasy of demolishing the most delicate device for keeping a creature alive. And then we recall: *LeDoux's early model of brain functioning is very likely to be wrong.*[13] As, too, was MacClean's. How many animals spent their lives isolated in cages—then had their brains surgically wrecked—to create paradigms that lasted barely a decade? This, I would suggest, is a perfect specimen of the kind of secular body created by the practice of animal experimentation. The suffragettes who infiltrated the UCL vivisection theater 100 years ago described the scientist on stage as a "priest of vivisection," casually smoking a pipe and cracking wise to his students even as his white surgical garments were smeared up and down with blood and an animal writhed on the table in front of him. This is the portrait of Man, in Wynter's sense—aloof from the severe violence required to produce his status, cloaked in a pretense of neutrality, but totally defined by his own set of affective determinations—exhilaration, fascination, fun.

One of the challenges in trying to police the tendency to waste and abuse in the animal experimentation industry is the simple fact that—much like the vertical integration of agricultural oversight bodies with the factory farm industry—the monitors and the actors come from the same culture, with the same set of dispositions and priorities. Not only that, much of the violence of wet labs is now outsourced, with scientists designing experiments in their laboratories and harvesting data from animal procedures that they themselves never see. And as scientific research intersects with the motives of for-profit entities (Big Pharma not least among them), this combination spirals up into a profoundly devastating hidden empire of animal misery, a sprawling network of joyless, windowless prisons.

How much pain is too much pain? How much misery can we justify creating to reduce pain somewhere else? Secular formations don't ask this question often enough. It's taken as given that the ends of the secular justify all kinds of violence. Asad points out that secularism doesn't even see the violence that underpins the maintenance of certain liberal orders as "violence" per se (2003, 11). Only religions can be violent, precisely because they are the domain of the *non*rational. *Secular* violence, in Asad's depiction, is more like cutting the grass, or exterminating rats. An invisible necessity.

As I've said, this chapter isn't designed to make an ethical intervention. Or maybe it is. But it does offer an account of how certain kinds of ethical problems crop up—and how those problems are rendered invisible to those enmeshed in them. They're obscured by desire. When we want something badly enough, our ethical reservations become easier to bury. Scheer, Fadil and Johansen note that one of the features of secular embodiment is that "an embodied, affective response to an offence to the secular sensibility argues for the truth of the secular and shuts down any deliberation over it" (2019, 6). Secular bodies build their own immune system, refusing challenges to their affective determination by defining themselves as essentially unaffected. But the secular body fantasized by Darwin—whisking away moral pain and frustration with the firm and bloodless hand of scientific necessity—is a sham. The voracious secular body is defined *not* by bland commitment to truth, but by an affectively rich fascination.

Voracious secularism isn't inhuman. It's all too human. Science doesn't create circumstances in which some bodies matter and others don't because it is bloodless. Like all that we do, science distributes value and meaning to some bodies and violently rips it from others through the operation of a dynamic interplay of feeling, including our intellectual feelings. As we face escalating crises that are centrally related to the status of shared knowledge—climate change, pandemics, conspiracism, and misinformation—the moral urgency of science is sharper than ever before. Precisely for this reason, science must present itself as a moral force, rather than as guided solely by a compassionless commitment to knowledge-production itself.

Notes

1 This chapter was previously published as Schaefer, Donovan O. "Voracious Secularism: Emotional Habitus and the Desire for Knowledge in Animal Experimentation." *Religion* 54.4: 700–723. DOI: 10.1080/0048721X.2023.2258710.

2 Lesley A. Sharp offers an extensive discussion not only of the controversy surrounding this monument (2019, 24–5), but a fascinating window into other practices of commemoration of dead laboratory animals commissioned by animal experimenters themselves (2019, 181–85).

3 Barbara and Karen Fields reframe "scientific racism" as "bioracism" to better highlight its status as a form of pseudo-knowledge (2014, 4). Although I will make mention of instances of bio-racism, my primary focus in this chapter will be on animal testing as a site of scientific violence. This is not to downplay the atrocities informed by bio-racism, let alone other zones of violence driven by science, like weapons production or eugenics. Whereas these atrocities are now discussed and universally condemned in a rearview mirror, animal testing remains a live topic and a live practice. But that's not to say we can't see the immune response of secular bodies as they face the invalidation of their intellectual pleasures elsewhere.

4 See, for instance: Agrama 2012; Asad 1993, 2003; Berger 1969; Calhoun, et al. 2011; Engelke 2013, 2014; Farman 2020; Fernando 2014; Gourgouris 2013; Hirschkind 2011; Lee 2015; Levine 2011; Mahmood 2016; Ogden 2018; Pelkmans 2017; Scott 2018; Smolkin 2018.

5 See, for instance: Blankholm 2018, 2022; Cady and Fessenden 2013; Jakobsen and Pellegrini 2008; Josephson-Storm 2017; Kahn and Lloyd 2016; McCrary 2022; Modern 2011; Pellegrini 2009; Schaefer 2022; Scheer, et al. 2019; Sullivan, M. 2020; Sullivan, W. 2005; Taves and Bender 2012; Thomas 2019; Wenger 2009.

6 See, for instance, Pellegrini 2009 and Scheer, Fadil and Johansen 2019.

7 See, for instance, Colombetti 2014; Maiese 2016; Schaefer 2022; Shah 2018; Thagard 2006.

8 In an 1871 letter, Darwin wrote that vivisection "is justifiable for real investigations on physiology; but not for mere damnable and detestable curiosity," adding that "It is a subject which makes me sick with horror, so I will not say another word about it, else I shall not sleep to-night" (Darwin 1871). A recent article suggested that large numbers of animal experimenters experience "insomnia, chronic physical ailments, zombielike lack of empathy, and, in extreme cases, severe depression, substance abuse, and thoughts of suicide" (Grimm 2023), which is consistent with what one of Lesley Sharp's interviewees (herself an animal experimenter) told her—that some of the studies designed by her

colleagues cause her to "lose sleep at night" (Sharp 2019, 67). While writing this chapter, I lost a lot of sleep, too.

9 Contemporary uses of the Forced Swim Test often rely on a variation in trials for antidepressant medications. In this version, the length of time the animals swim before allowing themselves is measured. Those that give up hope more quickly are considered to be more depressed, allowing a quantitative measure of the efficacy of antidepressants.

10 I am personally aware of at least one institution where the Forced Swim Test is part of undergraduate instruction.

11 One thing that becomes clear when reading Sharp's book is that standards of animal welfare vary a lot, and different lab cultures, experimental designs, and oversight protocols lend themselves to very different patterns of overall treatment. This is a caveat, then, that referring to animal experimentation contexts as a singularity papers over some important differences.

12 The history of racialized and vulnerable people abused and exploited for science's benefit—from Sarah Baartman, an African woman displayed in scientific and entertainment contexts in the nineteenth century, to the Tuskegee University syphilis experiments, which ran right into the 1970s—also illustrates voracious science.

13 LeDoux's early work emphasizes a split between "high road" and "low road" thinking, with the latter bypassing cognitive operations (LeDoux 1996). His more recent publications endorse an understanding of emotion that emphasizes the interplay between feeling and cognition (LeDoux 2014).

Bibliography

Agrama, Hussein Ali. 2012. *Questioning Secularism: Islam, Sovereignty, and the Rule of Law in Modern Egypt.* Chicago: University of Chicago Press.

Aristotle. 1933. *The Metaphysics.* Translated by Hugh Tredennick. London: William Heinemann.

Asad, Talal. 1993. *Genealogies of Religion: Discipline and Reasons of Power in Christianity and Islam.* Baltimore: Johns Hopkins University Press.

———. 2003. *Formations of the Secular: Christianity, Islam, Modernity.* Stanford: Stanford University Press.

Barrett, Lisa Feldman. 2019. "Zombie Ideas." *APS Observer*, October. www.psychologicalscience.org/observer/zombie-ideas.

Berger, Peter. 1969. *The Sacred Canopy: Elements of a Sociological Theory of Religion.* New York: Anchor Books.

Berkson, Gershon. 1973. "Social Responses to Abnormal Infant Monkeys." *American Journal of Physical Anthropology* 38: 583–86.

Blankholm, Joseph. 2018. "Secularism and Secular People." *Public Culture* 30, no. 2: 245–68.

———. 2022. *The Secular Paradox: On the Religiosity of the Not Religious.* New York: New York University Press.

Bourdieu, Pierre. 1990. *The Logic of Practice.* Translated by Richard Nice. Stanford: Stanford University Press.

Butler, Ann B. 2009. "Triune Brain Concept: A Comparative Evolutionary Perspective." In: *Encyclopedia of Neuroscience*, 1185–93. Cambridge: Academic Press.

Cady, Linnell E., and Tracy Fessenden, eds. 2013. *Religion, the Secular, and the Politics of Sexual Difference.* New York: Columbia University Press.

Calhoun, Craig, Mark Juergensmeyer, and Jonathan VanAntwerpen. 2011. "Introduction." In *Rethinking Secularism*, edited by Craig Calhoun, Mark Juergensmeyer and Jonathan VanAntwerpen, 3–30. New York: Oxford University Press.

Colombetti, Giovanna. 2014. *The Feeling Body: Affective Science Meets the Enactive Mind*. Cambridge: MIT Press.

Damasio, Antonio R. 1994. *Descartes' Error: Emotion, Reason, and the Human Brain*. New York: Putnam.

Damasio, Antonio R., and G.W. Van Hoesen. 1983. "Emotional Disturbances Associated with Focal Lesions of the Limbic Frontal Lobe." In *Neuropsychology of Human Emotion*, edited by Kenneth M. Heilman and Paul Satz, 85–110. New York: Guilford Press.

Darwin Correspondence Project. 1871. "Letter no. 7612." March 22. www.darwinproject.ac.uk/letter/?docId=letters/DCP-LETT-7612.xml

Darwin, Charles. 1882. *The Descent of Man and Selection in Relation to Sex. 2nd ed.* London: John Murray.

de Waal, Frans B. M. 1997. *Good Natured: The Origins of Right and Wrong in Humans and Other Animals*. Cambridge: Harvard University Press.

Duncan, Seth, and Lisa Feldman Barrett. 2007. "Affect Is a Form of Cognition: A Neurobiological Analysis." *Cogn Emot* 21, no. 6: 1184–211.

Engelke, Matthew. 2013. *God's Agents: Biblical Publicity in Contemporary England*. Berkeley: University of California Press.

———. 2014. "Christianity and the Anthropology of Secular Humanism." *Current Anthropology* 55.S10: S292–S301.

Farman, Abou. 2020. *On Not Dying: Secular Immortality in the Age of Technoscience*. Minneapolis: University of Minnesota Press.

Ferdowsian, Hope, Agustin Fuentes, L. Syd M. Johnson, Barbara J. King, and Jessica Pierce. 2022. "Toward an Anti-Maleficent Research Agenda." *Cambridge Quarterly of Healthcare Ethics* 31: 54–58.

Fernando, Mayanthi. 2014. *The Republic Unsettled: Muslim French and the Contradictions of Secularism*. Durham: Duke University Press.

Ferreira da Silva, Denise. 2007. *Toward a Global Idea of Race*. Minneapolis: University of Minnesota Press.

Fields, Karen E., and Barbara J. Fields. 2014. *Racecraft: The Soul of Inequality in American Life*. London: Verso Books.

Foucault, Michel. "Technologies of the Self." 1988. In *Technologies of the Self*, edited by Luther Martin, Huck Gutman, and Patrick Hutton, 16–49. Amherst: University of Massachusetts Press.

Gluck, John P. 2016. *Voracious Science and Vulnerable Animals: A Primate Scientist's Ethical Journey*. Chicago: University of Chicago Press.

Gould, Deborah. 2009. *Moving Politics: Emotion and ACT UP's Fight against AIDS*. Chicago: University of Chicago Press.

Gourgouris, Stathis. 2013. *Lessons in Secular Criticism*. New York: Fordham University Press.

Grimm, David. 2023. "Suffering in Silence." *Science*, March 9. www.science.org/content/article/suffering-silence-caring-research-animals-can-take-severe-mental-toll.

Gross, Aaron S. 2014. *The Question of the Animal and Religion: Theoretical Stakes, Practical Implications*. New York: Columbia University Press.

Guenther, Lisa. 2013. *Solitary Confinement: Social Death and Its Afterlives*. Minneapolis: University of Minnesota Press.

Halteman, Matthew C. 2011. "Varieties of Harm to Animals in Industrial Farming." *Journal of Animal Ethics* 1, no. 2: 122–31.

Haraway, Donna J. 2008. *When Species Meet.* Minneapolis: University of Minnesota Press.
Hirschkind, Charles. 2011. "Is There a Secular Body?" *Cultural Anthropology* 26, no. 4: 633–47.
Holden, Constance. 1979. "Paul MacLean and the Triune Brain." *Science* 204, no. 4397, June 8: 1066–68.
Hume, David. 2007. *An Enquiry Concerning Human Understanding*, edited by Peter Milligan. Oxford: Oxford University Press.
Jakobsen, Janet R., and Ann Pellegrini. 2008. "Introduction: Times Like These." In *Secularisms*, edited by Janet R. Jakobsen and Ann Pellegrini, 1–35. Durham: Duke University Press.
James, William. 1884. "What Is an Emotion?" *Mind* 9, no. 34: 188–205.
Josephson-Storm, Jason Ā. 2017. *The Myth of Disenchantment: Magic, Modernity, and the Birth of the Human Sciences*. Chicago: Chicago University Press.
Kahn, Jonathon S., and Vincent W. Lloyd, eds. 2016. *Race and Secularism in America*. New York: Columbia University Press.
Kalaska, John, and Bruce Pomeranz. 1979. "Chronic Paw Denervation Causes Age-Dependent Appearance of Novel Responses from Forearm in 'Paw Cortex' of Kittens and Adult Cats." *Journal of Neurophysiology* 42, no. 2: 618–33.
Kant, Immanuel. n.d. "What Is Enlightenment?" Translated by Mary C. Smith. www.columbia.edu/acis/ets/CCREAD/etscc/kant.html
Kean, Hilda. 1995. "'The Smooth Cool Men of Science': The Feminist and Socialist Response to Vivisection." *History Workshop Journal* 40 (Autumn): 16–38.
LeDoux, Joseph. 1996. *The Emotional Brain: The Mysterious Underpinnings of Emotional Life*. New York: Simon and Schuster.
———. February 25, 2014. "Coming to Terms with Fear." *PNAS* 111, no. 8: 2871–78.
Lee, Lois. 2015. *Recognizing the Non-Religious: Reimagining the Secular*. Oxford: Oxford University Press.
Levine, George. 2011. "Introduction." In *The Joy of Secularism: 11 Essays for How We Live Now*, edited by George Levine, 1–23. Princeton: Princeton University Press.
Lind af Hageby, Lizzy, and Leisa Katherine Schartau. 1903. *The Shambles of Science: Extracts from the Diary of Two Students of Physiology, 3rd ed.* London: Ernest Bell.
Livingstone, Margaret S. 2022. "Triggers for Mother Love." *PNAS* 119, no. 39: e2212224119.
Lloyd, Genevieve. 1993. *The Man of Reason: "Male" and "Female" in Western Philosophy, 2nd ed.* Minneapolis: University of Minnesota Press.
MacLean, Paul D. 1990. *The Triune Brain in Evolution: Role in Paleocerebral Functions*. New York: Plenum Press.
Mahmood, Saba. 2009. "Religious Reason and Secular Affect: An Incommensurable Divide?" *Critical Inquiry* 35 (Summer): 836–62.
———. 2016. *Religious Difference in a Secular Age: A Minority Report*. Princeton: Princeton University Press.
Maiese, Michelle. 2016. *Embodied Selves and Divided Minds*. Oxford: Oxford University Press.
McCrary, Charles. 2022. *Sincerely Held: American Secularism and Its Believers*. Chicago: University of Chicago Press.
Modern, John Lardas. 2011. *Secularism in Antebellum America*. Chicago: University of Chicago Press.
Ogden, Emily. 2018. *Credulity: A Cultural History of US Mesmerism*. Chicago: University of Chicago Press.

Pelkmans, Mathijs. 2017. *Fragile Conviction: Changing Ideological Landscapes in Urban Kyrgyzstan*. Ithaca: Cornell University Press.

Pellegrini, Ann. 2009. "Feeling Secular." *Women and Performance: A Journal of Feminist Theory* 19, no. 2: 205–18.

Pessoa, Luiz. 2008. "On the Relationship between Emotion and Cognition." *Nature Reviews Neuroscience* 9: 148–58.

Phelps, Elizabeth A. 2006. "Emotion and Cognition: Insights from Studies of the Human Amygdala." *Annual Review of Psychology* 57: 27–53.

Pogliano, Claudio. 2017. "Lucky Triune Brain: Chronicles of Paul D. MacLean's Neuro-Catchword." *Nuncius* 32: 330–75.

Polanyi, Michael. 1962. *Personal Knowledge: Towards a Post-Critical Philosophy*. London: Routledge.

Porsolt, R.D., M. Le Pichon, and M. Jalfre. 1977. "Depression: A New Animal Model Sensitive to Antidepressant Treatments." *Nature* 266 (April): 730–32.

Schaefer, Donovan O. 2015. *Religious Affects: Animality, Evolution, and Power*. Durham: Duke University Press.

———. 2022. *Wild Experiment: Feeling Science and Secularism after Darwin*. Durham: Duke University Press.

Scheer, Monique, Birgitte Schepelern Johansen, and Nadia Fadil. 2019. "Secular Embodiments: Mapping an Emergent Field." In *Secular Bodies, Affects and Emotions: European Configurations*, edited by Monique Scheer, Nadia Fadil, and Birgitte Schepelern Johansen, 1–14. London: Bloomsbury Academic.

Scheer, Monique. 2012. "Are Emotions a Kind of Practice (And Is That What Makes Them Have a History)? A Bourdieuian Approach to Understanding Emotion." *History and Theory* 51 (May): 193–220.

Scott, Joan Wallach. 2018. *Sex and Secularism*. Princeton: Princeton University Press.

Shah, Esha. 2018. *Who Is the Scientist-Subject? Affective History of the Gene*. New York: Routledge.

Sharp, Lesley A. 2019. *Animal Ethos: The Morality of Human-Animal Encounters in Experimental Lab Science*. Berkeley: University of California Press.

Smolkin, Victoria. 2018. *A Sacred Space Is Never Empty: A History of Soviet Atheism*. Princeton: Princeton University Press.

Sullivan, Marek. 2020. *Secular Assemblages: Affect, Orientalism, and Power*. London: Bloomsbury.

Sullivan, Winnifred Fallers. 2005. *The Impossibility of Religious Freedom*. Princeton: Princeton University Press.

Taves, Ann and Courtney Bender. 2012. "Introduction: Things of Value." In *What Matters? Ethnographies of Value in a Not-So-Secular Age*, edited by Ann Taves and Courtney Bender, 1–33. New York: Columbia University Press.

Taylor, Charles. 2007. *A Secular Age*. Cambridge: Belknap Press of Harvard University Press.

———. 2011. "Western Secularity." In *Rethinking Secularism*, edited by Craig Calhoun, Mark Juergensmeyer, and Jonathan VanAntwerpen, 31–53. New York: Oxford University Press.

Thagard, Paul. 2006. *Hot Thought: Mechanisms and Applications of Emotional Cognition*. Cambridge: MIT Press.

Thomas, Jolyon Baraka. 2019. *Faking Liberties: Religious Freedom in American-Occupied Japan*. Chicago: Chicago University Press.

Torrella, Kenny. 2022. “Neuralink Shows What Happens When You Bring ‘Move Fast and Break Things’ to Animal Research.” *Vox*, December 11. www.vox.com/future-perfect/2022/12/11/23500157/neuralink-animal-testing-elon-musk-usda-probe

Wenger, Tisa. 2009. *We Have a Religion: The 1920s Pueblo Indian Dance Controversy and American Religious Freedom*. Chapel Hill: University of North Carolina Press.

Wynter, Sylvia. 2003. “Unsettling the Coloniality of Being/Power/Truth/Freedom: Towards the Human, After Man, Its Overrepresentation—An Argument.” *CR: The New Centennial Review* 3, no. 3: 257–337.

4 Synchronicities of the Living and the Non-Living

Wahida Khandker

In order to articulate a process philosophical account of "nature" as we might understand it in today's context of accelerated climate change, this chapter will survey a range of biological accounts of the synchronization of living processes with their non-living milieus. My analysis spans the evolutionary synchronicity of species with environment, as the biologist Julian Huxley contemplates in his reflections on the effects of geological shifts and the forces of "denudation" exerted by the weather, by plants, and by animals, such as the chamois, a goat-like animal living in the mountains of Europe. I then consider the evolution of birdsong, in some speculative passages of Charles Hartshorne's work of philosophical ornithology, *Born to Sing*, and I compare it with Henri Bergson's account of the continuum of memory, perception, and action that operates within the malleable framework of virtual objectivity.

I want to highlight, in particular, the ways in which birdsong constructs a dynamic field of significance in unique, auditory landscapes, an account that aligns readily with Bergson's description of adaptive (or dissipative) virtual frameworks. Such human and more-than-human frameworks resonate with our endeavors, as theorists in Critical Animal Studies, to re-conceptualize our relationships with other species in order to find better ways to live with them. The latter part of the chapter considers the forces that arise contrary to the good of individuals or species. This includes not simply the processes of "natural selection" towards extinction if the conditions are unfavorable, but also the reported disruptions of synchrony as a consequence of climate change and a range of "excessive" human activities such as habitat destruction, overfishing, or hunting. In short, the chapter will take the popularized environmentalist ideal of achieving harmony with one's environment, and reconceive it within a process philosophical framework.

Synchronicity

Let us start with a relatively obscure and early piece on mountaineering from 1912, written by the biologist, and key architect of the Modern Synthesis, Julian Huxley. I also want to reference here, for drawing attention to this piece, Emily Herring's 2018 article, "'Great is Darwin and Bergson his poet': Julian Huxley's Other Evolutionary Synthesis" (2018) and her argument that, contrary to popular

DOI: 10.4324/9781032659985-6

historiographical tendencies, Huxley did not advance from an early fascination with the thought of Henri Bergson towards a more "mature" alignment with Darwin. Huxley, Herring continues, "operated a synthesis of both theories" and "when Huxley eventually encountered Bergson, he was already working within a Darwinian framework" (2018, 42).

Why start with this association between Huxley and Bergson? First, Herring's article usefully broaches the common problem that besets philosophers, especially the Process or Continental variety, who engage with the sciences: their work is seen either in conflict with, or inferior to, those sciences. Secondly, the resonance between the more processual passages in the early writings of Huxley (whether it is intentionally Bergsonian in inspiration or not) and his strictly scientific body of work, exemplifies and opens up the possibility of rethinking scientific observations of "nature" along the lines of process philosophy. And why should we wish to rethink nature along process philosophical lines? This speaks to the broader aim of the paper: to contribute to writings in Critical Animal Studies and the environmental humanities that seek to change the very ways in which we speak about our fellow species and the nature and depth of our interactions and interdependencies on this planet. Process philosophy is well placed to offer this change in vocabulary and perspective.

The relation of climber to mountain is an epistemological one: knowing how to solve "problems" (which, I am told, is climber's parlance) echoes the slow evolutionary acclimatization of mountain-dwelling species such as the chamois to their seemingly inhospitable milieus. Reflecting on the "transience of the hills" gives the contemplative climber an appreciation of the multiplicity of processes of weathering that are constantly, if imperceptibly to the human eye, altering our physical landscapes (Huxley 1955/1912, 308). We do not ordinarily classify mountains as living things, but as Huxley observes, they "live":

> not only in their form and individuality, but in the constant cycle of their changeableness. They approach to being closed systems, independent in some degree of the rest of the world; partial individuals, they have a share in determining their future selves. Once raised to mountains, they contain within themselves the germs of their own destiny; and if not possessing such power as true life possessed of blossoming into a predetermined form, scarcely to be altered by all the efforts of the outside world.
>
> (311)

Furthermore, amongst the numerous living, as well as non-living, forces of "denudation" (including the wind, rain, frost, the life cycles of lichens and fungi), even the chamois itself has a role to play in shaping the mountainous landscape. Huxley's brief account does two things here. First, it reiterates the Darwinian insight into the complex interactions of species with the land, taking a step towards a biogeographical understanding of the context or "milieu" of evolution, as Georges Canguilhem notes in his essay, "The Living and its Milieu":

> Darwin was looking for the appearance of new forms in the interplay of two mechanisms: a mechanism of production of differences that is variation, and a mechanism of reduction and criticism of the differences produced, that is, the struggle for existence [*la concurrence vitale*] and natural selection. The fundamental biological relationship, in Darwin's eyes, is a relationship between living things and other living things. It trumps the relationship between living and milieu, conceived of as a collection of physical forces. The primary milieu an organism lives in is the set of living things around it that are enemies or allies, prey or predators.
>
> (2001, 13)

Second, it provides a very concrete model of knowing that is always *immanent*. Each agent or actual entity (or "actant" to use Latour's term) is altering its surroundings simultaneously with "taking it in" or "knowing" it.

Bergson's model of virtual-to-real action *implies* that there is a distance (due to its resonance with Kant's transcendental framework) between knower and known, but the relation of virtual to actual is precisely intended to embed each knower, deeply and inextricably, into its environment. Our ways of knowing and acting, virtually and actually, are implicated in a continual process of exchange and refinement as we each learn and grow. In what follows, I want to use the example of birdsong and the auditory landscape, or soundscape, to expand upon Bergson's account of the virtual/actual relation and how we might use it to articulate our both creative and destructive relation to the "natural world."

In the opening chapter of his book, *Born to Sing* (1973), Charles Hartshorne summarizes a problem that is still being challenged by critical animal theorists today: the tendency to use behaviorism to explain the actions of species other than *Homo sapiens*. As Hartshorne explains, this application of the principle of parsimony, in which the simplest explanation is better than the more complex one, dictates that "whatever [more-than-human] animals 'do,' they move parts of their bodies and their bodies as wholes, and these movements are in principle intersubjectively discernible, whereas any sensations or emotions the animals might be thought to experience seem problematic inferences from the movements, the behavior" (Hartshorne 1973, 1). In other words, Hartshorne is highlighting that in the field of animal behavior, it is acceptable to speak about animals' physical or physiological characteristics and movements, but it is less acceptable to speak about their psychological or "inner" lives.

Today's theorists would point to historical examples of the importation of human norms and standards onto other species—for example, Donna Haraway's critique in *Primate Visions* (2013) of the use of human models of male and female behavior to interpret social interactions amongst groups of primates—in order to reveal the pitfalls of anthropomorphism. Conversely, the charge of anthropomorphism is often casually laid at the door of animal rights campaigners and environmentalists for attributing too much value to species outside the human sphere or the circle of moral consideration.

However, this importation of seemingly human traits beyond the human does not preclude the possibilities of reversing anthropomorphism such that human characteristics might be interpreted within an overarching natural, or ontological, framework. A generalization of concepts, such as the aesthetic phenomena of harmony or "unity in variety," to which Hartshorne points, might yield distinctive insights into our affinities with other species (1973, 2).

On this view, we can start to ask whether our complex actions and emotions, from musicality to mourning, are in fact common to a range of mammals, or to species that evolved within similar climatic and geographical conditions. Behaviorism, in any case, ought to be rejected, and, as Hartshorne continues, "in view of the evolutionary continuity of life, and the ideal of a unitary explanation of nature as a whole, it seems an unsatisfactory dualism to make man a mere exception" (1973, 1). Thus, Hartshorne rejects human exceptionalism whilst remaining alert to the risks of anthropomorphism. In this, he follows (and cites) Huxley's work, *Animal Language* (1939). There, Hartshorne finds the means to separate aesthetic *feeling* (universalizable across species) and aesthetic *thought* (characteristically human) (1973, 2). We can use the idea of aesthetic feeling to encompass the actions of other species that we recognize as fellow "emotional creatures" who are guided by the desire for "harmony" and the avoidance of "the twin evils of discord and monotony" (2). For example, "Although the chief function of bird song is to maintain territory, it does not follow that the chief, still less the sole, emotive meaning of singing for the bird is territorial 'hostility.' Why not say it is 'liking' for the territory? Or why not, part of the time, simple joy in the singing?" (3).

Considered from this perspective, animal utterances would not be understood as primitive forms of language, since the parts (of a song) cannot be read as bearing meaning in isolation in the way that parts of sentences can. Rather, a song, even a complex one, should be taken as an expressive whole, announcing a bird's presence to potential mates or to warn rivals away from their territory (Hartshorne 1973, 4–5).

Julian Huxley defends this way of interpreting animal utterances in a response to a book review of *Animal Language*. Huxley recalls the following:

> In passing, in compiling [*Animal Language*] I was astonished at the lack of any coherent body of scientific information on the comparative aspect either of sound-production or of hearing: here is a fruitful field for anybody with time and knowledge. As regards the point concerning the rattlesnake, Gadow, in the 'Cambridge Natural History,' writes: 'The rattle is decidedly useful to the snake as an instrument of warning off any approaching possible enemy, since no snake likes to bite unless in self-defence or in order to kill its prey.' The 'Encyclopaedia Britannica' (eleventh edition) expresses a similar idea. So does that eminent authority, Dr. Thomas Barbour, in his book 'Reptiles and Amphibians' (1926), but in considerably greater detail, and with the specific suggestion that the bison is the enemy to be warned off.
>
> (Huxley 1939, 725)[1]

Hartshorne goes on to review and speculate upon, with reference to numerous published studies, the factors that have favored the development of song. My discussion of them here is not intended to make factual statements about "why birds sing," but rather to underline the ontological implications of Hartshorne's study. In the first part of this chapter, I sought in Huxley's work an account of the synchronicity between geological landscapes and organisms evolved to survive in them through a certain resistance or resilience. Unstoppable forces negotiate immovable objects (at least to the human/animal eye), as it were. Hartshorne's work, here, allows us to reflect on the manipulation of what we might call "lived space" by organisms that have evolved to manipulate them.

We are wandering into Bergsonian territory here, echoing his analysis of the *creation,* rather than the discovery, of geometrical space as a means of navigating, through a form of "understanding," one's physical surroundings. In *Matter and Memory*, Bergson's study of the immersion of human perception within a constantly altering and meaningful environment offers a distinction between virtual and real action. We create our own demarcations between objects, rather than those boundaries being ready-made, according to the uses that we can make of them:

> That which is given is the totality of the images of the material world, with the totality of their internal elements. But, if we suppose centers of real, that is to say of spontaneous, activity, the rays which reach it, and which interest that activity, instead of passing through those centers, will appear to be reflected and thus to indicate the outlines of the object which emits them. There is nothing positive here, nothing added to the image, nothing new. The objects merely abandon something of their real action in order to manifest their virtual influence of the living being upon them.
>
> (Bergson 1991, 37)

Bergson's identification of the virtual-to-real relation has the potential to help us to understand the effects that birds might have on their surroundings.

In the case of song, the physical environment is one that is constantly created anew, musically rather than geometrically. Song, Hartshorne writes, transforms "an auditory emptiness into a world of meaningful sounds" (1973, 14). Consider the factors that Hartshorne surveys favoring the development of song. Some of the most elaborate songs come from ground-foraging bird species. The abundance of larger amounts of protein from seeds and small insects means that less energy is expended on the search for food, thus freeing up more energy for the enhancement of methods of communication (16). The purposes of communication appear to range from mere contact with others, to sounding an alarm, courtship, and to demarcate territorial boundaries (15). Regardless of whether this is sufficient to capture the complexity of the social existences of numerous bird species, we can derive some sense of the ways in which birds carve up the auditory milieu in rich and rapidly dynamic ways. Consider also the specificity of soundscapes in different parts of the world: in the U.K. it might combine blackbirds, sparrows, thrushes, and magpies; in the U.S., sparrows and thrushes might be heard amongst scarlet tanagers and ovenbirds; in Australia, the calls of noisy miners, kookaburras, and magpie larks might predominate. These dynamic auditory landscapes are invested with meaning for birds within and across species, even if they "add nothing" to the real environment and are thereby of negligent interest to the human ear. Yet they attest to the processes of harmonization of a broad array of species with "land" and "air." In more Bergsonian terms, the unanswered calls to meet, or the unheeded sounds of alarm, form a dynamic virtual interpretation of the surrounding forest fragment, whilst the convergence or scattering away of birds would see their fulfillment in real actions.

Disharmony

In this section, I turn to phenomena of disrupted harmonies between species and environment, which, in some cases, can even lead to species loss or extinction. The following set of examples have entered our common discourse on the effects of climate change on plants and animals across the globe, and I present them here as a means of complicating the accounts I have provided in the previous section in which "nature" itself is read as a striving for, and achievement of, harmony.

Radchuk, et al. (2019) explain the adaptive processes that come into play when an environment becomes challenging or perilous:

> Fitness losses (i.e. reductions in survival or reproductive rates) can be mitigated, however, if populations respond adaptively by undergoing morphological, physiological or behavioural changes that maintain an adequate match—or at least reduce the extent of mismatch—between phenotype and environment. Such adaptive phenotypic changes —which we call "adaptive response" (to climate change)—come about via phenotypic plasticity, microevolution or a combination of both, and can occur in tandem with geographic range shifts.
>
> (2)

The phenotypic plasticity of any species would only count as an "adaptive response" to climate change if that phenotypic change (a change in the external expression of an internal genetic modification) provided an advantage to the survival of that species. From a survey of the scientific literature on adaptive responses, they conclude that "advancement of phenology is adaptive in some bird species, but this response is not universal. Further, modeling suggests that even bird species responding adaptively to climate change may adapt too slowly to be able to persist in the longer term" (Radchuk et al. 2019, 3). In short, the indications are that the kinds of adaptive response articulated by Huxley (chamois adapting to mountain) cannot keep up with the pace of modern climate disruption.

The geographic, rather than phenotypic, range of responses to changing climate includes the redistribution of species, which occupy the majority of studies of animal responses to climate change (Radchuk et al. 2019). Ornithologist Humphrey Crick writes in particular of changes "due to El Niño events, when periodic warming alters oceanic currents along the west coast of the Americas, leading to crashes in fish abundance and catastrophic breeding failure or even adult mortality among seabirds in the Pacific" (2004, 48). Recent climate change, he continues, has resulted in "earlier breeding; changes in timing of migration; changes in breeding performance (egg size, nesting success); changes in population sizes; changes in population distributions; and changes in selection differentials between components of a population" (49). Rising sea-surface temperatures (SSTs), the discussion of which are increasingly commonplace today in the context of international "net zero" targets to stave off more than a 1.5 degree change in global average temperatures, have already been blamed for declines in species of Emperor Penguins in Antarctica since the 1970s, and Rockhopper Penguins on Campbell Island, New Zealand, which have "declined by 94% since the early 1940s, associated with rises in SSTs and declines in their fish prey" (52) Finally, Crick notes what he refers to as "phenological miscuing" in which higher average temperatures in the Spring have seen the early arrivals of migratory birds, such as the American Robin, and the emergence from hibernation of species such as the Yellow-bellied Marmot. Such phenomena are problematic, Crick explains, since

> despite higher temperatures, the date of snowmelt has not changed because of the greater volumes of snow present due to increases in winter precipitation with the higher temperatures. Thus, the interval between arrival or emergence and the first date of bare ground has actually grown and may cause problems for migratory and hibernating species.
>
> (2004, 53)

In July 2022, the Royal Society for the Protection of Birds (RSPB) in the U.K. reported on the presence of a small group of bee-eaters that nested and hatched chicks in North Norfolk, making far more frequent nesting attempts, and doing so farther north than in their usual favored nesting sites, in the last two decades. Whilst in August 2022, amid the widely reported record high temperatures experienced in the U.K., a phenomenon dubbed "false autumn" was witnessed in British deciduous tree populations in response to the unusually dry weather (even in the typically inclement Manchester). As Tng et al. (2018) explain of drought-response in trees:

> The occurrence of longer-than-usual drought periods or droughts over an extended number of years can result in widespread seedling mortality, reduced plant growth, and even large-scale tree dieback. Recurring droughts may occur at frequencies that do not give forests time to recover, leading to perpetually damaged ecosystems that can negatively affect the global carbon budget.
>
> (12479)

The most visible effects we witnessed in the U.K. included the premature shedding of leaves and the early emergence of seeds. Whilst the first would help the tree to "produce smaller, thinner, or 'cheaper' leaves (i.e., leaves with lower leaf mass per area)," the second poses a problem for many species of bird that rely on autumnal seeds and berries for their survival in the colder months of the year (Tng et al. 2018, 12480).

Now, at the extreme end of the range of responses to climate change is the risk of extinction. Here, we see some of the consequences of *disharmony* or imbalance. These are often attributable to human activity, although contemporary studies of indicators of extinction highlight the more complex interplay of environmental and specific factors that play a role in species loss. For example, it is not simply the numbers within a population that determine its extinction risk, although this is certainly an idea perpetuated and popularized through high-profile cases of critically endangered species (for example, large mammals such as rhino, panda, etc., as categorized by the International Union for Conservation of Nature [IUCN]) typically selected for breeding programs in zoos across the globe. In an article entitled, "The Sixth Mass Extinction: Fact, Fiction or Speculation?" (2022), Cowie et al. tackle the skepticism in some quarters towards the idea that the Earth is currently in its Sixth Mass Extinction event. On the one hand, the idea of "environmental shocks" provides us with powerful images of massive, destructive events such as meteorite collisions with the Earth, large volcanic eruptions, or earthquakes, which can lead to large-scale extinctions. These types of catastrophic events are "not normally encountered during the geological life spans of such species or groups, and the shock must be applied rapidly enough over a broad geographic area to prevent adaptation by natural selection or escape by migration" (Raup 1994, 6762). On the other hand, the dynamics of a sixth period of mass extinction might be less discernible to the unschooled observer, but Cowie et al. underline the following:

> If all human-caused extinctions (i.e. over a period of up to 200 [thousand years] so far) are to be included in the current event, while realizing that the rate has increased dramatically in more recent times, the time span is of a roughly similar order of magnitude as that over which extinctions took place in the five traditional previous mass extinction events.
>
> (2022, 643)

They go on to highlight the perils of over-emphasizing the effects of this process on "charismatic megafauna," that might detract from thousands of invertebrate species losses, which "vastly outweigh the small number of much better known vertebrate extinctions" (645). Furthermore, arguments over the current rate of extinctions and how it compares to the "background rate"—for example, De Vos et al. argue that extinctions are occurring at one thousand times faster than the background rate, which is slower than some popular estimates (2015, 460)[2]—ultimately bear out the conclusion that "regardless of the relative numbers, whether one accepts a background rate of 1.0 or 0.1 E/MSY[3], extinction rates assessed in many studies have increased within the last 500 years" (Cowie et al. 2022, 651).

If we return to our example of the effects of drought on forests (as mentioned above), we find that "large-scale tree dieback" entails the permanent loss of habitat for several surrounding plant and animal species. Hanski and Ovaskainen (2002) highlight the accumulation in such cases of what is termed "extinction debt." As they explain:

> it takes some time following habitat loss before the extinction will occur, which generates an extinction debt in a community of species. We show that extinction debt is especially great in a community in which many species are close to their extinction threshold following habitat loss because the metapopulation-dynamic time delay is especially long in such species.
>
> (Hanski and Ovaskainen 2002, 666)

The solution is to return "to near pre-industrial global temperatures as quickly as possible [that] could prevent much of the projected, but slower acting, climate-related extinction from being realized" (Thomas et al. 2004, 147). More direct impacts of human activity include "the well-known impacts on large animals of human hunting and fishing for meat and secondary products, such as ivory and horn, during both the Pleistocene and present-day extinctions" (Davidson et al. 2009, 10704). But it is also pointed out that there is a wider range of impacts that "include not only hunting, but also extermination programs, climate change, spread of diseases and exotic species, deforestation, habitat fragmentation, urbanization, desertification, and the conversion of landscapes to agriculture" (10704).

Alexander et al. (2016) show that the correlation between age and extinction (exemplified in Darwin's *Origin of Species*, that the longer a species exists, the greater its risk of extinction) has some merit as an explanation for, and indicator of, potential species losses, but that, again, a range of factors play a role. They argue that younger species also face distinct risks, and it is not necessarily age itself that decides fitness ("senescence"—or deterioration due to age), but rather the exposure to "ecological variables" that, over time, accumulate risk.[4] Finally, the effects of habitat loss in significant forested areas such as the Amazon, as Doughty et al. (2016) argue, might not yet be fully evident:

> we find some empirical and theoretical support in the Amazon region for the controversial 'anachronism' theory of Janzen and Martin (1982) that the Late Pleistocene–early Holocene megafauna extinctions led to loss of dispersal ability and subsequent range declines in large-seeded tree species adapted to dispersal by large-bodied frugivores. This not only has strong implications for ecosystem function and biodiversity conservation, but also may have reduced the ability of the Amazon to sequester carbon. Our model simulations suggest that we may not yet have seen the full impact of the past extinctions on these processes.
>
> (202)

In summary, the wide variety of impacts felt from the disruptions or fluctuations in climate and its often-imperceptible effects on plant and animal communities, represents a radical reconfiguration of the Earth's ecosystems. Such a radical alteration in our shared habitat requires us to alter (radically) both our ways of seeing and ways of saying if we are to become better co-habitants of the Earth.

Unchecked Growth

In *Braiding Sweetgrass* (2013), botanist Robin Wall Kimmerer evokes the Native American myth of the Windigo to reflect on contemporary and historical concepts of consumption and excess. The book interweaves narratives from the sciences, indigenous mythologies, and Kimmerer's personal experiences as a member of the Potawatomi Nation, and as a mother, the juxtaposition of which exposes the contrasts and commonalities between these seemingly incompatible worldviews (at least from the perspective of conventional scientific writing). Her insights into the processive tendencies of the Potawatami language start with the frustration of learning a language of such apparently excessive specificity:

> …the verb *wiikwegamaa*—to *be* a bay—releases the water from bondage and lets it live. 'To be a bay' holds the wonder that, for this moment, the living water has decided to shelter itself between these shores, conversing with cedar roots and a flock of baby mergansers. Because it could do otherwise—become a stream or an ocean or a waterfall, and there are verbs for that, too. To be a hill, to be a sandy beach, to be a Saturday, all are possible verbs in a world where everything is alive.
>
> (Kimmerer 2013, 55)

Grasping the complexities and subtleties of a language grounded in an animist worldview provides us with multiple perspectives from which to assess persisting problems. Mythological frameworks, storytelling to explain why the world works in the way that it does, are deeply embedded in all cultures, but the relevance of those creative, *fictional* frameworks has long been characterized, in the Western philosophical and scientific traditions, as a mark of primitive human culture or denigrated as pastimes or products of idle time (much like the foraging birds whose high-protein diets free them up to create beautiful songs). The image of the Windigo, as Kimmerer explains, itself serves as a cautionary tale to guard against a lifestyle that allows for the creation of beauty but, when taken to excess, can create the monstrous. The Windigo is

> a being in the shape of an outsized man, ten feet tall, with frost-white hair hanging from its shaking body. With arms like tree trunks, feet as big as snowshoes, it travels easily through the blizzards of the hungry time, stalking us… . Windigo stories were told around the fire to scare children into safe behavior lest this Ojibwe boogeyman make a meal of them. Or worse. This monster is no bear or howling wolf, no natural beast. Windigos are not born, they are made. The Windigo is a human being who has become a cannibal monster. Its bite will transform victims into cannibals too.
>
> (Kimmerer 2013, 304)

Kimmerer explains that whilst the experience of starvation would have been common to her ancestors, the relevance of Windigo mythology may have been boosted at the time of the fur trade "when overexploitation of game brought famine to the villages. The ever-present fear of winter famine is embodied in the icy hunger and gaping maw of the Windigo" (304). Reflecting further on our more recent concerns with the effects of climate change, and the acknowledgment both scientifically and politically of the need to reduce our use of fossil fuels, Kimmerer laments the role of multinational corporations whose ambitions have "spawned a new breed of Windigo that insatiably devours the Earth's resources 'not for need but for greed'" (306).

As Brianne Donaldson outlines in *Creaturely Cosmologies* (2015), combining metaphysical insights from Whitehead's process philosophy, the Jain tradition, and contemporary Critical Animal Studies, we need to expand our epistemological and ontological frames of reference that have hitherto excluded more-than-human forms of perception and feeling:

> Performative action is no longer limited to the safe confines of the humanist stage. Rather, it becomes an imaginative play between what we feel and perceive now and what we might yet perceive, or feel, in the future—two simultaneous directions of our own creaturely becoming that are both familiar and deeply strange, one and many simultaneously, a gradual expansion of our personal identity to include a more diffracted and impersonal scope.
>
> (18)

Like Kimmerer, Donaldson's study does not seek to extricate us from our "human perspective," as if such a thing ("the human") has ever been properly or comprehensively defined. Rather, her work helps to expose, through Process and Jain metaphysics and, with Foucault's "practices of freedom," the "contradictions and multiplicity that exist within our own experiences of self, and to extend that to the manifold aspects of other life forms" (2015, 38). As Donaldson explains further:

> By conceiving of *ahiṃsā* [nonviolence] as an experimental practice—between our conventional day-to-day experience of life and the transcendent possibility of coexistent understanding without limit or loss in the Jain ideal of omniscience—we can sidestep extremes that require us to either reify nonviolence as a dogmatic aim or to reject it as always insufficient. Nonharm and noninterference can be as lively as its practitioners' imaginations: as dynamic, active, and abundant as the many situations in which it finds embodied expression.
>
> (2015, 38)

In this endeavor, the variety and depth of responses being constructed across disciplines in Critical Animal Studies reflects an effort to think, or *conceptualize*, that is more like birdsong, at least in the sense I have borrowed from Bergson in his account of virtual-to-real action. The construction of dynamic systems of meaning across species lines offers an alternative to positivist or reductionist trajectories in the Western philosophical tradition that are inherently exclusive of more-than-human others.

Notes

1 And, as Emily Herring observes, this willingness to ascribe emotional significance to other animals seems to indicate a synthesis of Bergsonism and Darwinism in Huxley's thought: "In 1912, while he was developing his Bergsonian theory of progressive evolution, Huxley had become convinced that there was an element of choice in the birds' behaviour, more precisely in the female's attitude towards a potential mate. Huxley pushed the anthropomorphism as far as comparing the female Redshank's rejection of the male with the attitude of female humans" (Herring 2018, 48).

2 As they explain, "there is no widespread evidence for high recent extinction rates from either the fossil record or from molecular phylogenies; thus, in general extinction rates cannot exceed diversification rates. Combining the evidence from fossils, the separation of speciation and extinction rates from molecular phylogenies, and from overall diversification rates, we conclude that the benchmark estimate of the natural background rate of 1 E/MSY extinctions per million species per years is too high. A more defensible estimate should be closer to 0.1 E/MSY, whereupon current extinction rates are 1,000 times higher than the natural background and future rates of extinction are likely to be 10,000 times higher" (De Vos et al. 2015, 460).

3 E/MSY metric = number of extinctions (E) per million species-years (MSY) (Cowie et al. 2022, 651).

4 See also Paddle (2000, 7–8), for a discussion of the negative impacts of some of the nomenclature used to distinguish between different types of mammals. For example, the terms "eutherian" and "metatherian" suggest that the former are superior to the latter in terms of evolutionary development. Paddle favours instead the terms "placental" and "marsupial" which are more descriptive and carry less "cultural baggage."

Bibliography

Alexander, Helen K., Amaury Lambert, and Tanja Stadler. 2016. "Quantifying Age-dependent Extinction from Species Phylogenies." *Systematic Biology* 65, no. 1: 35–50.

Atkinson, Paul. 2020. *Henri Bergson and Visual Culture: A Philosophy for a New Aesthetic*. London: Bloomsbury Publishing.

Bergson, Henri. 1991. *Matter and Memory*. Translated by Nancy M. Paul and William Scott Palmer. New York: Zone Books.

Canguilhem, Georges. 2001. "The Living and Its Milieu." Translated by John Savage. *Grey Room*, Spring, no. 3: 6–31.

Cowie, Robert H., Philippe Bouchet, and Benoît Fontaine. 2022. "The Sixth Mass Extinction: Fact, Fiction or Speculation?." *Biological Reviews* 97: 640–63.

Crick, Humphrey Q. P. 2004. "The Impact of Climate Change on Birds." *Ibis* 146: 48–56.

Davidson, Ana D., Marcus J. Hamilton, Alison G. Boyer, James H. Brown, and Gerardo Ceballos. 2009. "Multiple Ecological Pathways to Extinction in Mammals." *Proceedings of the National Academy of Sciences* 106, no.2: 10702–05.

De Vos, Jurriaan M., Lucas N. Joppa, John L. Gittleman, Patrick R. Stephens, and Stuart L. Pimm. 2015. "Estimating the Normal Background Rate of Species Extinction." *Conservation Biology* 29, no. 2: 452–62.
Donaldson, Brianne. 2015. *Creaturely Cosmologies: Why Metaphysics Matters for Animal and Planetary Liberation*. Lanham: Lexington Books.
Doughty, Christopher E., Adam Wolf, Naia Morueta-Holme, Peter M. Jørgensen, Brody Sandel, Cyrille Violle, Brad Boyle, et al. 2016. "Megafauna Extinction, Tree Species Range Reduction, and Carbon Storage in Amazonian Forests." *Ecography* 39, no. 2: 194–203.
Glenn, Susan M., and Scott L. Collins. 1992. "Effects of Scale and Disturbance on Rates of Immigration and Extinction of Species in Prairies." *Oikos* 63, no. 2 (March): 273–80.
Hanski, Ilkka, and Otso Ovaskainen. 2002. "Extinction Debt at Extinction Threshold." *Conservation Biology* 16, no. 3: 666–73.
Haraway, Donna J. 2013. *Primate Visions: Gender, Race, and Nature in the World of Modern Science*. London: Routledge.
Hartshorne, Charles. 1973. *Born to Sing: An Interpretation and World Survey of Bird Song*. Bloomington: Indiana University Press.
Herring, Emily. 2018. "'Great Is Darwin and Bergson His poet': Julian Huxley's Other Evolutionary Synthesis." *Annals of Science*, 75, no. 1: 40–54.
Huxley, Julian. 1955/1912. "On the Behaviour of a Chamois: And Incidentally of Some Other Matters." In *The Book of the Mountains, Being a Collection of Writings about the Mountains in All Their Aspects: With 64 Pages of Illus. in Halftone and Gravure, and Many Line Drawings*, edited by Auguste C. Spectorsky, 306–12. New York: Appleton-Century-Crofts.
Huxley, Julian S. 1939. "Animal Language." *Nature* 143: 725.
Janzen, Daniel H., and Paul S. Martin. 1982. "Neotropical Anachronisms: The Fruits the Gomphotheres Ate." *Science* 215, no. 4528: 19–27.
Kimmerer, Robin. 2013. *Braiding Sweetgrass: Indigenous Wisdom, Scientific Knowledge and the Teachings of Plants*. Minneapolis: Milkweed Editions.
Low, Tim. 2016. *Where Song Began: Australia's Birds and How They Changed the World*. London: Yale University Press.
Paddle, Robert N. 2000. *The Last Tasmanian Tiger*. Cambridge: Cambridge University Press.
Radchuk, Viktoriia, Thomas Reed, Céline Teplitsky, Martjin van de Pol, Anne Charmantier, Christopher Hassall, Peter Adamík, et al. 2019. "Adaptive Responses of Animals to Climate Change are Most Likely Insufficient." *Nature Communications* 10, no. 1: 3109.
Raup, David M. 1994. "The Role of Extinction in Evolution." *Proceedings of the National Academy of Sciences* 91, no. 15: 6758–63.
Royal Society for the Protection of Birds. 2022. "Protecting Norfolk's First-ever Colony of Breeding Bee-eaters." *YouTube*. Accessed February 4, 2024. www.youtube.com/watch?v=TMCw5s12Hxw
Thomas, Chris D., Alison Cameron, Rhys E. Green, Michel Bakkenes, Linda J. Beaumont, Yvonne C. Collingham, Barend F. N. Erasmus, et al. 2004. "Extinction Risk from Climate Change." *Nature* 427, no. 6970: 145–48.
Tng, David Y. P., Deborah M. G. Apgaua, Yoko F. Ishida, Maurizo Mencuccini, Jon Lloyd, William F. Laurence, and Susan G. W. Laurance. 2018. "Rainforest Trees Respond to Drought by Modifying Their Hydraulic Architecture." *Ecology and Evolution* 8, no. 24: 12479–91.

5 Environmental Ethics, Climate Change, and African Social Epistemology

Kai Horsthemke

Considering Indigenous Knowledge for Sustainable Development

"Indigenous," "traditional," or "local knowledge," and "African ways of knowing," are among the buzz words used by African environmentalists over the last couple of decades, often in tandem with "sustainable development," in both postcolonial and environmentalist discourses.[1] Otto Dennis, for example, emphasizes the importance of recognizing traditional African knowledge as part of the solution of environmental challenges facing the planet and for paving the way towards "sustainable environmental development" (2022, 187–88). While the very mention of these terms together seems to facilitate acceptance of conference papers, publication of journal articles, funding for research and related projects by bodies like the South African National Research Foundation (NRF), etc., there have been few contributions that have engaged critically with the suitability and defensibility of the respective notions. Indeed, proponents of the ideas of indigenous knowledge and epistemic diversity invariably fail to elaborate on their respective understandings of "knowledge" and "epistemology." They are characteristically more concerned with explaining their use of the terms "indigenous" and "diversity."[2]

The present chapter attempts to redress the balance, or at least express some crucial misgivings. I argue that: (1) "indigenous knowledge" faces insurmountable logical and epistemological difficulties and can, at best, be understood either as "indigenous skills" or as "indigenous beliefs" and that this also applies to "African ways of knowing." More often than not, these phrases are used to describe either "African ways of doing" or "African opinions," neither of which reflects the epistemologically salient meaning of "knowing" or "knowledge." (2) "Sustainable development" is an ethically suspect term, since it reflects an unhealthy market-driven obsession with growth. Moreover, given the inherent anthropocentrism of "sustainability" as primarily sustaining the lives of humans, and the implicit prejudice *vis-à-vis* nonhuman nature this characteristically involves, the concept is also unlikely to contribute to a fundamental rethinking of our relationship with the natural environment and, therefore, to a long-term solution to our environmental and ecological predicament. Therefore, while "indigenous knowledge for sustainable development" looks attractive and perhaps even promising on paper, it is largely a

DOI: 10.4324/9781032659985-7

nonstarter conceptually, epistemologically, and ethically. Towards the end of this chapter, I examine how one might make sense of the ideas of "African ways of knowing" and "epistemological plurality" (Ryan and Ferreira 2019), namely in terms of different and distinctive epistemic priorities.

Relational Epistemology and Nature: Lessons from Africa?

In the words of Zimbabwean novelist Chenjerai Hove, Africans "have neither cataloged nature nor pinned it down and preserved it in formaldehyde. We see it differently and speak to and about it differently" (Grill 2003, 363).[3] Indeed, what is arguably characteristic of African conceptions of knowledge is a strongly relational element that is also found in African ontology and ethics (Horsthemke 2021, 27). Coming to know is understood as a process of persons developing insights in relation with one another and with all that exists. This indicates not only an intimate relationship between knower and known, or between what it is to know and what it is to be known, but in effect also a relational or communalist understanding of knowledge: *I know because we know.* In other words, a knower is a knower because of other knowers. These examples help to explain my reference to "social epistemology" in the title, by which I mean interpersonal practices that promote adequately justified true belief.

Regarding African ontology and ethics, the now-classic Nguni aphorism is *Muntu ngumuntu ngabantu*: "A human being (person) is human (a person) because of other humans (persons)." Or, as sometimes summarized, "I am because we are" (Crippen 2021, 236; Echekwube 2017, 112; Horsthemke 2015, 3–10; Topidi 2022, 51–52). This principle of *ubuntu*[4] is also captured in the southern African (Sotho) notion of *botho* (*Motho ke motho ka batho*; the meaning is identical to the Nguni phrase), the Shona terms *hunhu* (*Munhu munhu navhanhu*; ditto) and *ukama*, and the northern African concept *maat*, both meaning "relationality." Relevant ideas in indigenous American thought include the Cree term *wahkohtowin*, and Lakota *Mitakuye oyasin*, meaning *All is related*, or *We are all related*, ideas that express the interconnectedness of humans and other-than-human nature (Ong'eta 2013, 163). In other words, the self is defined relationally and best understood in relation.

How do these concepts of epistemological, ontological, and ethical relationality bear on environmental ethics and, especially, on the ongoing climate crisis? They express a sense of unity and harmony with all forms of life: with other people, other-than-human animals, plants, and also with rocks, rivers, mountains, and valleys. What we can learn from African and other indigenous environmental philosophy is that, simply and succinctly put, we are all in this together. But can these social aspects of African indigenous knowledge systems actually make significant contributions to local and global environmental problems and dilemmas?

Anthropogenic Global Climate Change

Anthropogenic climate change is probably the most urgent problem facing humanity and its policymakers, and indeed, all organic existents. The world (and

Africa in particular) is in a progressive state of environmental crisis, caused by global warming, loss of biodiversity, human overpopulation, pollution, massive deforestation and desertification, urbanization, and many other environmental problems and risk factors. Arguably, one of the most formidable challenges facing humankind at the beginning of the twenty-first century is the state of our planet and, coupled with this, our relationship with the natural environment. Most, if not all, other concerns, however significant, are necessarily secondary in this regard. Human population growth and advances in technological capacities and control continue producing previously non-existent environmental problems.

What is at stake here is nothing less than the survival of our planet in its present state as being inhabitable, and, therefore, also human survival. At the very least,[5] it is a matter of the quality and conditions of human lives, present and future. If this is correct, it follows that one of the greatest priorities—if not *the* greatest—of academic, scientific, philosophical, educational, and other kinds of research, should be how to arrest and possibly reverse the present decline.

There are difficulties, of course. The problem is not only climate change denialism—that is, doubt whether extant diagnoses and prognoses are correct. Substantial philosophical and pedagogical questions also concern the articulation and definition of the approaches we human beings should adopt in response to the environmental predicament(s). Further questions are raised about the metaphysical and ethical foundations of our concern for our planet, and about educational and pedagogical implications. Should educational policy and practice, for example, be informed by a concern for nature and the environment for our (human) purposes, i.e., as instrumentally valuable? Or should we teach and learn for the natural environment in and for itself, as having inherent value?

Philosophy, in all its forms and traditions, has always been centrally preoccupied with truth. However, in recent years truth appears to have become unfashionable. Climate change denialism, coronavirus denialism, and various conspiracy theories have all contributed to the current state of affairs in which truth and facts are treated with disdain. For example, the ideas that truth is not objective or universal and that there can be "alternative facts"[6] have taken hold not only of popular culture but also of politics, to the obvious detriment of the latter. The precedence given to assorted political and ideological (not to mention personal narcissistic) agendas, along with the rising popularity of pseudoscience, has led to a decline of truth both as a serious subject and as an intellectual tradition. Climate change denialism and coronavirus denialism are objectionable not only because they peddle blatant untruths but also because they are cynical and cruel. They demean the many millions, human and nonhuman, who have forfeited their health and/or their lives as a result of the changing climate and the recent pandemic. It remains an important task to unmask and undo the pernicious forms of cultural, moral, and epistemological relativism that have held impressionable people in their grip and that continue to thwart attempts to establish a sane and morally sustainable set of relationships between us, human beings, and other animals, as part of the animate and inanimate environment.

The "Hierarchy of Being": Animals and the Environment in African Thought and Practice

Quoted earlier, Hove is right about one thing: it is hardly Africans who are to blame for the consequences of the crass objectification of nature. On the contrary, Africans bear the brunt of centuries of Western (or Northern) overexploitation of natural resources. Africans have for the longest time been concerned about living *with* nature, rather than perceiving nature as something to be vanquished and dominated (see also Mutwa 1996). This is also connected with their struggle to regain the dignity lost after centuries of violent colonization and degradation, to combat poverty, malnutrition, and disease, and to maintain optimism in the face of threats presented to the African continent both by the changing climate and overpopulation.

That said, the communalist and relational orientation manifest in thought and practice throughout the African continent has little in common with the egalitarianism that has come to characterize contemporary animal rights and animal liberation movements, deep ecology, or radical environmentalism. While I am aware of the dangers of facile generalization, and while there exists no single unified "African ontology" or "African ethic," there are nevertheless certain core ideas that appear with astonishing regularity across African (especially sub-Saharan) societies and cultures, like the "hierarchy of being," totems and taboos, as well as rituals of sacrifice and slaughter. All these indicate that the relationships that exist between humans and other-than-human nature are determined essentially by human preoccupation with their own benefits, material or spiritual. Since humans are situated below the Creator-God, spirits, and ancestors (or "living-dead") in the ontological hierarchy—which is also an ethical hierarchy, since it yields norms of interaction—it is important that these superior beings remain favorably disposed. This is achieved not only by adhering to specific customs and rituals of sacrifice and slaughter but also by respecting divine creation. As Cyril Emeka Ejike has bluntly put it, "Inferior Forces…include animals, plants, and minerals. The living [human beings] exercise control over them. Inferior forces basically exist for the purpose of being utilized to increase the vital forces of the living" (2020, 27). Therefore, duties towards nonhuman nature, such as animals, plants, and inanimate objects that all occupy lower ranks in the hierarchical arrangement, are at best derivative duties; that is, they emanate from human obligations towards the Creator-God, the ancestors, and to living human beings.

In more secular thought and practice within African contexts, relations are also generally inegalitarian, as the following section will make clear. Other-than-human nature exists essentially for the purposes and use of human beings, and it is therefore in the interest of the latter to ensure sustainable interaction with natural resources—that is, not to overfish, over-harvest, hunt to extinction, or exploit beyond productivity (see Shava and Masuku 2019, on the "multiple roles" of "livestock" as "living currency" and "livelihood sustenance"). A survey of the various African philosophical views and claims pertaining to nonhuman nature demonstrates that tendencies vary quite broadly from: (1) more or less outspoken human-centeredness and more or less qualified endorsement of human moral superiority, by way of, (2) lip service toward environmental or animal-friendliness, while humans remain the measure of all things and at the center of ethical concern and deliberation, to (3) explicit pro-animal attitudes and/or the view that nature matters in and for itself, and outright rejection of moral anthropocentrism (Horsthemke 2015, 2018a, 2018b, 2019). It is the middle group of pro-animal/environmental lip service that often constitutes the greatest challenge to the critical commentator, in terms of having to pinpoint not only the deficiencies of the respective views and knowledge claims but also the reasons why an environmental-friendly view may not necessarily (indeed, often does not) incorporate any direct concern for individual nonhuman animals. For one thing, the critical commentator may not want to belittle or discourage the ethical progress and advancements, however small and tentative, that have been made by both theorists and activists on the African continent. For another, endorsing a radicalization in pro-animal and environmental thought and practice (associated, for example, with Western animal liberation and deep ecology movements) may seem to threaten or at least minimize the distinctly African contribution to the ethical and epistemological debates in question. This is arguably where a sustained focus on the last group of arguments—that reject moral anthropocentrism, develop pro-animal attitudes, and articulate the inherent value of nature—becomes all important, in attempting to determine what values and other conceptual and practical resources exist not merely to establish future-oriented environmental or ecological awareness but also to bring about substantial changes in the conditions under which the many billions of animals live (and die) on the African continent.

Problematizing "Indigenous Knowledge" and "Sustainability"

Prior to the 2012 United Nations Conference on Sustainable Development in Rio de Janeiro, the 2011 "Treaty on Environmental Education for Sustainable Societies and Global Responsibility" was published in the Rio+20 Portal. It contains several telling clauses, including:

> 14. Environmental education must recover, recognize, respect, reflect and utilize indigenous history and local cultures, as well as promote cultural, linguistic and ecological diversity. This implies acknowledging the historical perspective of native peoples as a way to change ethnocentric approaches, as well as the encouragement of bilingual education…
>
> 16. Environmental education values all different forms of knowledge. Knowledge is diverse, cumulative and socially produced and should not be patented or monopolized…
>
> 20. Environmental education must integrate knowledge, skills, values, attitudes and actions. It should convert every opportunity into an educational experience for sustainable societies.
>
> (Rio+20 2011)

These clauses are noteworthy not only for their emphasis, with regard to sustainable development, on indigenous cultural perspectives and perceptions and on knowledge diversity, but also on the difference between "knowledge" and "skills," or between factual knowledge ("knowledge-that") and practical knowledge ("knowledge-how"). (In what follows, a third sense of knowledge, knowledge as familiarity or acquaintance—knowing a person, a place, etc.—will only be considered insofar as it bears on factual and practical or skill-type knowledge). The latter acknowledgment is also contained in the UNESCO statement on education for sustainable development (ESD): "ESD gives learners of all ages the knowledge, skills, values, and agency to address interconnected global challenges including climate change, loss of biodiversity, unsustainable use of resources, and inequality" (UNESCO 2023).

Indeed, in many accounts of (African as well as other) "indigenous knowledge" the knowledge in question appears to be skill-type knowledge, like San[7] tracking abilities, protection of biodiversity (examples below), being able to find water in extremely arid surroundings, or predicting changes in weather patterns, seasonal climate, rainfall, etc. on the basis of astronomical and other observations (Horsthemke 2021, 97–98, 166). It is perfectly plausible how this knowledge could be considered "indigenous," "local," or "traditional." These skills may indeed be unique to individual (groups of) practitioners, like traditional healers, and also indigenous farmers, artisans, and craft workers. Similarly, acquaintance or familiarity (i.e., knowledge in the third sense distinguished above) with geographical terrain, tree phenology, or animal behavior may be a distinguishing feature of local agents.

However, the question is not only what makes indigenous knowledge "indigenous," but also, and importantly, what makes it "knowledge" in the theoretical (factual, or propositional) sense of knowledge-that. Knowledge, in this sense, has three necessary conditions (whether these are also jointly sufficient has been the subject of much philosophical debate but need not detain us here): belief, truth, and adequate justification. It should be noted that this is not a Eurocentric understanding of knowledge. In Yoruba (Nigeria) pertinent distinctions are made between *gbàgbó* (belief; the subjective, private, or personal component of knowledge) and *mò* (knowledge in the sense of "knowledge-that"). Barry Hallen and J. Olubi Sodipo observe that:

> *gbàgbó* that may be verified is *gbàgbó* that may become *mò*. *Gbàgbó* that is not open to verification and must therefore be evaluated on the basis of justification alone (*àlàyé*, *papò*, etc.) cannot become *mò* and consequently its *òótó* [truth] must remain indeterminate.
>
> (1997, 81)

Belief, Justification, and Truth

"Belief in the powers of the earth and its deity" (Apusigah 2011, 6) is just that: belief. In the absence of the additional conditions of justification and truth, it involves, at best, a knowledge *claim*. It is necessary but not a sufficient condition for knowledge. Similarly, "African belief systems" that deal with the special role played by ancestors, the supernatural, and the religious (Breidlid 2009, 141) may be "indigenous"—but they are hardly *knowledge* systems. Why is that? In order for something to constitute knowledge in the factual, propositional, or theoretical sense, it must incorporate more than mere belief, for obvious reasons. Beliefs may be false and/or wholly unjustified, or at least insufficiently justified. Knowledge that something is the case therefore necessarily involves an adequately justified true belief that it is so. It follows that to speak of "legitimate" or "valid knowledge" (Ryan and Ferreira 2019, 117) involves a tautology, as does "true knowledge." The qualifying adjectives imply that there could be "invalid" or "untrue" knowledge—but this is a conceptual error. Theoretical knowledge, *qua* knowledge-that (with justification and truth as its necessary components), could not be anything other than "legitimate," "valid," and "true."

Moreover, the plausibility problem of "indigenous knowledge"—the problem of both questionable practices and questionable knowledge claims—is starkly illustrated in the following observation by Lisa Ryan and Jo-Anne Ferreira:

> Not all indigenous knowledges can be conceived of as environmentally friendly or benign and many unsustainable traditional practices are based on deeply rooted beliefs. Practises such as the widespread killing of owls, snakes, and frogs (among other species) and the harvesting of animal parts for the *muthi* (traditional medicine) industry are all derived ostensibly from traditional indigenous knowledges.
>
> (2019, 111)

It is difficult to see how these practices derive from or involve "knowledge" as justified true belief. As stated by Ryan and Ferreira, "ecologically destructive behaviors [are] underpinned by traditional beliefs (such as the killing of frogs because they cause thunderstorms)" (2019, 109). Or more to the point, these practices are informed by superstition. They are also, as the authors note, "unsustainable" (111).[8]

As noted earlier, practices and skills can, frequently and correctly, be identified as "African," "indigenous," "traditional," and the like. The same goes for beliefs, so there is at least one component of knowledge-that to which the descriptor "indigenous" could be applied. What about the remaining necessary conditions of knowledge-that: justification and truth? Regarding justification, it is not clear at all what *African* evidence, *indigenous* proof, or *local* inductive and deductive reasoning would look like. But even if it were granted that something like "indigenous" justification is, in principle, not inconceivable, the same could not be maintained with regard to truth. Unlike belief, which constitutes the subjective condition of factual or propositional knowledge, truth is the objective condition (with justification acting as a kind of bridging condition between subjective and objective). To deny any such objective and universal grounding is to open the floodgates to a pernicious cultural and epistemological relativism. Why is relativism problematic? Relativism about factual knowledge and relativism about truth are both logically untenable (are knowledge claims and truths regarding the relativity of knowledge and truth also relative, or do they have universal purchase?) and practically unfeasible. The latter point is made clear by considering the possible implications of having no objective recourse to rational judgment, comparison, evaluation, or assessment. Conspiracy theories, lies, prejudice, biased reporting, racist, sexist, or speciesist practices, policies condoning or enabling environmental degradation, etc., could be questioned only (if at all) within (but not across) local or indigenous communities.

"Sustainability"

The promise, but also the problem, of the notion of sustainable development in question is expressed in the words attributed to an African tribal chief who defined sustainability as "always enough for everyone" (Schäfer 2016). The notion in question, first articulated in 1987 in the World Commission on Environment and Development Report, has received global support, being defined as "development that meets the needs of present generations without compromising the ability of future generations to meet their needs" (World Commission 1987, 43; see also Agbedahin 2019, 3). Tellingly, the spirit here is clearly anthropocentric. Only the needs of humans (present and future) are mentioned, not the needs of nonhuman beings or the value of ecosystems and the environment. A similar spirit informs the Kenyan proverb, "The world was not given to you by your parents, it was lent to you by your children" (Stewart 2004, 137). The characteristically relational and intergenerational concern with use and development of "the world" focuses essentially on human beings.

The following reference to the epistemic and ethical character of *ubuntu* renders this human-centered orientation explicit. "In the African context," writes Catherine Odora Hoppers:

> the relationship *with,* and *to nature*, *human agency*, and *human solidarity*, for instance, underpins the knowledge system and the human existence around it. Relationships between people *hold pride of place*, expressed in the various philosophies across Africa and best captured by the African concept of *Ubuntu*—a human-trophic philosophy…
>
> (2008, 30; emphasis added)

Kyriaki Topidi likewise asserts that "Ubuntu, as indigenous knowledge, contains consideration about the physical environment, especially as it relates to the social environment of humans and with regards also to the interaction between the two," before continuing, "This is because Ubuntu is a cosmological notion (among numerous other indigenous cosmologies) that is based on the ways that the spiritual and the material aspects of our existence are connected" (2022, 52).[9]

It should be noted, however, that this connection, this relationality, referred to in *ubuntu* is a hierarchical one. *Ubuntu* (like its semantic stablemates *botho* or *hunhu*; see above) expresses an explicitly anthropocentric idea. As shown earlier, the African dictum from which "*ubuntu*" is derived, *Muntu umuntu ngabantu* (and its semantic equivalents, *Motho ke motho ka batho*, and *Munhu munhu navhanhu*) means "A human being is human because of other human beings," and also "I am because we are"—where "we" signals not only the collective but also (and especially) membership of the human species. Therefore, the environmental and ecological concern expressed by appeals to *ubuntu* and semantically identical notions like *botho* and *hunhu* can, by definition, not be a concern for the environment (the other-than-human biosphere) *in and for itself*. The natural environment is valued only either because human beings are part of it, or because it constitutes a (set of) resource(s) for (present or future) human beings to draw on. In a telling example of deploying *ubuntu* in ways that are directly exploitative to the environment is employed (and endorsed) by Matthew Crippen, who writes, "African hunting operations…by and large, adopt Africapitalist and ubuntu attitudes to enhance community wellbeing, environmental sustainability, and long-term economic viability" (2021, 235).[10]

Some theorists recognize the human-centric limits of such notions and attempt to forge ecological expansions. Lesley Le Grange and Callie Loubser, for example, acknowledge that the meaning of sustainable development as "development that takes place in such a way that it does not compromise the needs of future generations" has been criticized for its anthropocentrism (2005, 114). Yet, they claim, "sustainable development is not a monolithic entity, and a more nuanced understanding of the concept could incorporate values such as interspecies equity, e.g., that all living organisms have the right to being treated decently and to be protected from cruelty" (114). After distinguishing between two alternative conceptions of sustainable development—conservative and radical—Le Grange and Loubser contend that these "should not necessarily be seen as discrete categories but rather as opposite ends of a continuum" (114). It is, they claim, this tension between conservation needs on the one hand and development needs on the other that constitutes the beginning of environmental education's association with sustainability (115). For some, the authors say, sustainable development is the ultimate goal of environmental education, hence the phrase "environmental education for sustainability," yet

> For others, sustainable development encompasses specific objectives that should be added to those of environmental education, thus the expression 'education for environment and sustainable development.' For still others, environmental education inherently includes education for sustainable development, thus rendering the distinction between the two meaningless.
>
> (116)

Regrettably, Le Grange and Loubser fail to critically interrogate these different conceptions of environmental education. There is, clearly, a distinction between "education for the environment" and "education for sustainable development." The former signals a concern for the environment in and for itself, while the latter commonly emanates from an anthropocentric perspective.[11] It comes as no surprise, then, that the authors should claim that "different conceptions of environment, education, and sustainability should coexist" (Le Grange and Loubser 2005, 117). While Le Grange and Loubser do acknowledge the "dangers in uncritically accepting a plethora of concepts that we use in environmental education," they ultimately reiterate their "view that there should be an appreciation of the diversity of meanings and usage of the concept of sustainability, but also that such meanings should be (re)clarified within specific sites and discourses of environmental education" (117). Sustainability, they claim, is a "polysemous…concept"; its "diversity of meanings should be celebrated and continuously (re)clarified. .so that sustainability (sustainable development) becomes a reflective social process rather than a fixed idea" (120). This strikes me as a somewhat anemic move, as trying to invoke an unfeasible compromise. Surely, the authors cannot mean that all meanings, uses, and applications are equally valid. Some, perhaps even most, can be legitimately considered worth rejecting, not least because of an underlying, unargued presumption in favor of anthropocentrism.

"Indigenous African Knowledge" and Environmental Ethics

How, then, could "indigenous knowledge" contribute towards environmental and ecological progress? A traditional healer's insight that one should only use a limited amount of bark from a given tree, or that one should harvest no more than one-tenth of a given natural resource (i.e., harvest a plant only if it is one of ten such plants growing in the vicinity), or San acquaintance with the appetite- and thirst-suppressing properties of the*!khoba* cactus (*hoodia gordonii*) all constitute insights that may not be shared by many, but they have universal value and application—and there is no reason why these insights should not be included in a comprehensive educational curriculum (Horsthemke 2021). Another formidable example of indigenous insight and problem-solving is the planting of chili pepper hedges by Ugandan farmers, to protect their crops against marauding elephants (Mukhwana 2019). Apart from proving highly effective, and thus offering a nonviolent resolution to an age-old conflict, the chili production venture has become a considerable commercial success. There is a surprising amount of common ground between cultures, not only in terms of factual knowledge but also in terms of values (the moral premium placed on truth-telling and preservation of life, other things being equal, comes to mind here). A *rapprochement* between so-called "indigenous" and "non-indigenous" insights is not only possible but desirable—on educational, ethical, as well as political grounds.

In conclusion to this section, "indigenous knowledge" is meaningful if reference is made to indigenous skills and practical knowledge, or knowledge-how.[12] Applied to theoretical (factual or propositional) knowledge, or knowledge-that, it is either not a matter of knowledge at all, but rather of belief—or the knowledge in question is not "indigenous." As a prospective knower, I do not want to believe or be able to say things that are then passed-off as "my truth," or that are considered to be "true for me," or "true for or within my society or culture." If I say something that I consider to be true, that I believe sincerely and intensely, but that contravenes basic common sense and reason, or that involves certain blind spots or blank spots on my part, this should be pointed out to me, criticized, and rejected on the basis of available evidence, and of norms and rules of argumentation that apply to all language users, all those participating in relevant discursive exchanges. On the same basis,

I want to be able to engage critically with what is believed or stated within and also outside of my society or culture, and not to have to accept it merely because of majoritarian consensus, or for reasons of "political correctness," however well-intended. When I abandon or change my beliefs, as a result of such criticism or rejection, this arguably constitutes a qualitative change, indeed progress—just like the transition from a geocentric to a heliocentric worldview constitutes progress, something a rigorous relativism would reject, at least implicitly. What I (have come to) know is different from faith, superstition, bias, and prejudice. My knowledge may have a particular historical and geographical source, a particular origin both in time and in space, but once I know something, once it qualifies as "knowledge," it is no longer "local" or "indigenous." It is, purely and simply, knowledge.

African Ethics and Social Epistemology: A Question of Priorities

Reiterating Chinua Achebe's description of diversity "as the reality of our planet," Milton Nkoane asserts that "the nurturing of diverse cultures, ideas, perspectives, interpretations, and worldviews are what enable the construction of knowing to reflect and relate to its society" (2015, 42). However, the plausible consideration that diversity is both an epistemic and an ethical virtue does not support epistemological, moral, or cultural relativism, alternative facts, or the notion of different "regimes of truth" (Ryan and Ferreira 2019, 105). The value of diversity for knowledge resides in the possibility of different individuals and groups having different experiences that lead them to pay attention to things that may go unnoticed by others. Thus, accounts of their experiences may function as research data that permit examination of the social system or structure in question from the researchers' particular location or standpoint.

While many African and decolonialist scholars arguably err in postulating the existence of diverse knowledges and truths, the diversity in question is conceivably generated by (characteristically) practical epistemic priorities—priorities that emanate from different lived experiences, individual as well as social and cultural. A plausible view appears to be that knowledge and truth do not fluctuate, that they remain invariant across individuals, societies, and cultures, but that there may well be distinctive sets of epistemic concerns that arise from particular personal, historical, and socio-political circumstances. If it is correct to assume that practical epistemic and educational priorities will emerge from life experiences and from the ways these are socially articulated, then one might assume that, given the different life experiences of people across the globe, the practical epistemic and educational priorities will also differ. For example, as Elizabeth Anderson has put it:

> No one disputes that personal knowledge of what it is like to be pregnant, undergo childbirth, suffer menstrual cramps, and have other experiences of a female body is specific to women. Gynecology has certainly progressed since women entered the field and have brought their personal knowledge to bear on misogynist medical practices. The claims get more controversial the more

> global they are in scope. Some people claim that women have gender-typical "ways of knowing," styles of thinking, methodologies, and ontologies that globally govern or characterize their cognitive activities across all subject matters. For instance, various feminist epistemologists have claimed that women think more intuitively and contextually, concern themselves more with particulars than with abstractions, emotionally engage themselves more with individual subjects of study, and frame their thoughts in relational rather than an atomistic ontology…There is little persuasive evidence for such global claims.
>
> (2002, 325)

Interestingly, too, Anderson "does not suppose that women theorists bring some shared feminine difference to all subjects of knowledge" (326).

To use another example, given the experience of "indigenous" Africans of a wide-ranging credibility deficit, it stands to reason that they would have as priorities matters of epistemic transformation and redress. If epistemic, ethical, and educational concerns and priorities arise from different forms of social life, then those that have emerged from a social system in which a particular race or group has been subordinate to another deserve special scrutiny. Given the (especially vicious) history of physical and psychological colonization, and of having to bear the brunt of Western (or Northern) climate sins, it is plausible that chief among the priorities will be to educate *against* development of a subordinate or inferior mindset, as well as *against* the obsession with growth and profit (and coupled with this, the instrumentalization of nature). While it does not follow that particular historical and socio-economic circumstances yield or bestow automatic validation or justification of (the content and objectives of) an "African epistemology," an idea like "decolonization of the African mind" has a particular resonance here (Ong'eta 2013, 171). Yet, rather than implying a relativist "post-truth" epistemology, it involves going back to more promising resources like epistemic and ethical relationality.

If what has been established above is cogent, it follows that so-called "epistemic diversity" or "epistemological plurality" (Ryan and Ferreira 2019) refer neither to a multitude of knowledges and truths nor to an "anything goes" conception of justification, but rather to different experiences connected to particular social locations, or to different social pathways to knowledge (note the singular!). In this sense, reference to "knowledges," "truths," or to indigenous, local, or alternative epistemologies is not only unhelpful but also misleading. While the protection of biodiversity constitutes a significant moral and political imperative, the protection of "truth diversity" involves an equally significant conceptual mistake.

In recent years there appears to be a growing ecological consciousness and awareness among African scholars—consider the Africa Network for Animal Welfare (ANAW), for example—of the untenability of a rigidly species-governed "us-against-them" thinking, that anthropocentrism shares many relevant features with ethnocentrism, and that speciesism is relevantly like racism—all of which fills me with (admittedly cautious) optimism about our relationships with nonhumans on the African continent and perhaps even on this planet. Our task, as philosophers and as human beings, is a formidable one that requires imagination as well as practical consistency and intellectual honesty.

In order to set a compelling example, African relationalism (the insight that "I am because we are") arguably needs to be more radical than previously intended in African thought and practice, and more radically egalitarian. The world was neither given to us by our parents, nor was it lent to us by our children. It exists independently of us. It does not need humans. On the contrary, it is highly likely to flourish in our absence. While bees, earthworms, and dung beetles are all important with regard to the diversity and integrity of nature, the removal of *Homo sapiens* from the world's ecosystems would have a positive effect. The realization that we are the one species that the world could well do without is something we ought to continue to bear in mind, something that should inspire in us a lasting sense of humility about our role within the relational web.

Notes

1 See Agbedahin 2019; Apusigah 2011; Breidlid 2009; Chanza and DeWit 2013; Darko 2014; Echekwube 2017; Eyong 2007; Kanu and Ndubisi 2020; Noyoo 2007; Onwuegbuzie 2010; Owuor 2007; Mafongoya and Ajayi 2017; Maila and Seroto 2013; Makinde and Shorunke 2013; Mandikonza 2019; Masoga and Shokane 2019; Mosimege and Holtman 2012; Ndlovu, James and Govender 2019; Ngcoza 2019; Nkoane 2015; Ong'eta 2013; Pesanayi, O'Donoghue and Shava 2019; Ryan and Ferreira 2019.

2 An exception is offered by Gbari (2020). However, although he provides a cogent analysis both of epistemology and of knowledge, he fails to bring the implications of this analysis to bear on the rather questionable notions of African epistemology or African knowledge and truth: "African epistemology has similarities and differences in the varying conceptions of knowledge and truth in different African cultures and tradition" (189).

3 This is my translation. I have not been able to find the original source of this quotation, so I am unable to say whether or not my translation accurately conveys Hove's intended meaning.

4 See also Odora-Hoppers 2008; Ryan and Ferreira 2019, 108.

5 Emphasis here is on "the very least," which will become clearer as the essay develops. There are numerous considerations that have nothing at all to do with human interests and benefits.

6 This is, of course, former White House counsellor Kellyanne Conway's memorable phrase in defense of the blatant untruth that the 2017 presidential inauguration had attracted the largest audience ever.

7 The San are southern Africa's oldest human inhabitants, having lived in the deserts and on the plains and mountains of the sub-continent for at least the past 20,000 years and possibly going back 40,000 years. Only about 100,000 remain, predominantly in Angola, Botswana, Namibia, South Africa, and Zambia.

8 For additional examples of unsustainable "indigenous knowledge," like "slash and burn agriculture, female circumcision, and overgrazing practises," see Ong'eta 2013, 167.

9 The reference to *ubuntu* as an example or "case of 'indigenous knowledge'" is repeated in Topidi 2022, 55.

10 Crippen later adds, in a section on hunting operations in South Africa and Zimbabwe, that "Ubuntu cosmologies and care hierarchies of course prioritize people over nonhumans, as do Africapitalist worldviews" (2021, 245).

11 See, for example, Dennis (2022), who argues that concern for "environmental and general sustainability" is a matter not only of "vital resources" of "living people" (181) but also of "obligation to [human] posterity" and "intergenerational justice" (181, 187–89).

12 Similarly, as illustrated above, one might plausibly refer to indigenous acquaintance- or familiarity-type knowledge.

Bibliography

Agbedahin, Adesuwa Vanessa. 2019. "Sustainable Development, Education for Sustainable Development, and the 2030 Agenda for Sustainable Development: Emergence, Efficacy, Eminence, and Future." *Sustainable Development* 27, no. 4: 1–12.

Africa Network for Animal Welfare (ANAW). 2021. "Annual Report." Accessed January 1, 2024. www.anaw.org/Reports_and_Publications/Africa_Network_for_Animal_Welfare-ANAW-Annual_Report-2021.pdf

Apusigah, A. Atia. 2011. "Indigenous Knowledge, Cultural Values and Sustainable Development in Africa." *Academia.edu*. Accessed August 31, 2022. www.academia.edu/1067839/Indigenous_Knowledge_Cultural_Values_and_Sustainble_Development_in_Africa

Breidlid, Anders. 2009. "Culture, Indigenous Knowledge Systems and Sustainable Development: A Critical View of Education in an African Context." *International Journal of Educational Development* 29, no. 2: 140–48.

Chanza, Nelson and Anton DeWit. 2013. "Epistemological and Methodological Framework for Indigenous Knowledge in Climate Science." *Indilinga: African Journal of Indigenous Knowledge Systems* 12, no. 2: 203–16.

Crippen, Matthew. 2021. "Africapitalism, Ubuntu, and Sustainability." *Environmental Ethics* 43, no. 3: 235–59.

Darko, Isaac Nortey. 2014. "Environmental Stewardship and Indigenous Education in Africa: Looking Beyond Euro-Centric Dominated Curricula." In *Emerging Perspectives on "African Development": Speaking Differently*, edited by George Jerry Sefa Dei and Paul Banahene Adjei, 179–206. New York: Peter Lang.

Dennis, Otto. 2022. "African Environmental Intuitionism and the Obligation to Posterity." In *African Philosophy in an Intercultural Perspective*, edited by A. Graness, E. Etieyibo, and Franz Gmainer-Pranzl, 181–90. Berlin: J.B. Metzler.

Earth Charter in Action. 2000. "Annual Report 2001." *Earth Charter International Secretariat*. Accessed September 4, 2022. https://earthcharter.org/library/the-earth-charter-in-action-2000-annual-report/.

Echekwube, Anthony O. 2017. "Traditional African Identity and Sustainable Development." *Ewanlen* 1, no. 1: 104–13.

Ejike, Cyril Emeka. 2020. "African Indigenous Knowledge Systems and Philosophy." In *African Indigenous Knowledge Systems: Problems and Perspectives*, edited by Ikechukwu Anthony Kanu and Ejikemeuwa .J.O. Ndubisi, 20–49. Yola: APAS.

Eyong, Charles Takoyoh. 2007. "Indigenous Knowledge and Sustainable Development in Africa: Case Study on Central Africa." *Tribes and Tribals* 1: 121–39.

Gbari, Usman Sylvester. 2020. "Understanding the Relevance of African Epistemology for Development." In *African Indigenous Knowledge Systems: Problems and Perspectives*, edited by Ikechukwu Anthony Kanu and Ejikemeuwa J.O. Ndubisi, 174–91. Yola: APAS.

Grill, Bartholomäus. 2003. *Ach, Afrika. Berichte aus dem Inneren eines Kontinents*. Berlin: Siedler.

Hallen, Barry and J. Olubi Sodipo. 1997. *Knowledge, Belief, and Witchcraft: Analytical Experiments in African Philosophy, 2nd edition*. Stanford: Stanford University Press.

Horsthemke, Kai. 2015. *Animals and African Ethics*. Basingstoke: Palgrave Macmillan.

———. 2018a. "*Isilwane*: The Animal—*Ubuntu*, *Ukama* and Environmental Justice." In *Africa and Her Animals: Philosophical and Practical Perspectives*, edited by Rainer Ebert and Anteneh Roba, 3–21. Pretoria: Unisa Press.

———. 2018b. "African Communalism, Persons, and the Case of Non-Human Animals." *Filosofia Theoretica* 7, no. 2: 60–79.

———. 2019. "Animal Rights and Environmental Ethics in Africa: From Anthropocentrism to Non-Speciesism?" In *African Environmental Ethics: A Critical Reader*, edited by Munamato Chemhuru, 239–53. Cham: Springer.

———. 2021. *Indigenous Knowledge: Philosophical and Educational Considerations*. Lanham: Lexington Books.

Kanu, Ikechukwu Anthony and Ejikemeuwa J.O. Ndubisi, ed. 2020. *African Indigenous Knowledge Systems: Problems and Perspectives*. Yola: APAS.

Le Grange, Lesley and Loubser, Callie P. 2005. "Education for Sustainability: A Critical Appraisal." In *Environmental Education: Some South African Perspectives*, edited by Callie P. Loubser, 113–26. Pretoria: Van Schaik.

Mafongoya, Paramu L. and Oluyede Olu Ajayi, ed. 2017. *Indigenous Knowledge Systems and Climate Change Management in Africa*. Wageningen: CTA.

Maila, Mago W. and Johannes Seroto. 2013. "Indigenous Ways of Knowing and Sustainable Development in Higher Education." *Journal of Human Ecology* 43, no. 2: 165–72.

Makinde, Omawumi O. and Oludare A. Shorunke. 2013. "Exploiting the Values of Indigenous Knowledge in Attaining Sustainable Development in Nigeria: The Place of the Library." *Library Philosophy and Practice* (Winter). Accessed August 31, 2022. http://digitalcommons.unl.edu/libphilprac/908/.

Mandikonza, Caleb. 2019. "Integrating Indigenous Knowledge Practices as Context and Concepts for Learning of Curriculum Science: A Methodological Exploration." *Southern African Journal of Environmental Education* 35: 145–60.

Masoga, Mogomme Alpheus and Allucia Lulu Shokane. 2019. "Indigenous Knowledge Systems and Environmental Social Work Education: Towards Environmental Sustainability." *Southern African Journal of Environmental Education* 35: 195–205.

Mosimege, Mogege David and Lorna B. Holtman. 2012. "Biodiversity and Indigenous Knowledge in South Africa: Developments and Ethical Challenges." *Biology International* 52: 24–34.

Mukhwana, Methodius. 2019. "Chili Peppers Are Helping Uganda's Elephants and Farmers Peacefully Coexist." *African Wildlife Foundation*, February 8, 2019. www.awf.org/blog/chili-peppers-are-helping-ugandas-elephants-and-farmers-peacefully-coexist#:~:text=Chili%20peppers%20are%20a%20simple,rugs%20hung%20on%20garden%20edges.

Mutwa, Credo. 1996. *Isilwane*: *The Animal. Tales and Fables of Africa*. Cape Town: Struik.

Ndlovu, Christopher, Angela James and Nadaraj Govender. 2019. "Towards an IK-SCIE Integrative Model: A Theoretical Reflection on the Agricultural College Curriculum in Zimbabwe." *Southern African Journal of Environmental Education* 35: 177–93.

Ngcoza, Kenneth Mlungisi. 2019. "Education for Sustainable Development at the Problem-Posing Nexus of Re-appropriated Heritage Practices and the Science Curriculum." *Southern African Journal of Environmental Education* 35: 135–43.

Nkoane, Milton N. 2015. "Sustainable Rural Learning Ecologies: A Pathway to Acknowledging African Knowledge Systems in the Arena of Mainstream of Knowledge Production?" *JHEA/RESA* 13, no. 1: 33–44.

Noyoo, Ndangwa. 2007. "Indigenous Knowledge Systems and Their Relevance for Sustainable Development: A Case of Southern Africa." *Tribes and Tribals* 1: 167–72.

Odora Hoppers, Catherine A. 2008. "Culture, Language, Indigenous Knowledge and the Role of Universities in Sustainable Rural Development." *Proceedings of the Conference Universities in Southern Africa as Catalysts for Sustainable Rural Development*, Centre for Education Policy Development at the Kopanong Conference Centre in Johannesburg, March 6 and 7, 2008. Johannesburg: Centre for Education Policy Development, 29–35.

Ong'eta, Wyclife. 2013. "Indigenous Knowledge Perspective on Holistic Education and Sustainable Development." *Msingi Journal* 1, no. 1: 148–75.

Onwuegbuzie, Henrietta N. 2010. "Sustainable Development Strategy Using Indigenous Knowledge and Entrepreneurship." *Social Science Research Network*, 1–23. Accessed August 2, 2013. http://papers.ssrn.com/sol3/papers.cfm?abstract_id=1841787.

Owuor, Jenipher A. 2007. "Integrating African Indigenous Knowledge in Kenya's Formal Education System: The Potential for Sustainable Development." *Journal of Contemporary Issues in Education* 2, no. 2: 21–37.

Pesanayi, Tichaona Victor, Rob O'Donoghue and Soul Shava. 2019. "Situating Education for Sustainable Development in Southern African Philosophy and Contexts of Social-Ecological Change to Enhance Curriculum Relevance and the Common Good." *Southern African Journal of Environmental Education* 35: 213–21.

Rio+20 Portal. 2011. "Treat on Environmental Education for Sustainable Societies and Global Responsibility." August 1. http://rio20.net/en/documentos/treaty-on-environmental-education-for-sustainable-societies-and-global-responsibility/.

Ryan, Lisa and Jo-Anne Ferreira. 2019. "Pursuing Epistemological Plurality in South Africa's Eco-Schools: Discursive Rules for Knowledge Legitimation." *Southern African Journal of Environmental Education* 35: 103–21.

Schäfer, Torsten 2016. "Nachhaltigkeit: Einige Gedanken zu Begriff und Bedeutung." (Sustainability: Some thoughts on the term and meaning) *UmweltDialog*. Accessed September 4, 2022. www.umweltdialog.de/de/management/wirtschaftsethik/2018/Nachhaltigkeit-einige-Gedanken-zu-Begriff-und-Bedeutung.php.

Shava, Soul and Sibongile Masuku. 2019. "Living Currency: The Multiple Roles of Livestock in Livelihood Sustenance and Exchange in the Context of Rural Indigenous Communities in Southern Africa." *Southern African Journal of Environmental Education* 35: 77–89.

Stewart, Julia, ed. 2004. *Stewart's Quotable Africa*. Johannesburg: Penguin.

Topidi, Kyriaki. 2022. "Ubuntu as a Normative Value in the New Environmental World Order." In *Environmental Constitutionalism in the Anthropocene: Values, Principles and Actions*, edited by Domenico Amirante and Silvia Bagni, 49–67. London: Routledge.

UNESCO. 2023. "What You Need to Know About Education for Sustainable Development." Updated November 17, 2023. www.unesco.org/en/education-sustainable-development/need-know.

World Commission on Environment and Development (WCED). 1987. *Our Common Future*. Oxford: Oxford University Press.

Section II

Growing Perceptions

PRAXIS B

THE OBLIGATIONS OF OUR ECOLOGICAL RELATIONS: A CHALLENGE FOR LAND ACKNOWLEDGMENTS

Angela Robinson and Laura Terrance

This short praxis offers reflective questions to help readers move beyond performative land acknowledgments toward concrete practices that reanimate land and shift from temporal to spatial thinking in order to interrupt patterns of multispecies and colonial violence.

Challenge 1: Reanimating the Lands We Acknowledge

Wider modes of knowing life, which is the topic of this collection, invite us to be experimental, creative, and expansive in our thinking around land acknowledgments. In our short reflection, we challenge readers to move beyond the performative and human-centered tendencies of acknowledgments toward responsible enactments of our obligations and ecological relations to land. This includes reconsidering the reciprocal giving *and receiving* of care from land as a living, knowing, and animated being, as well as from the Indigenous peoples and nonhuman entities who make a place home.

At their best, land acknowledgments are attempts to reveal settler colonial relations of power, the fact that we live on violently settled terrains, and the colonial erasure that has historically dominated the space and its human and nonhuman beings.

Land acknowledgments as they currently exist, however, are primarily performative, a way to demonstrate "wokeness." In that pursuit, they often work to assuage settler anxiety and white guilt at the expense of true anti-colonial thought. It is easy to stand in front of a crowd or post a statement that simply states the

DOI: 10.4324/9781032659985-8

current conditions that settler colonialism has constructed. Doing so provides an avenue of expression without requiring further action or accountability for either the speaker or the listener. It is far more challenging and time-consuming to work with Indigenous peoples and ecologies towards a future where land acknowledgments no longer need to exist.

Perhaps, as a consequence, this leads to land acknowledgments' second shortcoming, namely the human-centered approach they take in relationship to land. Often, land acknowledgments use a framework of stewardship to describe Indigenous peoples' roles and relationships to land. Furthermore, stewardship is often framed as the ideal form of relationality non-Native peoples should aspire to. This anthropocentric approach denies the complex and diverse ways nonhuman animals and beings care for land, and, importantly, how the land cares for us. Anthropocentric approaches to land acknowledgments are unable to properly conceptualize the eco-terror settler colonialism has wreaked upon the natural world, and how this has severed relations between land, beings, and peoples.

This leads us to what we believe is the most critical limitation of land acknowledgments: the inability to conceptualize land as anything beyond an object. By de-animating and sterilizing land, acknowledgments often obfuscate the relationality and reciprocity of land as a living being, and subsequently cannot ever fully confront the history of colonization, nor the "terrestrial intelligence that lies beyond our human knowing and grasping," as Chickasaw poet Linda Hogan writes (1995, 11). Across centuries, land has come to know the horrors of dispossession, genocide, and enslavement; it has been witness to the worst humanity has to offer. When they perpetuate the de-animation and objectification of land, land acknowledgments reinforce the processes of settler colonization that render such knowledge meaningless. Hogan's affirmation of "terrestrial intelligence" reminds us that as a living being, land remembers and retains beyond the human capacity for H/history. That is, land asserts the falsity of authoritative claims to the past at the same time that it transcends and exceeds the linear way of remembering that humans often embrace. Thus, land acknowledgments, in their de-animation and sterilization of land, can never allow us to mourn *with* the land the violence and terror it has been made to know, nor do they allow for the celebration of land's regeneration under such precarity.

Our first challenge for readers is to think beyond the human and begin noticing how the lands you live on are animate, vibrant, and full of life and life-giving power. Thinking beyond the human requires us to get dirty with land, in order to, as Anishinaabe scholar Melissa Nelson says, "embody an ethic of kinship so needed in the world today to address ecological and cultural challenges" (2017, 232). If we are to imagine a future that does not require land acknowledgments, we must begin to think otherwise. Kinship allows us to develop a relationship to land that embodies reciprocity. Kin are who we look out for and who look out for us.

Questions for Reanimation:

- How do the lands you live on care for you? How do you care for them?

To deepen our engagement with these questions, we will focus on spatiality to more fully understand land as genealogical kin and the obligations of our ecological relations.

Challenge 2: The Temporal Nature of Land Acknowledgments: Thinking Spatially Allows Us to Think Relationally and With Reciprocity

While land acknowledgments are often attempts to acknowledge physical space, we find that land acknowledgments are loaded with temporal implications that, on the one hand, deaden and objectify land, its beings, and sometimes even its peoples, and, on the other hand, assume a static settler future. This settler future becomes apparent in two primary ways that we touched upon in the first challenge: land acknowledgments rarely allow for a practice-based response, but are often solely performative, and they very rarely gesture towards ways to disrupt or challenge the present state of settler colonialism, assuming this state will continue into the future. We assert that thinking spatially can move us beyond a settler future.

To do this, we bring Standing Rock Sioux scholar and activist, Vine Deloria, Jr. into this space with us. Among his many important, field-forming insights, Deloria's critique of time and linear temporality is particularly profound in that he connects Eurocentric notions of linear, progressive temporality to a distorted relationship with land. He writes, "If time becomes our primary consideration, we never seem to arrive at the reality of our existence in places" (1994, 73). In other words, Deloria cautions against a singularly temporal framework because of the ways in which it can abstract the physical, material, tangible realities of place. By thinking past the idea that settler societies operate teleologically, wrongly asserting that society is moving within a framework of "progress," spatial thinking can productively engage histories of colonization and racialization. Consequently, we can, as Deloria suggests, come to understand ourselves in relation to the spaces and places around us. For example, storytelling prompted by a tree or other marker in a specific geographical location situates the storyteller and listener spatially through a process of place-making and memory. That triad—tree, storyteller, and listener—can only fully understand that story through the place that bore it. From such a grounded point of view, we can prioritize our ecological obligations with the purpose of recognizing and being responsible to the livingness of the land and other nonhuman relations.

As Deloria argues, "Spatial thinking requires that ethical systems be related to the physical world and real human situations, not abstract principles, and are believed to be valid at all times and under all circumstances. One could project, therefore, that space must in a certain sense precede time as a consideration for thought" (1994, 72). Thus, rather than emphasizing the temporal here and now of settler colonialism and assuming this will continue—as land acknowledgments often do—thinking spatially, in its emphasis on the physical, real world, can point us towards challenging a settler colonial future. Settler colonialism within a liberal multicultural society tends to view itself as moving through time toward an ever-perfecting future. Focusing on time and progress obfuscates the material realities of settler colonialism—its exploitation, its violence, and its destruction. A focus on temporal progress likewise obscures what we know about a settler colonial future, namely, the violence and terror settler colonialism has wrought upon the land, which would lead to an unlivable future. When land acknowledgments do not think spatially, such a future remains unchallenged and the movement through time drags us further into a world of colonial death and decay.

Thinking spatially, as a methodology of grounding and materializing relations of land, allows us to see how we are dependent upon land, literally formed from the land, and that, therefore, these relations with land are ongoing, reciprocal, and genealogical. Thinking spatially might mean that instead of acknowledging land, we tell grounded stories of place that account for complex histories, make clear our obligations and relations, and animate and enliven land and its beings, all while not assuming a settler future of further destruction. These spatial approaches can gesture toward a future when land acknowledgments are no longer necessary, because they work towards an Indigenous futurity that fulfills the aims of the landback movement where land is always already known to be a relation.

Questions for Spatialization:

- What story can you tell about a place to illuminate a complexity of its history?
- Do you have any experiences that awakened you to an obligation or relation with the land or its inhabitants?
- How can animating land in a new way challenge futures of ongoing destruction?

Ultimately, part of our obligations to ecological relations is making this way of spatial thinking a practice that goes beyond a chapter in a book or recited land acknowledgment. This can, and, we challenge, *should* be an ongoing responsibility to land as a genealogical relation. How have you been formed through land? We invite you to find these stories and share them. Through our stories, located in the place and space that situates our intimacy with land, we can come to understand and fulfill our obligations to ecologies.

Bibliography

Deloria, Jr., Vine. 1994. *God Is Red: A Native View of Religion*. Wheat Ridge, Colorado: Fulcrum Publishing.

Hogan, Linda. 1995. *Dwellings: A Spiritual History of the Living World*. New York: W.W. Norton and Company.

Nelson, Melissa K. 2017. "Getting Dirty: The Eco-Erotocism of Women in Indigenous Oral Literatures." In *Critically Sovereign: Indigenous Gender, Sexuality, and Feminist Studies*, edited by Joanne Barker, 229–59. Durham, North Carolina: Duke University Press.

6 Decolonizing Listening

From Grammars of *lo Inaudito* to Expansive Onto-Epistemic Thresholds

María del Rosario Acosta López and Juan Diego Pérez Moreno

This chapter started as a reflection on the possibility of putting two different questions into dialogue: how to connect our inquiry into memory and testimony to the question of the coloniality of being or, more specifically, to the question of the colonial distribution of *what counts as being* and the forms of epistemic and ontological violence this distribution entails. At the core of this essay lies an attempt to turn the question of listening to traumatic forms of violence into the question of listening to the trauma of coloniality and its aftermath to the very end, that is, to the point where it involves not only the frames that are produced and required to *make sense* of "the world," but also ways of being in and practices that *make* worlds. Ultimately, our question aimed at decolonizing listening altogether and inquiring about what it means to *render audible* not only the colonizing effects of trauma but also the trauma of coloniality and its hegemony. What would it mean to do justice to testimonies of traumatic experiences in colonial contexts? Is it possible to decolonize the frameworks or "grammars" of sense that determine in advance the conditions of audibility in the realms of memory-building and history-making without reenacting forms of colonial violence?[1] While thinking and writing together around these questions, we realized that questioning hegemonic forms of audibility might not be enough; that even remaining in the realm of the "audible" might still allow for the sense(s) of colonial trauma to be predetermined in ways that reproduce forms of violence inherent to the colonial regime as well as to its distribution of the sensible and, as we shall see, to its underlying ontology.

To be clear, the task at hand is not to imagine ourselves outside of this sensorial regime—if this is even possible—but to understand what would be required to subvert it and resist it from within. We conceive of this task, building on previous work, as one of decolonizing aesthetics, that is, ultimately, the search for a decolonial dis-organization of the sensible committed to radically subvert the criteria—inclusive of the aesthetic, epistemic, ethical, and ultimately ontological, as we shall see—that organize and distribute sense within a hegemonic colonial regime.[2] What follows is thus an exercise in self-critique.

We begin by sketching the questions that framed our earlier approach—namely, how to render audible traumatic memory in grammars that respond to the crisis of experience and meaning that trauma introduces—as in inquiry into the connections

DOI: 10.4324/9781032659985-9

between a decolonial approach to audibility and the task of listening to colonial trauma. We then shift our focus to a historical case of traumatic *and* colonial violence: the massacre of Bahía Portete against the Wayuu Indigenous community in Colombia. In doing so, we consider how the task of making audible the sense(s) of colonial trauma and its memory involves a critical challenge of listening to the testimony of human and nonhuman beings whose claim to existence exceeds, escapes, or resists the colonial grammars of Anthropocentrism. We ultimately theorize the consequent call to create radical grammars to listen to the memory of colonial trauma beyond anthropocentric audibility, that is, by opening sense and the senses to world-making processes that divergently contravene the colonial and racist division between humans and nonhumans, between the human subject and all other so-called "natural species." As a subversive and imaginative strategy against the erasure of colonial/traumatic violence, we propose *decolonial grammars of listening* that expand onto-epistemic thresholds in which the memory of divergent worlding practices can be heard, recalled, and reclaimed in their resistance to the ontological as well as epistemic violence inherent in the colonial regime.

Grammars of *lo Inaudito,* or On Radical Forms of Listening

We first offer a brief overview of the two questions and projects that this chapter is attempting to put in connection, even as we are also proposing ways in which we need to move beyond them, described in the second part of this essay. The first question, on which we have worked alongside one another for many years, concerns developing a philosophical perspective on memory in the aftermath of traumatic forms of violence, specifically considering the thousands of testimonies of Colombian survivors of paramilitary violence in the early 2000s, during some of the most intense years of the armed conflict. We especially reflect on the epistemic and aesthetic challenges that emerge when listening to testimony in these trauma-informed contexts.[3] Grounded in questions of epistemic violence and injustice, we have stressed the fact that testimonies emerging from traumatic forms of violence often remain unheard, not only because of a lack of disposition to listen to these testimonies on "their own terms"—without having to "translate" them into already available grammars—but also because of the unprecedented forms of violence they are attempting to communicate. We use the term "traumatic violence" to denote forms of violence that are not only directed against bodies and lives but also against sense and the worlds it gives place to; that is, forms of violence that introduce such unprecedented modes of harm and unimaginable realities that they leave their survivors with a lack of grammars to make and give sense to what has happened.

Consequently, listening to testimony, in these particular cases, requires more than a commitment to render audible what has been silenced, often both institutionally as well as politically, by the same actors who, in turn, exercise an aesthetic control of the means of representation.[4] There is also the need to inaugurate, and produce—perhaps each time anew—grammars of listening, frameworks of sense that will allow rendering audible, and therefore, believable, legible, and politically recognizable, what otherwise runs the risk of remaining inaudible, both because it will remain *unheard*, and because of its *unheard-of* character and the radical challenge this poses to its communicability. What is needed, we claim, is a radical form of listening that can open the possibility of sense where there is none available in advance, without having to speak for another, but instead taking on the radical responsibility of providing a grammar for the *unheard* and *unheard-of*. We refer to this as "grammars of *lo inaudito*." In Spanish the word *inaudito* refers to both the unheard and the unheard-of, namely, what has not (yet) been rendered audible and what emerges as ethically unacceptable. We claim there is a close and critical connection between the two. It is precisely because traumatic violence inaugurates unprecedented forms of harm absolutely unheard-of, which challenge the ethical imagination in such radical ways, that the survivors and their listeners do not *yet* have the grammars to approach them properly and make them audible, much less intelligible or even believable. And with this, we risk doubling the effect of forgetfulness and inaudibility that trauma usually imposes on its survivors.

The task is also therefore double, since rendering *lo inaudito* audible entails also denouncing the kinds of erasures that are at play in traumatic forms of violence. Accordingly, it is not enough to resist the criteria that control what is to be rendered (or not) audible. To listen in these cases also means denouncing the kind of erasures that are taking place and the structures that sustain them; it is as much an aesthetic as a political and an ethical task.[5] As we've argued elsewhere:

> A grammar of "lo inaudito" opens thus a site for listening to what does not "make sense" as meaningful within available and hegemonic frameworks of meaning. While signifying, and opening a site for audibility, it is also capable of denouncing erasure, concealment, and silencing. And in this capacity to denounce and remind us of the responsibility to listen to the unheard-of, it also finds ways of interrupting and subverting the structures that have made the erasure inaudible by having presented it as illegible, unbelievable, unrecognizable, and thus forgettable, not worthy of remembrance, not legitimate enough to be heard and documented as true.
>
> (Acosta López 2023c, 165)

The second question that we have previously explored concerns the connection between grammars of *lo inaudito* and decolonial grammars of listening, more exactly, the role that the grammars of *lo inaudito* can play in the task of decolonizing sense.[6] What we call "grammars of listening" refers to the search for sites that produce sense(s) capable of rendering audible, legible, and believable the lives that traumatic violence threatens to designate as impossible. This search responds to the demand that such sites be found, produced, or wagered. Indeed, traumatic violence threatens to inflict such impossibility and illegibility, not only because of the type of destruction to which it exposes these lives, but also because the destruction occurs at the limit, and even outside, of the frameworks of meaning that usually enable the working-through of reality, both psychically and historically. Consequently, traumatic violence is a form of violence that colonizes by erasing the trace of that which a grammar of listening should strive to render audible.

Grammars of *lo inaudito*, thus, open up a site and a praxis of signification capable of subverting the breakdown of meaning produced by traumatic violence and its colonizing effects. They do so, on the one hand, through their critical capacity to bring to light, denounce and make audible the very erasures that violence has enacted and reinforced at the level of both the psychic as well as the historical. To put it another way, grammars of listening seek to interrupt the further concealment of erasure itself—its inaudibility—which allows traumatic violence to perform its colonizing function by presenting the effects of its traces as illegible to a colonized form of listening. Grammars of *lo inaudito* are also, at the same time, creative forms of resistance. In their capacity to open a space for listening to what otherwise would remain designated as "illegible"—and sometimes, too, as "unbelievable," as in not even "possible"—they confront the "abominable originality"[7] of traumatic forms of violence with the inventive capacity of imagination to produce its own frameworks where there are none. Grammars of *lo inaudito* intend to make sense out of what is otherwise interpreted as "unrecognizable," and to render perceptible what otherwise runs the risk of disappearing, and being disappeared, in a regime of the sensible that structurally cannot listen beyond the deafening roar of its own colonial and colonizing hierarchies of sense.

Any task of decolonization must consequently be directed towards the radical reconfiguration of the aesthetic regime that serves as the foundation of the criteria for existence and/as legibility. This seems to be the only way out of a kind of epistemic violence that demands the reinvention of the very conditions of possibility of sense. It is the only escape when violence not only has dispossessed these conditions of possibility and made them obsolete but has also introduced something radically other to whom it seeks to deny access to comprehension by all available means. By decolonizing the frameworks that organize and determine colonial and colonizing distributions of sense, grammars of *lo inaudito* understood as decolonial grammars can contribute to resisting and disorganizing the criteria for legibility and audibility that colonizing forms of violence not only institute but constantly actualize in their attempt to perpetuate their silencing power.

The Memory of *Jepirachi,* or Considering the Wind as Witness[8]

Having established this conceptualization of "grammars of *lo inaudito*" as aesthethic strategies that seek to creatively counteract the silencing effects of colonial erasure, let us turn now to a specific case of traumatic and colonial violence that illuminates the epistemological and ontological challenges involved in the attempt to radically decolonize hegemonic frames of sense. Dwelling on these challenges is crucial to notice, and thus prevent, the unintentional perpetuation of the colonial grammars that might still be operating in practices of radical listening and the sense(s) they make audible. Only attending to the concrete, and thus, to the singularity of the voice that speaks from out of the traumatic wound—and to how this voice is forced to speak a double tongue, *within and in resistance to* colonial organizations of sense—can one start to imagine a way to make sense otherwise. Only then can a new grammar emerge as one tries to listen to this voice, to the singularity of a claim that is each time unique, each time uttered (and listened to) anew however ancestral it might be. In order to do so, the forms of history and memory that make this claim unheard, thereby silencing the *unheard-of* speaking through it, must be deconstructed and decolonized in the process.

On April 18, 2004, a paramilitary squad belonging to the Autodefensas Unidas de Colombia (AUC) violently entered Bahía Portete, a small town on Colombia's northern coast located in the middle of the Guajira desert. The squad invaded the town with guns in hand, murdered the locals who resisted them, looted their homes, dismembered their bodies, and desecrated their cemetery before sexually torturing four women in public whose bodily remains are yet to be found. The murdered women were prominent leaders of the Wayuu Indigenous community who had refused for years to form an allegiance with paramilitary groups that forcefully aimed to establish control over their territory due to its strategic location for drug trafficking. The massacre survivors were forcefully displaced; they hid in the mangroves before starting an aimless exodus across their desert. *Their desert*, indeed, because the Wayuu have lived in this region since the time before the Spaniards or any other *alijuna*, meaning "non-Wayuu people" in the Wayuu language of Wayuunaiki, stepped on their land. In fact, the Wayuu are one of the only Indigenous societies in Latin America who resisted colonial subjugation as they retreated into the dunes to protect their people, their language, and their worlding practices from any contact with *alijuna* settlers.[9]

When listening to the testimonies of Wayuu survivors in the process of elaborating a report on the Bahía Portete massacre, researchers from Colombia's National Center for Historical Memory (CNMH) were repeatedly challenged as they tried to reconstruct the memory of these traumatic events and the effects its violence had on the community.[10] Or better put, their listening was radically challenged; the grammars that were available to them to "make sense" of the stories and testimonies being shared seemed inadequate to render those stories and testimonies truly audible. In order to reconstruct the story of the massacre and its aftermath, for instance, researchers from the Center were told to also listen to the testimony of the "wind." "Wind" here denotes indeed something like the wind and something more than what we *alijuna* imagine when this word is uttered, since in Wayuunaiki and in the Wayuu world, the entity that is translated as "wind" is a manifold being. Each of its manifestations is linked to a specific mythological force which, in turn, becomes tangible through its actions on the Wayuu territory, that is, both on the desert's materiality and on the rhythm of the assemblages and worlding practices embedded in it.

The most influential and dominant of these manifestations is the trade wind of the Northeast also known as *jepirachi*.[11] The name of this wind-entity alludes to its movement throughout the peninsula: *jepirachi* blows toward the northern region of jepira, the sacred realm of the dead, whose point of entry is located at the northern cape called Cabo de la Vela. To accompany the spirits of the dead, or *yoluja*, in their journey to *jepira*, the Wayuu first bury the bodies and then exhume and rebury the bones in a second burial. The initial wake and burial take place in the days immediately after death, as every *yoluja* begins a journey toward this northern region under the guidance of *jepirachi*. Following the signs they receive in dreams, elderly women then prepare the bodies for their final burial. It is only then, when the bones are buried in the mother's territory years after their death, that the spirits reencounter their ancestors in the original time and space of *jepira*. Their life journey ends there: the balance between life and death is restored as the spirits transform into the rain of the new season (Finol 2007, 265–85).

The torture, humiliation, and disappearance of the bodies of Wayuu women in the macabre spectacle of Bahía Portete's massacre not only disturbs the life of the community through its unimaginable violence, but also interrupts Wayuu mourning rituals and, in consequence, disrupts the balance between life and death that sustains the community's social fabric as well as the fabric of the world that Wayuu worlding practices engender. The traumatic absence of bodies to be mourned, as well as the violence these bodies had to undergo as a social spectacle—intended as such by the paramilitaries, who knew very well the symbolic effect and disturbance such a spectacle would have on the community—forecloses the access of the *yoluja* to *jepira*. As stated in the CNMH's report cited above,

> as facilitators of the work of mourning and the rituals that make sure the dead undertake their journey towards *jepira*, Wayuu women play a fundamental role in healing the pain that violent events cause in their families and communities. They haven't been able to carry out this duty since the bodies have not been found" (Grupo de Memoria Historica 2010, 97; trans. Acosta López and Pérez Moreno).[12]

The fact that these *yoluja* can't reach the end of their journey not only means they cannot return to the cycle of life they belong to, a cycle that is one of the very foundations of the Wayuu world, it also means they can't be mourned, that their lives cannot be remembered nor retold since, strictly speaking, they have not yet ended. Being deprived of a proper burial, neither dead nor alive, they impossibly wander toward *jepira* somewhere between life and death in the obliterated space marked by their absent bodies. Living an all too sudden and yet belated death—that is, the (im)possible death of forced disappearance—the violence inflicted on these *disappeared* bodies speak of something radically unheard of surrounding these traumatic events. In the words of the CNMH's report on memory-making practices of Indigenous communities in Colombia, the massacre of Bahía Portete is "a fracture in the flow between life and death that implies a profound alteration in the way life is produced, circulated, and reproduced collectively in Wayuu society" (Grupo de Memoria Histórica and ONIC 2019, 186).[13]

What kind of attunement is needed to understand, make audible, or at least render communicable the kind of wound that is left here as an untraceable trace? How can one listen to this radical disturbance of sense altogether *and* acknowledge the worlding practices it attempts to fracture and erase? For the survivors of Bahía Portete, the trade wind of the North still bears witness and carries in its breath the memory of the *yoluja*'s truncated journeys to *jepira*. It is the testimony of *jepirachi*, then, that ought to be heard to reclaim the existence of the disappeared bodies in a present in which they are otherwise untraceable. The testimony of *jepirachi* introduces a grammar of the unheard-of as it bears witness to their sudden-yet-belated death, while also inscribing a form of memory and survival that resists and has resisted for centuries the erasures of colonial history. How do we respond to the surviving call of the spirits that speak through the memory of this wind-being? How to listen to the voices of *yoluja* as they blow in it? Are we still in the realm of colonial testimony? And are we still, can we still be, in the realm of listening *as we know it* given that this would be a testimony that, even before claiming the existence of the unburied bodies, and as a condition to do so, would first claim the existence of beings that speak from a more-than-human locus of enunciation? Indeed, what would it even mean to render the unheard-of audible when all that is left, all that is audible at least *for yoluja ears*, is the sound of the wind and the traces it waves on the sand of the desert? Perhaps our task here is not even, and no longer, about listening, especially if listening is reduced to the faculty of making sense audible *to human ears*. Perhaps listening here requires letting sense resonate in ways different from, and not entirely reducible to, the anthropocentric realm of the audible? In this task, one finds no longer (or not only) a grammar, but something closer to an ontology, given that listening can only take place here if we dwell in the ontological threshold—the threshold between worlds and worlding practices—that is opened at the edge of the (in)audible.[14]

Listening To/From Onto-Epistemic Thresholds: Towards the Anthropo-Not-Seen (Not-Heard)

How, then, to enact a form of listening that can render audible what escapes or resists the anthropocentric regime of audibility? How to listen to the traumatic memory that blows through *jepirachi as a witness*? In order to approach these questions in all their complexity, it is not enough then to rethink the available grammars that allow an entity to count, to be audible, and to *make sense* as a witness. The question, ultimately, is not only about "acknowledging" the ontological status of *jepirachi* as a nonhuman "person." The risk, if we remain only with that kind of task at hand, could be that of still being trapped within the realm of the "human" as the only way to listen to a testimony, the only way to count as audible.[15] To be sure, the question is not to extend humanity to "nonhuman" beings; this apparently inclusive move would reiterate the ontological division between the human and the nonhuman that deprives all beings that are deemed to be not (or not completely) human of the "human" capacity to make sense audible. The question, as suggested, might be rather about critically reframing the criteria that connect sense to audibility altogether. This includes, but is not reduced to, questioning the criteria that: (1) interpret the voice as an exclusively human attribute; (2) requires any form of audible sense *to be interpreted as a voice* in order to *make sense*; and (3) assumes all forms of sense and all forms of making sense take place in the realm of audibility. Such a critical take could expose us to forms of sense that can be rendered meaningful without being translated into a "grammar of listening," even if by listening we mean, indeed, listening to *lo inaudito*.

We must take seriously the radical character of this claim. Its underlying demand is, first, not to grant or extend humanity *as an attribute* to "nonhuman" entities through specular projection, posing the existence of objectified beings in "nature" as subjects whose alleged subjectivity would mirror, and thus leave unquestioned, the ontological primacy of the human subject. If *jepirachi* is to be taken as a witness, we must first admit that "the common point of reference for all beings of 'nature' is not humankind as a species but rather humanity as a condition," that is, the condition of having a point of view or a *perspective* (Descola 1986, 120). This condition is extended to more-than-human entities such as *jepirachi*, whose existence *disobeys* the colonial distinction between "human subject" and "non-human nature," but this extension does not make them merely a mirrorlike iteration or projection of human species. On the contrary, the condition of having a perspective is extendable *as an always singular and plural one* precisely because it is epistemologically and ontologically irreducible to the allegedly universal perspective of the human species as *the one and only* reference of personhood. Viveiros de Castro theorizes this turn to "perspectivism" in his study of the challenges that Amazonian metaphysics pose to the foundational dualisms of Western ontology, particularly to the human/nonhuman binary. In his words,

> All animals and cosmic constituents in Amazonian cosmologies are intensively and virtually persons, because all of them, it does not matter which, can reveal themselves to be (transformed? into) a person. This is not a simple logical possibility but an ontological potentiality. "Personhood" and "perspectiveness"—the capacity to occupy a point of view—is a question of degree, context and position rather than a property distinct to specific species. Certain "non-humans" actualize this potential more fully than others, and some, moreover, manifest it with a superior intensity than does our species and are, in this sense, "more human than humans".
>
> (Viveiros de Castro 2013, 24)

The fact that this "perspectiveness" is not reducible to a single epistemological and ontological position does not foreclose, however, the possibility of finding a "common ground" between entities whose diverging perspectives thus resist the logic of sameness and difference. Indeed, perspectivism asks us precisely to reframe what is common among perspectives, and to do so by resisting the idea of a universal quality—a single position or referent—that all perspectives would share to count as such. The "common ground" cannot be a quality or a given status—for instance, the status of being human—that predetermines which perspectives can be included in the relationship, implicitly excluding those who do not share that quality by othering them, that is, *by making this different*. The common ground, we argue expanding on Viveiros de Castro, is *the space of the relationship itself:* the ever-changing, multiple, and varied entanglements in which the common is (re)produced, as Raquel Gutiérrez Aguilar claims, through a constant renegotiation that acknowledges its irreducibility to a single, unaltered, and totalizing perspective (2016, 79–92).

The demand to listen to *jepirachi*'s testimony is, at its core, the demand to listen to a witness whose memory speaks from and through the onto-epistemic distance this irreducibility entails. The testimony of this more-than-human entity provides a form of memory that acknowledges the effects of traumatic violence by simultaneously reclaiming the existence of entities, and of the worlding practices they emerge from/with such entities, that colonial culturalism (and its reiteration in the neutralizing politics of inclusion in neoliberal multiculturalism) reduces to a set of "cultural beliefs" belonging to the colonial other.[16] Asserting *jepirachi* as a witness is, first and foremost, asserting that it is not the manifestation of a "belief" pertaining to a cultural worldview as opposed to its "true" or "real" status as the wind, that is, as a part of objectified nonhuman nature. Neither human nor nonhuman, neither subject nor nature, *jepirachi is a witness* insofar as its existence speaks of, and emerges with, worlding practices whose divergence is grounded on perspectivism as an ontological principle: one that opens the way to radically variable and coexisting, multispecies ontologies that are irreducible to the same referent or, more precisely, *to sameness as the point of reference*, to which we will return shortly.

The claim to listen—or, rather, to open our ears and our grammars even beyond anthropocentric audibility—to a divergent testimony like *jepirachi*'s is, therefore, one that both underscores and questions the *othering* operations through which the colonial world order naturalizes the criteria that determine what's "real" and what is "belief," and, in consequence, what counts as "audible" in order to be intelligible, that is, in order to *make sense*. Anything—any being, practice, idea, or thing—that exceeds and disobeys this colonial hegemony is consequently deprived of any claim to have the audibility that grants legitimate ontological status.

Listening, or opening ourselves beyond listening, to *jepirachi*'s testimony confronts us with an excess which, in fact, is the inextricable overlapping of two excesses. On the one hand, we face the *epistemological excess* of traumatic experience, that is, both the absence and excess of a form of memory that cannot be uttered into experience with the available grammars and frames of sense and which, in fact, often hinder its audibility. On the other, we confront the *ontological excess* of a witness, neither human nor nonhuman, such as *jepirachi*, that speaks of the memory and the actual existence of world-making processes that actively disobey and disarray the ontological separation between human "culture" and nonhuman "nature." *Jepirachi*'s testimony exposes us, then, to an *onto-epistemic excess*: the excess of a perspective whose voice—or perhaps another utterance different from the anthropocentric notion of a "voice"—speaks from outside the limit of colonial ontology and its nature and culture divide.

Like other complex entities inhabiting Amerindian cosmologies and worldings, *jepirachi* is not—that is, "does not exist"—within hegemonic colonial grammars as it is irreducible both to its subject of knowledge (the Anthropos) and to its objectified *other* (nature).[17] Neither human nor nonhuman, *jepirachi is not* because its excess is only partially reducible to the classificatory terms of the anthropocentric speciesist ontology it therefore defies. We should thus rephrase the question posed before: How can we listen to the memory of bodies, beings, and worlds—such as *jepirachi* and the Wayuu world—whose existence exceeds, escapes, or resists the ontological divide between nature and culture, that is, the founding principle of Anthropocentrism? How can we tune our ears to a voice (is it still a voice?), to the sounding of a memory that is *not only* human nor *not only* nonhuman, a memory that speaks before and beyond, from within and from without, colonial audibility? How can we make audible a form of sense that diverges from the voice and voicing effects of the colonial subject? In short, how can *we listeners* resist the colonizing frameworks of a "grammar of listening" whose "radical ear" may still only make audible a memory uttered in the voice of the Anthropos, the colonial and colonizing "human?"

The task is, then, to emancipate our grammars and our ears—our experience of sense and our senses altogether—from this onto-epistemic mandate; a mandate that depends on the obliteration and disavowal of the multiplicity of existents, assemblages, and worlds that are reduced to its dialectical *other*, namely, the mute, objectified and racialized monolith of "nature." Marisol de la Cadena conceptualizes the structural obliteration of such non-beings, and the ontological excess that emerges through their erasure, as the "anthropo-not-seen." In her words,

> The anthropo-not-seen mentions existents that are within a historically formulated hegemonic condition of impossibility: they simply cannot be—therefore they are not-seen, not-heard, not-felt, not-known. Importantly, they exceed radical conceptualizations like that of the human as more-than-human, or the history of human nature as the history of interspecies relationship: neither *geos* nor *bios*, the anthropo-not-seen that is all entities that are through relations that "cannot be"...may be dangerously denied being (or benevolently ignored as unimportant) if their possibility is not considered in a "complex we" grounded in divergence (rather than in sameness).
>
> (2019, 483)

As De la Cadena points out, it is not enough to listen to what has not previously been seen, heard, perceived, felt, and, consequently, known; what is needed is a disarray of the ontological regime that radically obliterates the possibility of being otherwise, perhaps even outside of and incommensurable with, the aesthetic regime that forecloses in advance not only what we hear but what it means to be rendered audible. In order to radically open ourselves to truly listen to the violence of coloniality and the unprecedented meanings it has introduced and continues actualizing in the present, the task must perhaps be reconsidered beyond that of *making audible the unheard-of*. If what we attempt to listen to is the roar of the onto-epistemic excesses the Anthropocene structurally *does not see*, per De la Cadena's words, then grammars of *lo inaudito* might turn out to be insufficient to contravene the colonial aesthetic, epistemological, and ontological regime.

Colonial obliteration not only occurs at the level of the grammars and/or frameworks that organize our sense but also (should we say even beyond?) at the level of what it means *to be* in terms of being-seen (perceived/known/heard). Disturbing the criteria that determine in advance what deserves to be (or not be) audible is a critical step, yet an insufficient one, if the task is not only to subvert the available grammars but, even further, to question whether audibility is indeed adequate to make and let the sounding of sense that emerges from colonial wounds circulate. Indeed, beyond the anthropocentric regime of audibility and the inaudible, there is sense; *jepirachi*'s testimony is not another kind of voice, it is a sounding that may have to *take over* the regime of the audible altogether to let itself, its testimony, and its memory be "heard." The inscription of this divergent testimony, like the traces it leaves on the sand, is not to be rendered legible; its existence is rather unquestionable and irreducible to any aesth-ethic that attempts to listen to its sound *as the voice of logos*. Strictly speaking, listening to *jepirachi*'s testimony is not, or not only, about listening; it is rather about opening ourselves to the sense(s) that circulate among a plurality of beings and worlds. Blaser and De la Cadena (2018), as well as Escobar (2020), among others, have conceptualized this plurality as the pluriverse: "the practice of a world of many worlds is what we call a pluriverse: heterogeneous worldlings coming together as a political ecology of practices, negotiating their difficult being together in heterogeneity" (Blaser and De la Cadena 2018, 4). Before anything else, *jepirachi*'s testimony bears witness to this pluriversal heterogeneity.

Let's then rephrase our question one last time: How does one listen to the memory of anthropo-not-seen existents in the present—a present that would disobey the implicit ontological "agreement" that only makes audible the voice of human existents in (their monologic) history? How can one make audible the reverberation of a sense here and now as it exceeds, and thus dismantles the "hegemonic condition of impossibility" that silences the very claim to existence of the world that emerges with it? And ultimately, how does one listen to the resilient soundings of its memory *before and beyond* the silencing effects of the anthropocentric I/eye and its colonial, racist, speciesist history? Learning to see, hear, feel, and know the "oxymoronic emergence" (De la Cadena 2019, 483) of entities like *jepirachi* demands not only a rearrangement of sense and the senses—namely, an aesthetic reconfiguration—but a radical decolonizing ontological revolution. The threshold between the former and the latter, between a subversion of the grammars that organize sense, and a decolonial and/or decolonized ontology, is what *jepirachi* opens up with its testimony.

How to cross from one to the other might still be a question of how to listen to a sound that is not a human voice, how to inscribe a testimony that cannot be reduced to the story of a human subject, and how to rethink history altogether in the light of this incommensurability: that of a time and a world that are not yet—or not entirely—meant to be turned into the "past," or even into a "memory," precisely because they can only emerge *as such* in their resistance to do so. Indeed, what the event of this emergence demands of any grammar attempting to render audible its *neither human nor nonhuman* roar, is a radical *interruption-as-redistribution* of the colonial sensorium. This kind of interruption could provoke, in De la Cadena's words, "an ontological disagreement among those who [intentionally or inadvertently] share sameness, thus inaugurating an altogether different practice of politics: one across divergence" (2015b, 65), one that takes place not between "equals" but rather across incommensurable worlds.[18]

If we want to listen anew to the diverging memory of existents like *jepirachi*, we must first open ourselves to feeling, sensing, and making sense otherwise.

"Listening" here involves not only our sense of hearing, but a reorganization of the hegemonic and hierarchical partition of the senses and, even more so, a radical departure from the anthropocentric ways of feeling and being. The task is therefore to open our bodies to being-with otherness beyond the colonial paradigm of difference, which always returns to the sameness of Anthropos as the measure of all that is and can be. We must search for ways to embrace the mutual excess of radical divergence and thus interrupt the colonial, speciesist, and universal imperative of *othering* that pervasively permeates our senses, our perception, and our grammars in a present that must be decolonized. We must render audible—that is, render perceptible—the power of radical divergence to subvert the colonial violence that *others* everything that is not "common" to what John Law has called the "one-world world," which assimilates all other worlding practices by presenting itself as the absolute measure of everything that is and can be, thus eliminating possibilities for what exceeds its limits (2015, 126–39).[19]

Such an aesth-ethic commitment to a decolonial, multispecies, and pluriversal sensorium entails creative strategies against both the silencing effects of traumatic colonial violence and the voicing effects of colonial and anthropocentric ventriloquism. As such, decolonial grammars of *lo inaudito* aim not to render audible the "impossible" others, but rather to open what we could call onto-epistemic thresholds. Building upon Mary Louise Pratt's foundational definition of contact zones as "social spaces where cultures meet, clash, and grapple with each other, often in contexts of highly asymmetrical relations of power" (1991, 36), we propose onto-epistemic thresholds as contested and unpredictable contact zones between divergent worlds where the multilayered sense of colonial trauma can be *sensed yet never completely understood.* Indeed, understanding grammars of *lo inaudito* as expansive onto-epistemic thresholds ultimately implies that the task of listening here is not that of producing sense but rather of opening a site where sense can be rendered audible, circulate, and resonate *in its critical, ontological incommensurability*. Listening to the resisting and unsettling call of this incommensurable sense—that is, to its unprecedented emergence *in between* worlding practices that "meet, clash and grapple with each other," defying the hegemonic grammars that make such emergence "impossible"—is perhaps the first step in recalling and reclaiming the liberatory power of decolonial memory. Such a memory renders the unheard-of audible as it multiplies and "un-commons" the textures of the present world, opening it to the plural worlds of the plural present that we—an ever-complex "we"—divergently share.[20]

Notes

1 For an earlier essay that begins to address some of these questions, as well as the conceptual background that explains in what sense we speak here of practices of "memory-building" and "history-making," see Acosta López 2022a.

2 Our common inquiry on decolonial aesthetics started in the panel "Estéticas Decoloniales: Resistencia, Desciframiento, Posibilidad" that we co-organized with Miguel Gualdrón for the *Philosophies of Liberation Encuentro II* in September 2020.

We first conceptualized decolonial aesthetics as practices that configure modalities of aesthetic experience that expose, subvert, and resist spatiotemporal structures inherent to the logic of coloniality, which includes epistemic operations, perceptual grammars, and how they over-determine and thus deprive bodies of any possibility of sense that unsettles the colonial regime. Decolonial aesthetics, therefore, can be understood as an aesthetics of resistance, emerging from resistance while also producing its (aesthetic) conditions of possibility, which entails a critique of coloniality rooted in the commitment to claim other ways of feeling, knowing, and being. See Acosta López 2021 and 2022b.

3 Our work began at a time (2010–14) when the Historical Memory Center in Colombia (CNMH) had started to recover survivor testimonies, organize them as public reports, and institutionalize them as political memory. For a detailed description of the Center's work, its role in institutionalizing a "politics of memory" in Colombia, and the effect its work had on current discussions of memory-building initiatives in the country, see Stern 2018 and Acosta López 2023a.

4 What is particularly hard about state-sponsored forms of violence is that they are not only devastating in their effects, but they also produce a narrative of legitimacy that normalizes and invisibilizes violence *as* violence, making it radically inaudible at the structural level as well as in its more "visible" effects. See Acosta López 2023b.

5 This is the reason why, in other places, we've referred to the task at hand as *aesth-ethic*, to highlight the fact that the ways in which we organize sense are intrinsically connected to an epistemic and political responsibility, to the responsibility to listen to the *unheard-of*; and thus, ultimately, to an ethical task. See Acosta López 2019.

6 Some of the following passages are taken from Acosta López 2022a, 213–14.

7 The expression is Hannah Arendt's (2004, 309); she uses it to refer to the kind of violence inaugurated by the Nazi death camps.

8 Parts of this section are in close dialogue with the catalog entry we wrote for the artwork installation *Deserere* by Colombian artist Clemencia Echeverri (see Acosta López and Pérez Moreno 2022, 16–44). That work, as is the case with our the present essay, is an attempt to approach the memories of Bahía Portete's massacre and its aftermath without translating them into the available grammars of sense that might reproduce the colonial forms of erasure underlying the massacre itself. We thank Echeverri for the generous conversations that foregrounded our contribution to the catalog as well as our ongoing work on this subject. Many of the reflections included in this section of our essay are a direct result of a dialogue with the artist that started around her work.

9 Anthropologists like Mario Blaser, Marisol de la Cadena, Arturo Escobar, Tim Ingold, and Eduardo Viveiros de Castro, among others, describe "worlding practices" as the quotidian actions that embody the network of relationships between human and more-than-human entities and the relational existence these practices *give place to*: their situated world. See, for example, Escobar 2021, 67–83.

10 For a detailed reflection on the methodological challenges that the testimonies of Wayuu survivors posed to the CNMH's previous exercises in reconstructing "historical truth," see the opening chapters of the official report on the case (Grupo de Memoria Histórica 2010, 15–35.)

11 For a comprehensive description of the ontological status and the role *jepirachi* plays in Wayuu worlding practices, see Guerra 2019, 249–54.

12 All subsequent translations of reports from Grupo de Memoria Histórica are also provided by the authors.

13 It is noteworthy that this specific report, which is co-authored with members of indigenous communities directly affected by the Colombian armed conflict, explicitly

acknowledges the traumatic violence of the conflict as an iteration of colonial forms of violence that started with the so-called "conquest" of their ancestral territories.

14 Our reflections here are in close dialogue with the recently produced work through Colombia's Truth Commission's final report, as part of the Testimonial Volume, and specifically within the interactive part of the volume collected under the title "Sounds and Memories." As Commissioner Alejandro Castillejo states in the brief introduction to this initiative uploaded on their website, the Commission has interpreted the mandate to collect memories of pain produced by the war as extended beyond human testimonies. The Commission acknowledges there are more-than-human subjectivities imbued with the agency to give testimony. The volume includes at least three hours total of recordings of "sonorous landscapes," and starts with an introduction that asks, first, what is and what counts as a "voice." Some of those sonorous landscapes were recorded in Maicao, in La Guajira region, and may possibly include, for those who can hear it, the sound of *jepirachi*. See Comisión de la Verdad de Colombia 2022, 9–15.

15 Kate Jenckes (2017) expands on Jacques Derrida's philosophy of difference and alterity in her examination of testimony within post-dictatorship literature and visual art in Chile and Argentina, challenging the conventional notion of the human voice as *the voice* of the witness. Some of the valuable questions she poses on the rethinking of sense from a non-anthropocentric perspective intersect with our questions here, yet our approach addresses Anthropocentrism in the context of colonial frameworks, thus departing from metaphysical inquiries into the connection between representation, traumatic violence, and death.

16 In Viveiros de Castro's words, "Cultural relativism imagines a diversity of subjective and partial representations (*cultures*) referring to an objective and universal nature, exterior to representation. Amerindians, on the other hand, propose a representative or phenomenological unity (the capacity to have a point of view) that is purely pronominal in kind applied to a real radical diversity" (2015, 25).

17 See, for example, Marisol de la Cadena's comprehensive ethnography of Andean worlding practices around/with Ausangate, an entity that emerges as mountain (*geos*) and *tirakuna* (an "earth being") in Cuzco, Peru (2015a), as well as Eduardo Kohn's engagement with beings like *runa puma*, a jaguar-person neither human nor animal, during his extensive fieldwork in Ecuador's Upper Forests (2013, 3–27).

18 De la Cadena's argument on the need to reconfigure hegemonic ontological "agreements" as an opening toward radically transformative politics—that is, towards a praxis of politics that questions the given distribution of the sensible, its neutralizing power effects, its *normalizing and normalized* grammars—resonates with poststructuralist approaches to the intersections between ontology, community, and politics, however arguably anthropocentric as they might be. See Acosta López 2017.

19 For a larger discussion on the concept in the context of political ontology, and the pluriverse in particular, see the introduction of Blaser and De la Cadena 2018.

20 We thank all the participants of our reading group "Lecturas decoloniales" for their generous reading of a previous version of this essay, as well as for previous conversations that have nurtured our collaborative work around decolonial ontologies.

Bibliography

Acosta López, María del Rosario. 2017. "Ontology as Critique: On Jean-Luc Nancy's Inoperative Community." *Research in Phenomenology* 47: 108–23.

———. 2019. "*One Hundred Years of Forgotteness*: Aesth-Ethics of Memory in Latin America." *Philosophical Readings, Special Issue on Philosophy in Colombia* 11, no. 3: 163–71.

———. 2021. "From Aesthetics as Critique to Grammars of Listening: On Reconfiguring Sensibility as a Political Project." *Journal of World Philosophies* 6: 139–56.

———. 2022a. "*Gramáticas de lo inaudito* as Decolonial Grammars: Notes for a Decolonization of Memory." *Research in Phenomenology* 52, no. 2: 203–22.

———. 2022b. "Hamaca Paraguaya de Paz Encina: Notas Sobre la Resistencia *de* la Memoria." *Huellas* 110: 100–9.

———. 2023a. "'Making Memory': Historical Memory in Colombia and Its Legacies." In *Memory Work in Colombia: Past and Present Experiences, Legacies for the Future, bilingual edition*. World Humanities Report CHCI–South America. Edited by Maria del Rosario Acosta López. Accessed November 1, 2023. https://worldhumanitiesreport.org/region/americas/

———. 2023b. "Storytelling as Grammars of Listening: On Trauma, Violence, and History in Benjamin's 'The Storyteller.'" *Journal of Continental Philosophy* 3, no. 1: 135–57.

———. 2023c. "Grammars of Listening: Or on the Difficulty of Rendering Trauma Audible." *Royal Institute of Philosophy Supplement* 93: 153–70.

Acosta López, María del Rosario and Juan Diego Pérez Moreno. 2022 "La Memoria de Jepirachi/The Memory of Jepirachi." In *Deserere*, edited by Clemencia Echeverri, 16–44. Bogotá: Espacio Continuo.

Arendt, Hannah. 2004. *Essays in Understanding, 1930–1954: Formation, Exile, and Totalitarianism.* New York: Schocken.

Blaser, Mario and Marisol de la Cadena, ed. 2018. *A World of Many Worlds*. Durham: Duke University Press.

De la Cadena, Marisol. 2015a. *Earth Beings: Ecologies of Practice Across Andean Worlds.* Durham: Duke Univresity Press.

———. 2015b. "Uncommoning Nature." *e-flux journal* 65. Accessed November 1, 2023. www.e-flux.com/journal/65/336365/uncommoning-nature/

———. 2019. "An Invitation to Live Together: Making the 'Complex We.'" *Environmental Humanities* 11, no. 2: 477–84.

Comisión de la Verdad de Colombia. 2022. *Cuando los Pájaros No Cantaban. Volumen Testimonial del Informe Final de la Comisión de la Verdad* (When Birds Did Not Sing: A Testimonial Volume of the the Colombian Truth Commission's Final Report). Bogotá: Puntoaparte.

Descola, Phillipe. 1986. *La Nature Domestique: Symbolisme et Praxis dans l'écologie des Achuar* (Domestic Nature: Symbolism and Praxis in Achuar Ecology). Paris: Maison des Sciences de L'Homme.

Escobar, Arturo. 2021. *Pluriversal Politics: The Real and the Possible.* Durham: Duke University Press.

Finol, José Enrique. 2007. *Mito y Cultura Guajira: Contribución a la Semiótica del Mito Guajiro* (Myth and Culture from the Guajira: Contributions to the Semictics of the Guajiro Myth). Maracaibo: Ediciones Universidad de Zulia.

Grupo de Memoria Histórica and ONIC (National Indigenous Organization of Colombia). 2019. *Tiempos de Vida y Muerte. Memorias y Luchas de los Pueblos Indígenas de Colombia* (Times of Life and Death: Memories and Struggles of the Indigenous Peoples of Colombia). Bogotá: Taurus.

Grupo de Memoria Histórica. 2010. *La Masacre de Bahía Portete: Mujeres Wayuu en la Mira* (The Bahía Portete Massacre: Wayuu Women in the Crosshairs). Bogotá: Taurus.

Guerra, Weidler. 2019. *Ontología Wayuu: Categorización, Identificación y Relaciones de los Seres en la Sociedad Indígena de la Península de La Guajira, Colombia* (Wayuu Ontology: Categorization, Identification, and Relationships of Beings in the Indigenous Society of the La Guajira Peninsula, Colombia). Bogotá: Universidad de los Andes.
Gutiérrez Aguilar, Raquel, Lucia Linsalata, and Mina Lorena Navarro Trujillo. 2016. "Producing the Common and Reproducing Life: Keys Towards Rethinking the Political." In *Social Sciences for an Other Politics: Women Theorizing Without Parachutes*, edited by Ana Cecilia Dinerstein, 79–92. New York: Palgrave Macmillan.
Jenckes, Kate. 2017. *Witnessing Beyond the Human: Addressing the Alterity of the Other in Post-coup Chile and Argentina.* New York: SUNY University Press.
Kohn, Eduardo. 2013. *How Forests Think: Toward an Anthropology Beyond the Human.* Berkeley: University of California Press.
Law, John. 2015. "What's Wrong with a OneWorld World?" *Distinktion: Scandinavian Journal of Social Theory* 16, no. 1: 126–39.
Pratt, Mary Louise. 1991. "Arts of the Contact Zone." *Profession*: 33–40.
Stern, Steve. 2018. *La Memoria Nos Abre Camino: Balance Metodológico del CNHM para el Esclarecimiento Histórico* (Memory Opens the Way for Us: A Methodological Balance of the CNHM for Historical Clarification). Bogotá: Centro de Memoria Histórica. www.jstor.org/stable/25595469
Viveiros de Castro, Eduardo. 2013. "Cannibal Metaphysics: Amerindian Perspectivism." *Radical Philosophy* 182 (November/December): 15–28.
———. 2015. *The Relative Native: Essays on Indigenous Conceptual Worlds.* Chicago: Hau Books.

7 The Tree of Life-Death

On the Vegetal Wisdom of Life in the *Book of Zohar*

Michael Marder

The separation of knowledge from life has long and far-reaching roots in the West. The Biblical tradition mentions two trees growing in the middle of the Garden of Eden, the tree of life and the tree of knowledge of good and evil (Genesis 2:9), as well as God's prohibition to eat the fruit of the latter tree. The original sin both results from and produces multiple splits and separations: of the forbidden fruit from the branch, on which it ripens; of fallen humanity from the place where the tree of life grows; and, ultimately, of knowledge from life itself. Henceforth, knowledge pertains to death, and it deadens everyone who is living by converting them into its objects. Outside the Edenic condition, knowledge and life become mutually exclusive. That is why attempts to recover a living knowledge, the knowledge of and for life as its source and destination, appear fanciful.

In other traditions, the cosmic tree often does not evince such a separation. In the *Bhagavad Gītā*, the eternal *aśvattha* tree is a tree at once of life and of knowledge. Resembling a fig or a banyan, it is imperishable (*avyayaṁ*), "having its roots above and branches below, / its leaves are the (Vedic) hymns. /" (15.1, trans. Sargeant). Immediately, though, *aśvattha*'s hierarchical arrangement is put in question: "Below and above [*adhaścordhvaṁ*] its branches spread, / nourished by the qualities, with objects of the senses as sprouts; / and below its roots stretch forth / engendering action in the world of men" (15.2). The realms above and below merge into a single domain above-below, *adhaścordhvaṁ*, just as living and knowing, sensing-thinking and acting, are different parts of the same vegetal being.

The stark contrast between "Biblical" and "Eastern" worldviews grows more complicated as soon as one takes into account the mystical strands of Judaism, particularly the *Book of Zohar*, the centerpiece of the medieval Jewish Kabbalah. Here, the tree of life, in the shape of the divine emanation *tif'eret* also associated with the Torah, continues to exert its influence on the world and, in turn, to be influenced by human actions, whether sinful or righteous. Indeed, the entire dynamic structure of ramified divine emanations (the Judaic version of the cosmic tree) is understood as a tree of life, suggesting a metonymic relation between a part (*tif'eret*/Torah) and the whole. And it is the appreciation of the tree, comprised of divine emanations, in its inherent unity that spells out the wisdom of *Zohar*, which is of a piece with life

DOI: 10.4324/9781032659985-10

itself, as opposed to knowledge that separates, analyzes, and so deadens everything that is, including the lower parts of this tree, detaching *šeḵīnah* from *tif'eret*.

Eating/Knowing Toward Ecological Receptivity

Jewish mystical tradition interprets the prohibition of eating from the Tree of Knowledge of good and evil with what I would call strategic minimalism. *Zohar* cuts the injunction halfway—"From every tree of the garden you are free to eat"—disrupting the semantic and hermeneutical flow between Genesis 2:16 and the following verses.[1] Separated from the rest of the text, the prohibition morphs into a permission, according to which "everything was permitted to him [to Adam] if eaten in oneness [*be-yiḥudah*)]" (1:35b; v. 1, 222).[2] On the face of it, the exegetical thrust of *Zohar* does the opposite of what it recommends, partitioning the Biblical narrative, while advocating a holistic understanding of the command. The point, though, is that Adam and Eve are forewarned not about *what* cannot be taken and appropriated from their environs, but about the *how* of the taking, namely a seizing that disrespects the unity of the taken with life and creation. Knowledge, which is the sublimation of digestion and which is literally present in its original form in eating, must strive toward the unity (*yiḥudah*) of existence: of life as such, of life and death, of life and knowledge. In other words, mystical knowledge must be, of necessity, ecological.

The wisdom of the forefathers (Abraham, Isaac, and Jacob) consists in the special know-how of abiding by the flow of divine emanations received as a whole, knowing without analyzing reality into bits: "they all ate [from the tree of knowledge of good and evil] and endured [*aḵalu ve-ḥayu*]" (1:35b; v. 1, 222). Literally: "they ate and lived [*ve-ḥayu*]," knowing in a way that is consistent with life (*ḥayyim*), the way that maintains and dispenses life to the knower and the known, where the two may refer to the same living being in the case of self-knowledge, of "going toward oneself," as the kabbalistic construal of *Leḵ leḵa*, God's appeal to Abraham, then called Abram, to leave his native land will indicate. Their gift, however, is rare. If Adam and Eve received death sentences as their punishment for the original sin, it is because of *how* they took the knowledge that the tree yielded, internally transforming it into a "tree of death": "But this tree is a tree of death; whoever takes it alone dies, for he has taken deadly poison. So, *on the day you eat from it, you will surely die*, for he separates saplings [*pareiš neti'in*]" (1:35b; v. 1, 222). The punishment meted out by God, thus, merely reiterates the consequences of dicing reality into manageable chunks to be eaten up and assimilated.

Zohar's reference to the separation of saplings alludes to the Babylonian Talmud (*Ḥagigah* 14b), where of the four rabbis who enter an orchard (*pardes*: recalling the garden of Eden, *paradise*), only Rabbi Akiva survives by virtue of not engaging in the practice of severing the saplings, *qitsets ba-neti'ot*. Although plant saplings can, under the right conditions, lead a life independent from the mother plant, the Talmudic expression implies the killing of human children (Poorthuis 2018, 81). In turn, the sapling at stake in *Zohar* is *šeḵīnah*, the last and the lowest of the *sefirot*, or emanations of divine energies, representing the tree of knowledge of good and evil, "a branch of those branches planted by the blessed Holy One when He created the world" (1:216b; v. 3, 303).[3] This "branch of those branches" contains nothing but emptiness and death when cleaved from the flow of emanations it is meant to receive. The formality of abstract knowledge is a reflection of *šeḵīnah*'s emptiness, endowed with an autonomous existence, which leads to the perishing of the lowest part of the Judaic cosmic tree.[4]

So, who is *šeḵīnah*? Though an emanation of God, she is a figure of emptiness and of poverty; she is the withdrawal of God which is still (in) God and which makes existence possible. In commentary on Noah, *šeḵīnah* appears in the guise of a cistern or a well, of emptiness and potential fullness that belong together, that co-belong in the indeterminate nexus of giving and taking: "A *cistern* [*reyqan'ya*] is empty, unflowing, while a *well* [*be'er*] bubbles with water. Yet, all is one site. The site embracing the poor is called *cistern*, possessing nothing of its own, only what is deposited there. Who is that? *Dalet*" (1:60a, v. 1; 345–46). When sefirotic dynamics work as they should, emptiness and fullness, a well and a cistern, are one and the same, "all is one site [*atar ḥad*]." But when *šeḵīnah* is grasped through dualistic knowledge (of good and evil) apart from the flow of emanations she receives from the upper *sefirot*, knowers come to possess something or someone "possessing nothing of its own." These knowers are not simply empty-handed; they eat up nothingness and, therefore, assimilate emptiness, devastating poverty, and death. The letter *dalet* signifies *šeḵīnah* because she is *d'la*, poor.

The text presents its readers with two disparate options, two approaches to poverty and emptiness that cardinally affect the relation of knowledge and life. On the one hand, empty knowledge receives the receiving alone, while rejecting the precious content of life, which only persists in a unity with the forms it is received by. On the other hand, it is possible to treat the withdrawal, or the inner voiding of *šeḵīnah*, as a welcome extended to life itself, as an ecological receptivity that makes room for the living.[5] The choice is between a denying and a giving withdrawal. Poverty can be as much the poverty of abstraction, implying the dearth of concreteness, as that of an absenting that makes room for existence It is the second option that resonates with the Platonic *khora* (χώρα), the receptive embrace of a place, not to be mistaken for the arid abstraction of space. Everything *Zohar* will convey on the subject of the welcoming opening of *šeḵīnah*, with its ineluctable absence that allows for the presence of beings, belongs within this embrace.

Seeking a Living Knowledge of Unity, Rather than Deadening Knowledge of Division

A living, vivifying knowledge is not sundered from the source of life, whereas a dead and deadening knowledge erects a dam across the river of life and marks its own domain on the arid terrain of abstraction that comes through in the aftermath. The sefirotic tree channels the flux of vitality to *šeḵīnah*, which, in her fully feminine divine form, ought to be taken as a part of a vast unfolding, rather than as a free-standing product. The flux, in turn, waters the entire cosmic tree, or the tree of life, and the tree of knowledge of good and evil (*šeḵīnah*) in particular:

> The tree of which we have spoken was watered, flourished, and rejoiced, as is said: *A river issues from Eden to water the garden* (Genesis 2:10). *The garden*—this is woman, the *river* entering Her, watering Her, and all was one [*ḥad*], while from there below is division [*piruda*], as is written: *From there it divides* [*yipared*].
>
> (1:35b; v. 1, 223–24)[6]

Initially, the tree of life and that of knowledge are one and the same tree. The division of primordial unity is as much an upshot of human behavior (taking and eating the fruit from the latter tree in such a way that the growth and development leading up to its ripening are obscured) as of the cosmico-ontological scheme, in keeping with which "whoever takes it [this tree] by itself [*deaḥidan*], takes it together with hordes below embraced by it, takes idolatry, murder, and exposing nudity" (1:36a, v. 1, 226). Remarkably, the ungluing of *šeḵīnah* from the other, upper *sefirot* emanations—separating knowledge from life, etc.—coincides with the mix, not only of good and evil but also of the multitudes below *šeḵīnah* unaware of their interdependence with each other and with the ever-coming flow of vitality, which they, to some extent, also express. And, on the contrary, a close integration of the cosmico-ontological scheme with ethical conduct is a staple feature of the kabbalistic worldview: the world is not firmly established without righteousness, as symbolized by the *sefirah yesod*, which is directly above *šeḵīnah*.

To sum up, the rift between *šeḵīnah* and the other *sefirot* that opens up as a result of the "unwise" eating from the tree of knowledge of good and evil means that: (a) the flow of emanations or cosmic energies is interrupted, (b) the lower branches of the world tree are no longer in touch with its celestial roots, (c) the symmetrical, bilateral, top-down and bottom-up causality—"as it is above, so it is below"—no longer works as it should, and (d) what is grasped by means of knowledge disconnected from life is pure emptiness and death. The four aspects of the rift reinforce one another, while also deepening it and repeating the initial separation. Among a slew of its effects, there are not only the ontological or the epistemological consequences, but also the ethical ones. The formal and abstract knowledge that remains after this traumatic event is cut off from life *and* from righteousness is the kind of knowledge that is both dead and unethical, dead because unethical, not to mention deadening as far as life is concerned.

Despite the negativity of knowledge unfastened from life, the unfastening also has positive repercussions, some of them mentioned in *Zohar* itself. For all its apparent sobriety and seriousness, lifeless knowledge derives from inebriation, not least inebriation with power over life: "She [Eve] squeezed grapes [*'anavim*] and offered them to him [Adam], and they inflicted death upon the whole world [*u-grimu mota le-ḵol' 'olma*], for death abides in this tree, a tree ruling by night [*d'shal'ta b'leilia*]" (1:36a; v. 1, 228).[7] The reign of the night, nocturnal knowledge, is an alternative and an alternating element with respect to diurnal knowledge, the knowledge of and for life.

Rather than a stray occurrence, alternation is a necessary event in accord with the symmetrical arrangement of the sublime body of God made of *sefirot* that are placed above and below, to the right and to the left of each other, with the upper part located in the cosmic east, while the lower (that is, *šeḵīnah*) is in the west, the land of sunset, decline, night, and death already in Egyptian antiquity. The oscillations of *šeḵīnah* are ideally rhythmic, detaching and adhering again to *yesod*, or foundation, depending on the actions of the creatures below and the determinations above. It is the heartbeat of the world. Catastrophe looms when distinctions between different realms of being wax absolute, disrupting the diurnal and nocturnal cycles of being and knowing. It is then that the west, harboring the night, is on the ascendence. The Western tree of knowledge becomes *itself*, acquires its distinct identity, by being chopped off from the tree of life.[8]

The reign of the night that awakens with the original sin participates in a chain of associations that involve empty knowing, inebriation, sleep, and death. "When She [*šeḵīnah*] rules, all creatures of the world taste a taste of death [*kol' bni 'alma t'aamu t'aama d'mota*]" (1:36b; v. 1, 228–29). The "taste of death" refers to sleep, as per *Beraḵot* 57b in the Babylonian Talmud ("Sleep is one-sixtieth of death"). Apparently translucid and ethereal, abstract knowledge is dense, obscure, intoxicated with itself, and asleep to life. The light of the Enlightenment is darkness that has claimed luminosity exclusively for itself. It is the darkness that has either projected impenetrable dimness onto matter,[9] life, finite existence, or has become convinced that dimness that may be altogether dispelled as soon as the ray of pure reason shines over obscure regions of being. In the night, at least, there is intimacy, a cover of darkness that protects. Yet, the knowledge incompatible with life belongs in the reign of the night *and* coincides with sheer exposure, with the nakedness of creatures rendered fully available to its anonymous, scrutinizing, unblinking gaze. Hence, the nakedness, which Adam and Eve discovered in the wake of the original sin and in which kabbalistic authors see the loss of "the supernal radiance enveloping them" (1:36b; v. 1, 229). That shine, that envelope of supernal radiance, was the sense of life itself that shimmered in the flux of emanations and clung to the creatures, whether human or not, who were one with it. When the night of abstract knowledge encroaches, creatures are stripped of their radiance and are subject to the external assignation of meaning (or, more often than not, meaninglessness). However deficient, this epistemico-axiological mechanism is an outcome of knowledge's unilateral and unidirectional detachment from life.

The other positive feature of negative, death-bearing knowledge disconnected from life is the God-like creativity it bestows upon whoever eats of its fruit. Seducing Eve, the serpent utters, according to Rabbi Yose's reading of the Hebrew Bible (and of the midrash *Berešit Rabbah*): "With this tree, the blessed Holy One created the world. Eat from it and you will really be like *Elohim*, God knowing good and evil. For so it is: *Elohim* is its name, the tree of knowledge of good and evil"; and Rabbi Yehuda adds: "Eat from it and you will be creating worlds! [*Yiḵlu minei ve-atun tehon be-ran 'olmin*]" (1:36a; v. 1, 225).

If the tree is *šeḵīnah*, who bears one of the names of God (*Elohim*), then eating her fruit is eating something of God. Eating the fruit of this tree is ingesting divine essence, and, moreover, separating one name of God (*YHVH*) from another (*Elohim*), the living God from a polysemic emanation/fruit (ZH 18c; v. 10, 191).

In the spirit of Neoplatonism, the serpent insinuates that, partaking of God, Adam and Eve would also become creators, that is, their knowledge and world-making, their contemplation and doing, theory and practice would be one and the same. But plucking the fruit from the tree, segregating knowledge from life, culminates in the reverse of what the serpent promises: the act drives a wedge between being and knowing. To treat knowledge as a tool (even of world-creation), to instrumentalize it, is already to fall far short of God in an attempt to digest, to break down and assimilate the fruit. Assuming that Adam and Eve were endowed with God's world-creating potency, the world or the worlds they would make would be unlivable and ephemeral, formal models in line with empty and abstract knowledge. They would be pale reflections of divine experimentation, God's pre-creation play with the letters spelling out the contents and the structure of possible worlds: "When the blessed Holy One wished to fashion the world, all the letters were hidden away. For two thousand years before creating the world, the blessed Holy One contemplated them and played with them. As He verged on creating the world, all the letters presented themselves before Him, from last to first" (1:2b; v. 1, 11).

Objectifying Knowledge as Exile From Life and Wisdom

The discrepancy between knowledge and life is not, for all that, a mere division; it is the *exile* of humanity and of the rest of creation from life. Exile is understood in *Zohar* as the muteness of the living: when Eve took the fruit, she peeled life away from death, the voice from speech, "for voice is never separate from speech, and whoever separates voice from speech turns dumb, speechless…Who is *silent*? *Praise, praise of David,* who turns *silent* in exile [*be-gluta*], voiceless [*u-štika be-lo qol'*]" (1:36a-b; v. 1, 228). Paradoxical as it may sound, the separation of life from death deadens life all the faster. In the sefirotic paradigm, *tif'eret* stands for voice, while *šeḵīnah* is articulated speech. Their separation leaves behind voiceless speech, speech exiled from a living voice, the voice of life. A dead speech, verging on silence, then. The loss of the shine that is the inner meaning of the living is accompanied by the deprivation of their voice, ruining in this double gesture the very mystical phenomeno-logy, which the authors of *Zohar* wish to recover. The culprit here is not so much Eve herself as a particular action: "and she took," *va-tiqaḥ*, the fruit. At fault is the kind of knowledge that seizes and appropriates, that objectifies the living and turns them into property. It is by *taking* the fruit, by removing knowledge from life, that humans dim down the light proper to the living and silence their voice. Their own light and voice, above all.

The newly acquired knowledge exiles human beings from life, still before their expulsion from the Garden of Eden and definitely well before the political exiles that have peppered the history of the Jewish people and have been assigned theological, if not metaphysical, significance. Exiled from life, we go on living, surviving ourselves. In the same breath, God is exiled from herself or from himself when the *sefirah šeḵīnah* comes unglued from the other nine *sefirot*. This event in divine history, this rupture in the sefirotic economy of God, is caused from below, by human actions militating against life.

The meaning of exile also needs to be assessed in the context of what the Aramaic *gluta* and the Hebrew *galut* are saying: the verb, from which it is derived (*legalot*) means, simultaneously, to exile and to expose. Just as the objectifying knowledge untethered to life mercilessly exposes beings in its neutered light, so the human subject is exposed (above all, to herself and to himself), made aware of her and his nakedness, stripped of the inner glow of meaning and of the "garments of light [*kotnot or*]" that are replaced with the nearly homonymous "garments of skin [*kotnot 'or*]" (1:36b; v. 1, 230).[10] Exposure is an exile from oneself, a self-imposed exile emblazoned on one's skin, *as* one's skin, which, in addition to its vulnerability signifies the absence of significations (the enveloping light) that would be inherent to the living. At the same time, and more radically still, life is exiled from life, exposed to the other, lacking anything like a self-identity, but this immemorial exile, the excess of vivacity over itself, is what, at bottom, renders it living. Divine emanations are rooted in this insight, lending depth and irreducible ambiguity to exile, viewed in dark mystical light.

The first exile occasioned by eating the fruit from the tree of knowledge of good and evil is an expulsion from the milieu of life, from the precise middle of the garden, where the tree of life is flourishing: "The tree of life is precisely in the middle of the garden [*'eẓ ha-ḥayim bimẓi'ut d'ginta mamaš*]" (1:35a; v. 1, 220). The middle where the tree of life grows is the place and the time of life, whereas the tree of knowledge is displaced with respect to the middle—it is evidently "not in the middle [*lo ihu b'emẓ'ita*]," though it "suckles on two sides and knows them as one suckling sweet and bitter, since it suckles on two sides, knowing them and shuttling among them, it is so called" (1:35a; v. 1, 220). The two sides are good and evil—the principal objects of the knowledge, which the tree yields. Here, a livable middle, the milieu of life, is replaced by a mix of the two, known as the sweet and the bitter and nourishing *šeḵīnah*. The mix will necessitate a patient and ongoing discernment between the good and evil that reach the knower as one.[11]

Later on in *Zohar*, the garden, in the middle of which the tree of life grows,[12] is converted into the "house of the world" (*beita d'olama*, or, to resort to Greek, the *oikos* of ecology), likewise hosting the tree in its center. "This house stands in the center of all [*hai beita ihu go emẓa'ita de-ḵola*], countless doors and chambers round and round—supernal sacred sites, where birds of heaven nest, each according to its species. Within emerges an immense, mighty tree, its branches and fruit nourishing all. That tree climbs to the clouds of heaven… . This house is saturated by it; within, it secretes supernal hidden treasures, unknown. Thereby the house is constructed and decorated" (1:172a; v. 3, 40–41).

The house of the world constructed around the tree of life *de*constructs the distinction between a cultural artifact and natural phenomena; it is a house that is livable only to the extent that it is thrust open, from within, by the tree that grows in the middle of it. In *beita d'olama*, creation swings back to the middle of life, but it does not, thereby, overcome the exilic condition: the house is a dwelling primordially broken into, indwelt by the uncontrollable and unobjectifiable, overwhelmed ("saturated") by the life that flows through all its chambers and spills over beyond them. Time, with its rhythmic drifts and vacillations, is not done away with. Seeing that the house of the world is *šeḵīnah* herself, hosting *tif'eret* in the middle, the alterations and alternations of day and night proceed in it unimpeded, in a dance of figure and ground: "That tree is revealed by day, concealed by night; this house rules by night, is concealed by day" (1:172a; v. 3, 41).[13]

The acquisition of knowledge by eating the forbidden fruit is, in equal measure, an exile from wisdom. "Until Adam sinned, he ascended and abided in supernal radiant wisdom [*b'ḥoḵmah din'hiru 'ilaah*], never parting from the tree of life" (1:52a; v. 1, 290). Now, *ḥoḵmah* (wisdom) is one of the upper *sefirot*, which would explain Adam's ascent and the supernal radiance he was enveloped in when he abided in it. In contrast to knowledge, wisdom is inseparable from life (*ve-lo... mitpareš*) and, more importantly, it is not something one "takes," has, or possesses, but the abode, a dwelling place, the milieu of life. In this sense, the abstract knowledge that weighs the values of good and evil is economic, but wisdom is ecological.[14]

The weights and other measurements, with which abstract knowledge operates, apply primordially to time, the limits of time and the time of the limit—*mo'ed*, "term," "season," slated or appointed time. "At first, extending long life eternally, so death would not prevail... . Now, time is allotted [*mikan ve-lehalaa ohel' mo'ed*], life rationed to the world" (1:52b; v. 1, 292). Time is not only allotted; it *is* allotment, the setting of a term that entails termination. The measure of time, its limit, that up until which (*'ad*) an event unfolds, belongs to a rhythmic sequence of alterations and alternations between life and death, good and evil, the sun and the moon "at first, joining and coupling...unceasingly," *tif'eret* and *šeḵīnah*.

Within these limits, *ex*istence has an *in*sistent character. It announces, wordlessly, almost breathlessly: "I am still here." The cosmico-ontological scheme itself changes after the original sin, exhibiting rhythmicality there where there was but an uninterrupted flow of emanations and divine energies. What has at first appeared to be an integral part of mundane and divine realities (a rhythmic alternation of opposites) is due to the bottom-up determination resulting from the act of eating the fruit from the tree of knowledge.

That said, the rhythm that dictates time is not, in and of itself, as worrisome as one-sidedness in privileging a single alternating element, where sin is understood in terms of "clinging to the tree where death abides" (1:52a; v. 1, 289)—to death, to the night, to the west. For, as soon as rhythms commence, they lapse into the arrhythmias of existence with excessive separations of night and death from day and life. The privileged element is then indefinitely extended in the parodic imitation of a life that does not know the distinction between life and death, precisely because it abides in and with wisdom, rather than knowledge.[15] The disruption of time sequences and terms, which is evident nowadays in the upsetting of the circadian rhythms of all living beings or in the manufacturing of non-biodegradable materials, is not a recovery of the tree of life, but a further distancing from it.

Not Grasping, but Grafting Onto the Tree of Life

The removal of humanity from the tree of life is not an irreparable incident. It can be deepened in a repetition of the inceptual knowing grasp, which is conventionally called "the original sin," or mended in line with the practice of *tiqun 'olam*, or "world repair." So, on the one hand, the character of wickedness is zombie-like, "dead while remaining alive" (2:106b; v. 5, 119). Repentance (literally, "return" or "response": *t'šuvah*) restores life: "Now he is really alive—alive on all sides, for he reunites with the tree of life [*aḥid b'ilana d'ḥayey*]. And, reuniting with the tree of life, he is called *ba'al t'šuvah*, master of returning" (1:106b; v. 5, 120).[16] On the other hand, though he enters into a covenant with God, Noah repeats the sin of Adam and Eve, separating knowledge from life: "When he sought to probe [*lemivdaq*] the sin probed by Adam—not to cling but to know, to mend the world [*leminda'u-latqana 'olama*]—he was incapable. He squeezed grapes to probe that vineyard, but as soon as he arrived, *he became drunk and exposed himself*" (1:73a; v. 1, 433). Like Adam and Eve, Noah is exposed, becomes naked as a consequence of eating, or probing, the fruit. (Incidentally, it is now clear why *Zohar* insists on the forbidden fruit being grapes: this specification makes the analogy with Noah complete).[17] The choice is the same in both instances: to probe and to know, to the detriment of life and its wisdom. Perhaps, our authors suggest, Noah tried something slightly different from Adam and Eve's experiment, notably to mend the world by knowing it better, *leminda'u-latkana 'olama*. But the end result is the same: inebriated with the power of knowledge and himself exposed in his knowing exposure of reality, Noah fails in his endeavor.

Since Noah's intent differs from that of Adam and Eve, his nakedness, too, assumes a faintly distinct shape. "So, *he exposed himself*, exposing the breach of the world that had been closed [*va-itgal'galei pirẓah d'olama*]" (1:73a; v. 1, 434). The breach of the world, *pirẓah d'olama*, is multifaceted, ranging from the loss of distinctions between the upper waters of the sky and the waters of the earth in a flood to the split between *tif'eret* and *šek̲īnah* or between life and knowledge. It was this breach that, supposedly, Noah's covenant with God mended (hence, "that had been closed"). Nevertheless, reliance on knowledge in the endless work of bridging the gap between knowledge and life is doomed to tear the fabric of existence further.[18] Exposed, Noah is exiled from life otherwise, after he approaches *šek̲īnah*, which here metamorphoses into a vineyard, with the tools of knowing, of probing, and extraction—squeezing the juice of her grapes, obtaining the essence of what is known. *This* is a new breach of the world that gapes open when essence is extracted from vegetal existence no longer in a mouth that bites into the fruit, as in the case of Adam and Eve, but with the assistance of the agricultural apparatus—"Noah arrived and invented spades and multitoothed files" (1:97b; v. 2, 117)—that relates to *šek̲īnah* in an instrumental fashion as arable soil, plantations, and their yield. (Needless to say, earning one's bread with the sweat of the brow is among the punishments meted out for the original sin, the punishment that, thanks to the technological improvements he has introduced, Noah views as the road to salvation).

The re-founding of the covenant with God at Mount Sinai undoes the separation between life and knowledge, *tif'eret* and *šeḵīnah*. "When Israel stood at Mount Sinai, before sinning, the slime of this serpent was eliminated from them, for the evil impulse was repulsed by them, universally abolished. So, they grafted to the tree of life [*itaḥidu b'ilana d'ḥayey*], ascending, and not descending; knowing [*yad'in*], seeing supernal specula; eyes glistening, delighting in discovery, in listening" (1:52b; v. 1, 290). Although they received the law on tablets of stone, those assembled at Sinai were bound to life (to the flow of emanations) by being grafted onto, or united with (*itaḥidu*), the tree of life, which is the Torah received there. In the same vein, Rabbi Ḥizkiyah affirms that "Torah is a tree of life, a tree of life for all those occupied with her [*ilana d'ḥayey ihi le-ḵol' inun d'askin be-yah*], extending them life in this world and in the world that is coming" (1:106b – 107a; v. 2, 149).[19] And again: "So long as Torah is found within her [within Jerusalem], she endures, since Torah is the tree of life, standing over her. As long as Torah is aroused below, the tree of life does not depart above. If Torah ceases below, the tree of life withdraws from her" (1:151a; v. 2, 344).[20]

The knowledge imparted by Torah is not exactly knowledge, but wisdom; it is another kind of approach that does not take, grasp, or appropriate reality, but, in a reversal of the active and possessive impulse, is grafted onto the tree of life. Instead of taking hold of the fruit of knowledge and squeezing them dry, extracting value or essence from them, one joins the living trunk, the trunk of life that extends from *tif'eret* to *yesod*, and one is nourished by it. That is why those versed in Torah do not "grasp" wisdom, as the Stanford translation suggests, but "occupy" themselves with it—'*askin be-yah*—preoccupied by, without intending to lay hold of, it. Their preoccupation (their care, even) is far from superfluous within *Zohar's* cosmico-ontological scheme: in line with the non-metaphysical, bottom-up causality developed in this body of thought, the "arousing" of Torah below affects the flourishing of the tree of life above, in particular its flourishing in continuity and in contiguity with the lowest *sefirah*, *šeḵīnah*, which is the tree's intricate crown, branching out toward the rest of the world. The flow of life, imagined as sap circulating in the cosmic tree, is then assured and not interrupted.

Similarly, prayer helps a community that voices it "with perfect intention" to overcome its own separation from the tree of life, as well as the separation carved into the tree itself. The most important Jewish prayer, *Shema Yisra'el*, *Hear, O Israel!* "enacts unification [*yiḥudah*]" by "striking a spark of darkness [*buẓina d'qardinuta*], which scatters into seventy lights, and those seventy flash into seventy branches of the tree of life" (2:133b; v. 5, 237). The seventy branches of the tree are the seven lower *sefirot*, from *ḥesed* to *šeḵīnah*, that flourish together thanks to the prayer and its animating intention. But their source, tapped by *Shema*, is the mystical spark of dark light, *buẓina d'qardinuta*, which is nearly impossible to translate[21] and, which refers to "a kind of primordial cosmic stylus that initiates creation" (Berman 2018, 148). Prayer is, therefore, responsible not only for a reunification with the tree of life as a whole, but for the tree's regeneration each time afresh, for a new creation or recreation of the world, in a performative gesture supported by the full cosmological and cosmogonic scope allotted to it in the "Terumah" section of *Zohar*.[22]

From Exilic Ecology to Replanting the World

With a return to the tree of life, the mending of the breach attributed to the effects of objectifying knowledge, is accomplished in vegetal terms: *tiqun 'olam*, the repair of the world, becomes a replanting of the world, as in the image of the immense tree growing in the middle of the "house of the world." Such a replanting, though, is not the agricultural operation that is Noah's solution to the problem of knowledge and life by means of knowledge, but the grafting of the human back onto the tree of life. It hinges on the ethical task of "nourishing and supporting someone in need" and thus "embracing the tree of life [*leitaḥada b'ilana d'ḥayey*], engendering life for himself and his children" (1:208b; v. 3, 280). To sustain the others in need, to nourish others, is to play the role of the tree of life vis-à-vis them. The "ground" for the act of supporting the vitality of the other is not knowledge but the good, the *sefirah* literally called "foundation," *yesod*, which mediates between *tif'eret* above and *šeḵīnah* below. And so, the path back to the tree of life winds through ethics, rather than knowledge. The mystical ecology of *Zohar* is ethical.

Lest it appear that the identification of Torah with *tif'eret* renders the tree of life highly idiosyncratic in cultural and religious terms (much like the letters of the Hebrew alphabet viewed as building blocks for the creation of the world), it is worth remembering that the entire sefirotic tree is also considered as the tree of life: "What is *a dew of lights? Lights*, literally! From those supernal lights [*orot*], through which He will pour forth life upon the world, for the tree of life will pour forth incessant life" (1:130b–131a; v. 2, 232). The "supernal lights" are the *sefirot*, through which cosmic energies are channeled, and, taken in their interconnectedness, they form the tree of life. Meaning "brilliance," the word *zohar*, which is adopted into the title of kabbalistic work, refers to the glow or the afterglow of these lights. As "Parašat Mi-Qeẓ" states: "The tree of life embraces many rungs distinct from one another, yet all is one." There are trees within the tree—trees that correspond to the rungs of "branches and leaves, bark, tree trunk, roots—and all is the tree [*ve-ḵola hu ilana*]" (1:193b; v. 3, 181). It follows that *tif'eret* stands in a synecdochic relation to the sefirotic tree: a part of the tree, it is also the reflection of the whole, precisely as the tree of life.[23] In the same way, the singular history of Jewish exile becomes a synecdoche for the human condition and for what I would like to call "exilic ecology," a mode of dwelling in and with the uprooting and displacement of humanity and many other species from a livable milieu and of life from itself.

Severing humanity from the tree of life is tantamount to abandoning the source of ensoulment. The "souls of the living being [*nefeš ḥaya*]…all enter beneath the wings of *šeḵīnah*," but they "emerge from the trunk of that tree [*migo gufa d'hahu ilana*], whence souls fly into earth, into Her womb, deep within" (1:13a; v. 1, 93). The Aramaic word *gufa* (Hebrew *guf*) signifies a tree trunk, as much as the body, instigating, once again, the interplay of a part and of the whole. It is the vegetal rendition of *tif'eret* and of the Torah scroll, the psychogenic body of the tree of life, the sublime body that yields souls. Dead, deadening, objectifying knowledge is deaf and blind to the souls of creatures that plop down to earth (to *šeḵīnah*, which is a vestige of chthonic, telluric, or earth-bound divinity) as ripe fruits, falling from the tree. In effect, according to Jewish mysticism, souls undergo a long process of ripening both on the sefirotic tree and on the part that stands in the relation of synecdoche to it, namely *tif'eret*: they metamorphose from *nišmeta* (the breath of life) that arises in one of the highest *sefirot*, *binah*, or the upper mother, to *ruḥa*, or the spirit-soul nurtured in *tif'eret*, and further to *nefeš*, a creaturely individuated soul, which passes from *tif'eret* to *šeḵīnah*.[24] Along the process of maturation, the fruit that is the soul is not available for the taking in the manner of the one growing on the tree of knowledge of good and evil, and, if it is, it immediately slips away from the objectifying grasp. This fruit remains faithful to *binah*, whence it issued, the *sefirah* responding to the question *who?* (*mi*) and unavailable for any determination as a *what* (*mah*) (1:2a; v. 1, 8).

What does it mean, finally, that every fruit granted to Adam and Eve is to be eaten "in oneness," *be-yiḥudah*? Is it not an eating without eating, a knowing without knowing, that is to say, without biting into, partitioning, analyzing, and assimilating external reality? That is what wisdom and a knowledge propitious to life would look and feel like. Nonetheless, *yiḥudah* does not refer to the undifferentiated One, but to a unification of differences. *Zohar* implies that the analytical apparatus of understanding is too crude, that finer discernments are needed to appreciate the flux of emanations all the way down to their emptying and branching out in *šeḵīnah*. The image of the soul as a fruit is one of these discernments, linking creaturely individuation to a shared root in the breath of life. Born from *binah*, it is nourished by the trunk of the tree, represented by *tif'eret*, in which "all worlds [*ḵol' 'olamin*] exist" and from which "all holy powers are nourished and deployed" (1:6b; v. 1, 39). To eat in oneness is to respect all the worlds sustaining the fragile and broken twigs of the cosmic tree that form "our" world.

Notes

1 Let it be mentioned in passing that, within a hermeneutical-exegetical tradition that pays close attention even to single letters and parts of letters, it is not a matter of sheer indifference where a verse of the Biblical text begins and where it ends.

2 All citations from *Zohar* refer to *The Zohar*, Pritzker Edition (12 volumes), translated by Daniel C. Matt (Stanford: Stanford University Press, 2018). Page references are to the volumes/pages in the Pritzker edition, as well as to the conventional page numbers of *Zohar*. All emphases within passages are original. Transcriptions from the original Aramaic and Hebrew are mine.

3 *Šeḵīnah* is explicitly identified with the tree of knowledge of good and evil in 1:7b–8a: "Why is it called *awe of YHVH*? Because it is the tree of good and evil. If a person is deserving, it is good; if not, evil. So, awe abides at this site, the gateway to all goodness of the world" (1: 7b–8a; v. 1, 50).

4 In the writings attributed to him, Moshe de Leon also associates the tree of knowledge of good and evil with *šeḵīnah*. See Berman 2018, 94.

5 This possibility parallels that of Deism, with the important proviso that the giving withdrawal is enacted by the feminine portion of God, or *šeḵīnah*.

6 On the tree of life as the Judaic version of the cosmic tree, growing between the celestial Eden and terrestrial paradise, see Scholem 1987, 2. Already in an early kabbalistic work, *Bahir*, the tree of life is imagined as a cosmic tree, growing from the earth up to heaven. See Scholem 1976, 95.

7 See also: "The tree with which Adam sinned was of grapes . . . and these are black grapes. For there are grapes, and then there are grapes! White grapes are good, since they derive from the side of life; black grapes require mercy, since they derive from the side of death" (2:144a; v. 5., 317). Now, the judgment itself, passed on the different kinds of grapes as good and evil, corresponding to life and death respectively, is traceable to the tree of knowledge of good and evil, which is said to be a grapevine!

8 There is also another night, the mystical night of the world, that points to the world that is coming (*ha-'olam ha-ba*) and its sublime knowledge. The midnight study of the Kabbalah is illuminated by such a night: "Just as this day has illuminated us in this world, rendering us worthy of the world that is coming, so too the night illuminates us, entitling us to the world that is coming, to adorn words of day and night in the presence of the Ancient of Days" (2:143a; v. 5, 308).

9 Already in *The Enneads* of Plotinus, matter is associated with darkness, shadows, and depth: *bathos*.

10 Between *or* and '*or*, light and skin, only one silent letter changes, from *alef* to *ayn*, while bearing the same diacritical mark, *o*).

11 If the principle of oneness applies, then the tree of life and the tree of knowledge are one and the same tree—an interpretation that has been present throughout the history of Jewish exegesis. Scholem cites Rabbi Ezra ben Solomon of Gerona (1160–1238), who holds that the tree of life and the tree of knowledge are one tree below and two trees above (Scholem 1976, 65). This metaphysical explanation, however, (1) contravenes the kabbalistic principle of symmetry (as it is above, so it is below), and (2) goes against the attribution of unity to the higher levels of reality and of division to the lower.

12 Midrash *Berešit Rabbah* states that the tree of life "was a tree which spread over all living things" (15.6).

13 Nevertheless, exceptional wisdom can retract time to an ecstatic state where the difference between day and night evaporates. See, for example, the kabbalistic description of a reaction to the words of Rabbi Abba: "When they departed, they did not know if it was day or night" (1:94b; v. 2, 98), or, again, to the words of an anonymous "old man." "The companions were amazed; they did not know if was day or night, if they were standing there or not" (2:105b; v.5, 81).

14 The connection between the tree of life and wisdom is already palpable in the earliest kabbalistic document, *The Book of Creation*, according to which, leading up to the tree of life are the thirty-two wondrous paths of *sophia* (*ḥoḵmah*, or divine wisdom) (Scholem 1987, 79, 153).

15 Perhaps, this excess of wisdom over knowledge is what Franz Rosenzweig means when he writes that "[t]he philosopher must be more than philosophy," invoking, at the same time, the fruit of the tree of life, rather than that of knowledge of good and evil: "The philosopher must be more than philosophy. We have heard: he has to be a man, flesh and blood. But it is not enough that he is merely that. As flesh and blood that he is, he must pray the prayer of creatures, the prayer to his own destiny where precisely the creature unknowingly acknowledges himself as creature. The wisdom that dwells in him, in his

flesh and blood—God makes it innate in him; now it hangs like a ripe fruit on the tree of life" (Rosenzweig 2005, 314).

16 Translation slightly modified.

17 "It [the grapevine] was banished from the Garden of Eden, and he [Noah] planted it here. . . . It was in the earth and he uprooted it, then transplanted it. The same day it yielded fruit, blossoming into grapes; he squeezed them, drank of the wine and became drunk" (1:73a; v. 1, 433). Midrash *Berešit Rabbah* attributes the view that the forbidden fruits are grapes to Rabbi Judah ben R. Ila'i. Other suggestions include wheat, *etrog* (citron), figs, and fruits from an unknown (unrevealed) tree (1.15.7).

18 In contemporary terms, such a solution would involve geoengineering and terraforming.

19 Translation slightly modified.

20 Similarly: "If one departs from Torah or separates from her, it is like separating oneself from the tree of life" (1:11a; v. 1, 74).

21 On the difficulties of translating or even transcribing this term, see Dan 2002, 141.

22 Note that unification is not only the effect but the avowed content of *Shema*, its first sentence ending with the words "YHVH is one [*eḥad*]."

23 Also: "In the trunk [*u-vegufa*: also, "in the body"] of this [sefirotic] tree all worlds [*ḵol' 'olamin*] exist; from it, all holy powers are nourished and deployed" (1:6b; v. 1, 39). We will have a chance to comment on this later. Generally speaking, synecdoche is the rhetorical device that allows those who resort to it to follow the injunction of taking everything in the unity of being, as one. After all, through synecdoche, a part is always reconnected to the whole it actually condenses in itself.

24 Thus, "holy soul, drawn from a supernal place, enters the recess of the tree of life. When the breath of the supernal priest blows and infuses this tree, those souls fly from there and enter a certain treasure-house. Woe to the world, for humans do not know to be careful!" (2:95a; v. 5, 6).

Bibliography

The Bhagavad Gītā, 25th Anniversary Edition. 2009. Translated by Winthrop Sargeant, edited by Christopher Key Chapple. Albany: SUNY Press.

The Zohar, Pritzker Edition (12 volumes). 2018. Translated by Daniel C. Matt. Stanford: Stanford University Press.

Berman, Nathaniel. 2018. *Divine and Demonic in the Poetic Mythology of the Zohar: The 'Other Side' of Kabbalah*. Leiden: Brill.

Dan, Joseph. 2002. *The Heart and the Fountain: An Anthology of Jewish Mystical Experiences*. Oxford: Oxford University Press.

Poorthuis, Marcel. 2018. "The Redemption of an Arch-Heretic: Elisha ben Abuya and the Flames of Gehenna." In *The Apostles' Creed: 'He Descended into Hell'*, edited by Archibald Van Wieringen and Marcel Sarot, 79–92. Leiden: Brill.

Rosenzweig, Franz. 2005. *The Star of Redemption*. Translated by Barbara E. Galli. Madison: University of Wisconsin Press.

Scholem, Gershom. 1976. *On the Mystical Shape of the Godhead: Basic Concepts in the Kabbalah*. New York: Schocken Books.

———. 1987. *Origins of the Kabbalah*. Edited by R.J. Zwi Werblowsky. Princeton: Princeton University Press.

8 Expansive Modes of Multispecies Knowing Toward Nonviolence in the Early Jain Canon

Brianne Donaldson

The ancient Jain tradition from India—now global—offers one of the most radically expansive multispecies ethics in existence today based on its commitment to nonviolence, or *ahiṃsā*, toward all living beings. In the face of an ever-increasing array of unchecked creative technologies to destroy life in the present—whether through habitat and species loss, the mass destruction of animals for food, use for science and technology advancement, hunting, clothing, and entertainment, or recent "depopulation" methods related to zoonotic outbreaks or pandemic slaughterhouse closures such as ventilation shutdown or foam-spray suffocation to dispatch millions of factory farmed animals—any thorough-going call toward multispecies nonviolence remains rare and essential.

Present in historical records for over 2,500 years, Jainism refers to those who emulate the teaching of 24 human teachers, called Jinas, or "victors" (from which the word "Jain" derives) who exemplify a path of nonviolent restraint in every epoch of time. The 24th and last teacher of the current era, called Mahāvīra, or "great hero," lived approximately in the fourth-fifth century BCE and is considered a contemporary of the Buddha. Mahāvīra led a fourfold community of monks and nuns ("mendicants"), and laymen and laywomen centered on particular vows related to nonviolence toward plants, animals, and even minute living beings in earth, fire, water, and air. In fact, many people are only familiar with the Jain tradition through images of its iconic practices of monks and nuns wearing mouth shields to prevent disturbing air-bodied beings, walking barefoot and eschewing all transportation so as not to injure invisible earth-bodied beings, using hand brooms to sweep the ground clear of insects when walking and sitting, and eating once-daily simple vegetarian food prepared by others that is further free of root vegetables due to the higher number of lives impacted in their harvest.

On one hand, such an expansive ethics can strike an outsider as an impractical idealism. Scholar Jeffery Long describes the greater resistance his university students display toward the Jain tradition—even more than related South Asian views of Hinduism or Buddhism—since, as Long puts it, "the idea that the life of an insect is every bit as valuable as my own is, to say the least, foreign" (2009, xix). On the other hand, the expansive quality of Jain ethics prefigures widely engaged contemporary multispecies approaches such as Donna Haraway's invite

DOI: 10.4324/9781032659985-11

of "Making Kin" with more-than-human others and assemblages in our time of planetary precarity (2016) or Jacques Derrida's assertion that "responsibility is excessive or it is not a responsibility" (1991, 118).

In this chapter, I explore varieties of knowledge in the earliest layers of the Jain canon, namely the first book of the *Ācārāṅga-sūtra* or "Book of Conduct." While this text is not a systematic account of multispecies existence, knowing, or ethics, it nevertheless has a capacious account of multispecies knowing that exceeds even contemporary animal rights or ecological politics. I will demonstrate how: (1) early Jain knowledge of universal sentiency prefigures Jeremy Bentham's utility calculations of pleasure and pain, and recognizes sensory capacities in every existing entity which, per the early Jain sages, can best be known through analogy with the self; (2) Jain knowledge of interchangeable life forms through innumerable rebirths offers a meaningful parallel to John Rawls' "veil of ignorance" by which reciprocal suffering is a transformative mode of (un)knowing oneself as separate and secure in order to produce justice; (3) how ethical restraints of carefulness arise from, contribute to, and function as right comprehension and right knowing of a living multiplicity, resonating with feminist physicist Karen Barad's claim that quantum world-making is best understood as the mutual-constitution of "onto-ethico-epistemology" in every becoming.

Knowing Universal Sentiency Through Bentham's Utility and Jain Analogy With the Self

In his exploration of "utility," or the primacy of pleasure and pain as the twin masters justifying legal protections, Jeremy Bentham writes, "The time will come when humanity will extend its mantle over everything which breathes" (1843, 1016). This wide-ranging approach to sentiency—or ascribing pleasure and pain to "everything which breathes"—as the foundation for ethical consideration extends, for Bentham, not only to those people with different skin color, such that "The French have already discovered that the blackness of the skin is no reason why a human being should be abandoned without redress to the whims of a tormentor," but also to more-than-human beings (2000/1781, 143). "Perhaps it will some day be recognized," Bentham writes, "that the number of legs, the hairiness of the skin, or the possession of a tail are equally insufficient reasons for abandoning to the same fate a creature that can *feel*" (143–44; emphasis original). This clear-eyed observation of animals' genuine feeling—counter to Descartes' 16th-century assertions of animals-as-machines lacking proper feeling, judgment, speech, and adaptable action—led to one of Bentham's most oft-cited sentiments, "The question is not *Can they reason*? Or *Can they talk*? But *Can they suffer*?" (144; emphasis original). This aspirational expansion of suffering reflects Bentham's acknowledgment that other global forms of jurisprudence (he references Hindu and Islamic law) had given more and different concern for animals than ancient Western legal opinions in which "other animals…are downgraded in a class of *things*" (143; emphasis original). This downgrade was merely the continuation of a long tradition running through Aristotle, Christian Scholasticism, and Cartesian dualism (though there were always Western counter voices[1]) which denied animals, respectively, the rational soul and/or morally significant sensory feeling.

Bentham's legal and moral insistence that (at least some) animals could suffer has fused his philosophy of "utility" in the contemporary Western imagination to the modern animal rights movement, even though his view did not actually preclude anyone eating animals or killing them in self-defense. "There is very good reason why we should be allowed to eat such non-human animals as we like to eat," writes Bentham, "we are the better for it and they are never the worse" (2000/1781, 143). Rather, Bentham primarily aimed—not least of which through inflicting a death "speedier and thus less painful than what would await them in the inevitable course of nature"—at protecting animals from torment and cruelty, though not from killing as such (143).

The Jain view of sentiency is much more comprehensive—since it extends even to insects, plants, and elemental beings—as are its resultant ethical practices such as vegetarianism that also eschews honey, eggs, root vegetables, and for some Jains, dairy, the modern production of which is exceedingly violent (Miller and Dickstein 2021). According to Jain cosmology, the universe is packed with infinite *jīva*s—also called *ātman*—referring to the core life force in every living entity, each on its own difficult journey across multiple lifetimes, or *saṃsāra*, toward the possibility of a state of liberation known as *mokṣa* or *nirvāṇa*. The four primary qualities of each individual *jīva* are pure consciousness (Skt: *upayoga*, *caitanya*), sometimes translated as *sentience* (TS 2.8, trans. Tatia), consisting of two aspects of (1) pure perception (*darśana*) and (2) pure knowledge/awareness (*jñāna*). The remaining qualities are (3) energy/strength (*vīrya*) enlivening the levels of perception and knowledge, and (4) bliss (*sukha*) describing the degree to which a *jīva* orients self-ward rather than outward (Jaini 2001/1979, 104). This Jain account, per scholar Anne Vallely, affirms "the world in its entirety as alive, purposeful, intelligent, sensory, and responsive" (2014, 39).

Moreover, counter to Bentham's view that only humans fear death (a Western supposition Beatrice Marovich neatly sketches in her chapter within this volume), universal sentiency in the Jain view is synonymous with a fear of injury and death since, as stated in the *Ācārāṅga-sūtra*, "All beings are fond of life, like pleasure, hate pain, shun destruction, like life, long to live. To all life is dear" (ĀS 1.2.3.4, trans. Jacobi[2]). At base, this means that every existent entity—regardless of its birth state—is characterized by a radical ontological equality of sentiency—an immense "sensory communitas" per Vallely (2014, 43)—including both perceptive abilities and a universal resistance to injury and death. However, because these four qualities of knowledge, perception, energy, and bliss are obscured due to karma—which the Jain tradition would come to theorize as subtle dust-like matter that adheres to each *jīva* through damaging mental, verbal, and bodily actions—the birth form and duration of life each *jīva* takes will differ, as will its family, feelings, and environmental surrounds (Bronkhorst 2011, 73).

Consequently, karmically differentiated beings are understood as fully perceiving and knowing entities whose capacities are dimmed by karma in each birth form across multiple lifetimes. Moreover, these beings are vulnerable to injury and death in a material universe in which they also cannot escape from committing karma-producing harm (*kamma-samārambha)* upon others (ĀS 1.1.1.5).

***Forms of Knowledge in Ācārāṅga I*[3]**

But how does one know these truths of the universal sentiency of *jīva*, its qualities, or its vulnerability to injury or death through the effects of karmic action? The *Ācārāṅga-sūtra* (Pkt: *Āyāraṁga-sutta*), or "Book of Conduct"—which is considered authoritative to the larger Śvetāmbara sect of the Jain community—details various modes of knowing, applicable to mendicants. The earliest layers of the first book of the *Ācārāṅga*, which will be my primary focus in this chapter, are believed to date back to the sixth century BCE, with later interpolations accruing over the centuries (Ohira 1994, 19). *Ācārāṅga I* states three overt modes of knowing: instruction from the liberated, omniscient teachers, or Jinas, such as Mahāvīra (Pkt: *paravāgaraṇeṇaṃ*); having heard it from other knowledgeable sages (and evaluated it alongside other views) (*aṇṇesiṃ aṃtie cā soccā*); or from one's own memory or knowledge (*sahasammaiyāe*), the latter of which takes various forms I will explore herein. (ĀS 1.2.3.1). Scholar-monk Shri Amar Muni clarifies that *Ācārāṅga I* does not merely emphasize the passive receipt of knowledge simply because it was stated by a Jina; rather "examine it with your own wisdom and sagacity before accepting" (Bothara 1999, 14) such that "He who sees by himself, needs no instruction" (ĀS 1.2.3.6).

The co-existence of received knowledge (such as testimony, teaching, or memory) that must be verified by sensory-based experiential knowledge is a common feature of Indian naturalist traditions such as Buddhism and Jainism, which, in various ways, challenged the ritual- or idealism-centric aspects of Vedic thought and priestly dominance. Philosopher Dale Riepe (1918–2006) illuminates several features of Indian naturalist traditions, such as a world that is objectively real and not merely an idea, a world that is self-making and manifesting order and not created by a supernatural force, a world that can be known primarily by the senses and not only the mind, and that the world's objectivity, self-making, and order do not preclude individual responsibility (1961, 6). While scholars argue that the *Ācārāṅga-sūtra* is better considered a fragmentary collection rather than offering any formal ontology or epistemology that would come in later postcanonical writing (Balcerowicz 2021, 659–60), the first book nonetheless provides valuable insights into the Jain account of knowing life. This account of universal sentiency—and a related fear of injury and death among all living beings—not only exceeds Bentham's capacious understanding of feeling but, nearly three millennia before Bentham, generates much more stringent and far-reaching ethical restraints against killing.

Comprehension and Renunciation: Knowing (Multiply) by Analogy

The first chapter of *Ācārāṅga I*, titled "Knowledge of the Weapon" (Pkt: *sattha-parinnā*), refers to a dual form of "comprehension and renunciation" (Pkt: *parijāṇiyavvā*; ĀS I.1.1.4, fn 2) of the many "weapons"—physical actions, implements, and unrestraints of mind, speech, and body—by which one harms oneself through injuring other beings that are "individually embodied…not one all-soul," in contrast with other rival philosophies of the time, such as the Advaita ("non-dual") Vedāntin view, that apparent multiplicity is merely an emanation from a singular monist source such as *Brahmā* (ĀS 1.1.2.2).

Ācārāṅga I lays out a true sentient multiplicity comprised of six kinds of living beings (Pkt: *chajjīvanikāya*)—animals, plants, and minute beings in earth, fire, water, and wind. Correctly comprehending the universal sentience of this multiplicity is intended to refute rival views and remedy what the text describes as a common lack of awareness about the true nature of self as it relates to all life. The text describes how "many do not remember" from where they have come, "whether their soul is born again and again or not; nor what they were formerly [before their current life], nor what they will become after having died" (ĀS 1.1.1.2–3). Understanding the self in this regard is essential to understanding others. As summarized by Jain studies scholar Piotr Balcerowicz "As Jaina thinkers would sometimes state it, to know one thing means to know everything, inasmuch as everything is interrelated" (2017, 79).

Right comprehension means knowing oneself (and by analogy, every other existent being) through four simultaneous views described in the opening lesson of *Ācārāṅga*: "*ātma-vāda, loka-vāda, karma-vāda*, and *kriyā-vāda*,"[4] or as Jacobi translates:

> He believes in soul [*ātman*], believes in the world [*loka*], believes in reward [*karma*], believes in action [*kriyā*] (acknowledged to be our own doing in such judgments as these): "I did it;" "I shall cause another to do it;" "I shall allow another to do it." In the world these are all the causes of [blamable action] (*kamma-samārambha*), which must be comprehended and renounced.
>
> (ĀS 1.1.4–5; see also fn 5)

These four perspectives are applicable to the self as well as other beings, each of whom is enlivened by a radically individual and sentient *jīva* perpetually vulnerable to injury as they undergo innumerable rebirths. "This is called the *Samsara*..." states *Ācārāṅga I*, "Having well considered it, having well looked at it, I [Mahāvīra] say thus: all beings, those with two, three, four senses, [one-sense] plants, those with five senses, and the rest of creation, (experience) individually pleasure or displeasure, pain, great terror, and unhappiness. Beings are filled with alarm from all directions and in all directions" (ĀS 1.1.6.2). The one- through five-sense framework would become a hallmark of Jain taxonomy—from the single sense of touch in plants and elemental beings, to two- to four-sense insects that also have taste, smell, and sight, respectively, and five-sense mammals, fish, birds, and people who possess hearing and, for some, a mind.

Two primary modes of comprehending the shared, though differentiated, multisensory reality of living beings accurately is through the teaching of the Jinas, including Mahāvīra, and other knowledgeable sages, as well as through analogy with oneself. Even minute earth-bodied beings, for example, whose feeling may be unrecognizable or unintelligible to human sensory observations, may yet be known in these ways:

> As somebody may cut or strike a blind man (who cannot see the wound), as somebody may cut or strike the foot, the ankle, the knee, the thigh, the hip, the navel, the belly, the flank, the back, the bosom, the heart, the breast, the neck, the arm, the finger, the nail, the eye, the brow, the forehead, the head, as some kill (openly), as some extirpate (secretly), (thus the earth bodies are cut, struck, and killed *though their feeling is not manifest*).
>
> (ĀS 1.1.2.5; emphasis added)

The text offers analogical knowing as an imaginative form of comprehension when sensory knowing may be inadequate. Just because a blind man cannot see his wound doesn't mean he is not injured. Likewise, just because earth bodies' feelings are not "manifest," or are unintelligible to gross human recognition, does not mean they are not alive, sentient, desiring to live, and fearing death. Such vulnerable sentiency also applies to minute earth-, air-, fire-, and water-bodied beings, motivating the renunciation of harmful actions:

> There are beings living in water, many lives...[T]o the monks water has been declared to be a living matter...He who injures these (water-bodies) does not *comprehend and renounce* the sinful acts; he who does not injure these, *comprehends and renounces* the sinful acts. Knowing them, a wise man should not act sinfully towards water, nor cause others to act so, nor allow others to act so. He who knows these causes of sin relating to water, is called a reward-knowing sage.
>
> (ĀS 1.1.3.7; emphasis added)

This analogical frame also extends to plants:

> As the nature of this (i.e., men) is to be born and to grow old, so is the nature of that (i.e., plants) to be born and to grow old; as this has reason, so that has reason; as this falls sick when cut, so that falls sick when cut; as this needs food, so that needs food; as this will decay, so that will decay; as this takes increment, so that takes increment; as this is changing, so that is changing.
>
> (ĀS 1.1.5.6)

These arguments are pointedly and repeatedly directed at rivals who, in the Jain view, fail to acknowledge the universal sentiency made evident through analogical knowing of one's own pain: "Ye professors! Is pain pleasant to you, or unpleasant? If they give the right answer, reply: For all sorts of living beings pain is unpleasant, disagreeable, and greatly feared" (ĀS 1.4.2.6).

The suffering of five-sensed animals is the easiest to perceive with the senses, such that one need not use analogy but only a willingness to employ observational attention. *Ācārāṅga I* details specific socially-normalized harms toward animals that are seemingly as relevant today as they were in Bentham's time and in Indian antiquity:

> Some slay (animals) for sacrificial purposes, some kill (animals) for the sake of their skin, some kill (them) for the sake of their flesh, some kill them for the sake of their blood; thus for the sake of their heart, their bile, the feathers of their tail, their tail, their big or small horns, their teeth, their tusks, their nails, their sinews, their bones; with a purpose or without a purpose. Some kill animals because they have been wounded by them, or are wounded, or will be wounded. (ĀS 1.1.6.5)

Such visceral descriptions remind seekers that the very tasks of living, including eating, medicine, and accumulating material goods, come at a tangible cost to living beings because of their universal sentiency and desire to live. As Vallely summarizes, "Jainism insists that we are embedded in a world of competing subjectivities—some of which we see, hear, and touch, and none of which we can blithely ignore" (2014, 46).

Nearly 2500 years prior to Bentham's hope that the "mantle" of legal and ethical consideration would extend to all that breathes, the earliest layer of the Jain canon affirmed and expanded this claim to all existing entities—including those with the single sense of touch—and, not just those most like the "human." On the contrary, universal sentiency provides a sufficient ontological similarity such that understanding the reality and experience of one's own self allows empathic and analogical knowing of others' pleasure, pain, fear of injury and death, and longing for life.

(Un)Knowing Self Through Reciprocal Suffering and Memory: Jain Interchangeability and Rawls' "Veil of Ignorance"

The accepted view of rebirth within the Jain account takes the recognition of universal sentiency one step further toward the knowledge that one will also be reborn in innumerable bodily forms with various sense capacities, and different families, feelings, and environmental conditions. Not only does right comprehension mean knowing other sentient beings through analogy with the self, but also that the self—as *jīva* or *ātman*—will actually be born *in and as* other sentient forms, though precisely which bodies one will inhabit cannot be anticipated nor easily remembered, on which more in a moment.

In *Ācārāṅga I*, the diagnosed, and/or willful, ignorance of where one comes from or to where one is going in rebirth is answered through a multispecies, multiclass, and multi-embodiment ancestry—a comprehension of one's susceptibility to karmic effects resulting in a multiplicity of birth states and conditions, which, when comprehended, shapes ethical orientation:

> Frequently (I have been born) in a high family, frequently in a low one; I am not mean, nor noble nor do I desire (social preferment). Thus reflecting, who would

> brag about his family or about his glory, or for what should he long? Therefore a wise man should neither be glad nor angry (about his lot): thou shouldst know and consider the happiness of living creatures. *Carefully conducting himself* he should mind this: blindness, deafness, dumbness, one-eyedness, hunchbackedness, [karmic] blackness, variety of color (he will always experience)[5]; because of his *carelessness* he is born in many births, he experiences various feelings.... There is nothing inaccessible for death. *All beings are fond of life, like pleasure, hate pain, shun destruction, like life, long to live. To all life is dear.*
>
> (ĀS 1.2.3.1–4; emphases added)

Noted here in *Ācārāṅga I*, and persistent up to the present, the Jain metaphysical account requires one to comprehend their countless birth forms and conditions, all characterized by a desire to live. Not only is one vulnerable to the carelessness of others in these embodied forms, but one's own carelessness is like a weapon of karmic accrual wielded against self and others, leading one to ever-more vulnerable and deluded birth states.

The ethico-political significance of imagining oneself into alternate birth states for the sake of improving moral action has precedence in the work of American political philosopher John Rawls (1921–2002), whose "veil of ignorance" provides a valuable, if culturally distinct, Western parallel to the Jain view of karmic rebirth, interchangeability, and reciprocal suffering. Rawls' "veil of ignorance," also called the initial (or original) position, is a hypothetical state of *un*knowing the state of one's birth, which plays a critical role in designing a social contract based on imagining a certain interchangeability with other birth states, especially those characterized by more precarious conditions (1999/1971, 11–12). The initial position, states Rawls, "excludes the knowledge of those contingencies which sets men at odds and allows them to be guided by their prejudices" (17). For example, "if a man knows that he was wealthy he might find it rational to advance the principle that various taxes for welfare measures be counted unjust," explains Rawls, "if he knew that he was poor, he would most likely propose the contrary principle" (17). The better aim was to move beyond knowledge of one's race, class, status, or abilities so that, as Rawls argues, a society would develop principles of equality and fairness that would benefit everyone regardless of the social and economic condition they are born into. Specifically, he developed two principles of (1) equality of basic rights and duties, and (2) that any inequalities of wealth and status must have a collective benefit, especially for the most vulnerable of society (13).

The veil of ignorance thought experiment invites people to imagine birth in states "mutually disinterested" and even opposed to one other spiritually or economically (1999/1971, 12). Contra utilitarian ethics in which the rights of the few could be sacrificed for the pleasure of the many—which Rawls observed as central to Civil Rights violations and the Vietnam War, among other oppressive conflicts—Rawls writes, "it hardly seems likely that persons who view themselves as equals, entitled to press their claims upon one another, would agree to a principle which may require lesser life prospects for some simply for the sake of a greater sum of advantages enjoyed by others" (13).

Explicit in Rawls' critique of utility is the justification of harms against certain individuals for the sake of others' benefits. The "veil of ignorance" invites one to imagine that they are born into just such a vulnerable state in order to recognize the universal desire to live and refusal of injury rooted in a principle of equality. Of course, Rawls clarifies that this is a "purely hypothetical" thought experiment (1999/1971, 14); it is not a cosmological or karmically earned state as in the Jain view. Nevertheless, the "veil of ignorance" aims toward societal and institutional policies, practices, and reforms based on "equality between human beings as moral persons, as creatures having a conception of their good and capable of a sense of justice" (17). The imaginative possibility of being interchangeably born into an alternate state is an essential aspect of developing an ethical contract of true social justice.

The Jain views offers a much more expansive view of interchangeability that *Ācārāṅga I* (and later Jain texts) extends to the six classes of all living beings, far beyond humans alone. While some recent theorists have argued that animals are entitled to (or participate) in Rawlsian justice and/or social contracts (Garner 2013), Rawls himself excluded nonhuman animals—to say nothing of insects and elemental beings included in the Jain taxonomy—as moral persons capable of relationships of giving and receiving justice (Baldwin 2023).

In the Jain view, however, analogical knowing through universal sentiency reveals a secondary knowledge of reciprocal suffering, such that "Knowing what is bad, [one] who knows it with regard to himself, knows it with regard to (the world) outside; and [one]who knows it with regard to (the world) outside, knows it with regard to himself: this reciprocity (between himself and) others (one should mind)" (ĀS 1.1.7.1). To put it another way, awareness of reciprocal suffering through the interchangeability of birth states—either hypothetically in the "veil of ignorance" or in the Jain view of countless karma-induced rebirths—means, what I do to you, I do to me.

Like Rawls' claim that social reciprocity undermines a utilitarian willingness to sacrifice the interests of some for the happiness of a majority, *Ācārāṅga I* rebels against heretical views of rival groups who justify harming others for their own ease. The text issues polemical reproaches against Buddhist, Ājīvika, and Shaivite mendicants who "pretend only to be houseless" and follow unworthy doctrines (ĀS 1.1.2.2[6]): "Many and several in this world, *brahmanas* or *sramanas*" claim that "all sorts of living beings may be slain, or treated with violence, or abused, or tormented, or driven away…there is no wrong in it" (ĀS 1.4.2.3–4). The text refutes this accepted view, because knowledge of reciprocal suffering emerges from analogical knowing: "[One] who denies the world (of water bodies), denies the self; and [one] who denies the self, denies the world of (water bodies)" (ĀS 1.2.3.3), and similarly stated for fire- and air-bodies, etc. Surendra Bothara further explains that this denial is not simply one of disavowing previous birth states but also "amounts to false accusation" since "to deny existence of something, to establish a lie as truth, and a truth as false, to show a being as non-being…are all [a] misstatement" (1999, 34).

Not only is one more likely to hurt other beings when one does not understand their universal sentiency as analogous to one's own, but one will reciprocally injure *themselves* through the karmic inflow caused by harmful actions toward others which further binds one in ignorance, violence, and difficult rebirths. *Ācārāṅga I* states:

> As it would be unto thee, so it is with him who though intendest to kill. As it would be unto thee, so it is with him whom thou intendest to tyrannize over. As it would be to thee, so it is with him who though intendest to torment… to punish, and to drive away. The [one] who lives up to these sentiments, does therefore neither kill, Nor cause others to kill (living beings). *He should not intentionally cause the same punishment for himself.*
>
> (ĀS 1.5.6.4; emphasis added)

The sense of "self," while not in any way meaningless or lost in the Jain view, is multiplied through the imaginative recognition of rebirth and reciprocal suffering in vastly different—even opposed—bodily forms. Indeed, this recognition is understood as having "a mind different (from that of common people)" (ĀS 1.2.5.3) who seek to satisfy pleasures, acquire property and wealth, or have professions, that come at tangible costs to other bodies. "A heretic professes to cure (the love of pleasure) while he kills, cuts, strikes, destroys, chases away" others who not only have universal sentiency, but in whose forms one may have previously existed or yet be reborn (ĀS 1.2.5.6).

Transformative Multispecies Knowing Through Memory

While *Ācārāṅga I* begins by addressing a lack of knowledge and recollection of one's true identity and path through multiple rebirths, some texts describe rare forms of knowing in which rebirth need not only be imagined but can be remembered. Another early Śvetāmbara canonical text called the *Uttarādhyayana-sūtra*, said to include Mahāvīra's final teachings, describe Prince Mṛga-putra's transformative experience of past-life recollection. Beloved by his parents and wives, the prince's lavish life of ease was disrupted one day when he observed a monk practicing restraints, startling the prince into a memory of his own previous life as a mendicant dedicated to the Jain vows and knowing the deep suffering of *saṃsāra* through impermanence, aging, illness, and death.[7] Trying to convince his parents to release him from his princely role in order to recommit to the mendicant path, the prince recounts memories of undergoing great suffering and violence, both in the many lifetimes his *jīva* was reborn as a hell being due to his own cruelty and ignorance towards others, and also his rebirth as animals, plants, and even iron:

> As an antelope I have, against my will, been caught, bound, and fastened in snares and traps, and frequently I have been killed.

> As a fish I have, against my will, been caught with hooks and in bow-nets; I have therein been scraped, slit, and killed, an infinite number of times.
>
> As a bird I have been caught by hawks, trapped in nets, and bound with bird-lime, and I have been killed, an infinite number of times.
>
> As a tree I have been felled, slit, sawn into planks, and stripped of the bark by carpenters with axes, hatchets…an infinite number of times.
>
> As iron I have been malleated, cut, torn, and filed by blacksmiths, an infinite number of times…
>
> Always frightened, trembling, distressed, and suffering, I have experienced the most exquisite pain and misery…In every kind of existence I have undergone suffering which was not interrupted by a moment's reprieve.
>
> (US 19.63–74)

Though Jain literature suggests that only the most advanced beings, such as Jinas, are able to recall *all* previous existences, many stories depict ordinary people or heretics recalling one or several past lives through which one comes to correctly understand the universal suffering of existence, the violent effects of ignorance, enjoining one to pursue the mendicant path toward liberation (Appleton 2014, 11).

Prince Mṛga-putra directly recalls being prey due to his interchangeability with others through rebirth. He is not learning *about* other beings' suffering nor seeing analogous similarities with his own. Rather he *un*knows his egoistic sense of self by knowing multiply his interchangeability in other forms through a profound memory of reciprocal suffering: "*I* have been killed," he attests, or caught, trapped, bound, slit, scraped, bound, sawn, or cut. Rebirth recollection shocks the prince's identity out of his comfortable life of ease into the direct perception and awareness of another's suffering *as his own*. While being born into a human birth state is always considered preferable in the Jain view, the knowledge of reciprocal suffering through the interchangeability of birth forms, according to Vallely, "takes for granted the experience of life as always and unavoidably in a state of being with others. The animate and intelligent life that encompasses us, that perpetually bombards us, and leaves its innumerable impressions on us every moment of every day, is, for Jains, taken as a given" (2014, 52). She continues, "This all-encompassing life—along with the myriad beings it contains—while certainly a given, is also the foundational source for an understanding of life's existential purpose"…so that one "engages with all senses open…fully attuned to life's complex stratifications…deeply cognizant of the suffering nature of the simplest of beings most especially" (52).

Ethical Knowing in the "Onto-Ethico-Epistemology" of Karen Barad's Material Becoming and the Jain "Three Jewels"

The *Ācārāṅga-sūtra* contains elements of knowing that are closely connected to right comprehension of the living multiplicity that exists as self and others and renunciation of harmful action towards all of these. This three-fold connection between what exists (ontology), knowing (epistemology), and restrained action (ethics) would be systematized in later Jain texts, forming the opening line of the *Tattvārtha-sūtra* (a title meaning "That which exists")—an early-centuries CE text considered authoritative by all Jain sects—"Right view (of reality), right knowledge, and right conduct constitute the path to liberation" (TS 1.1) As we have already seen above, rightly comprehending that plants, animals, and elemental beings experience pain, pleasure, fear, and a desire to live analogous to one's own, should naturally impact one toward careful action. One who knows rightly will be "[c]arefully conducting himself" while one who lacks right view, "because of his carelessness he is born in many births, he experiences various feelings" (ĀS 1.2.3.2). That is, one's view, knowledge, and ultimately harmful action has a decisive impact on one's immediate and future state.

Feminist physicist Karen Barad also suggests in her book *Meeting the Universe Halfway* (2007) that reality—primarily in its quantum entanglements—can be best understood through a fusion she calls "onto-ethico-epistemology" (381). An adequate account of what Barad calls "agential realism" reflects that what a quantum is, does, and knows emerges simultaneously as the "intra-active" becoming of what we consider space, time, and matter (381).

Barad's account is not a true parallel to the "three jewels" of Jain right view, knowledge, and conduct which are, in some ways, associated with the subjectivity of multiple perceptive and responsive *jīva*s. Barad is challenging any pre-existent agents, mediums, or containers within which being, doing, or knowing happen. Instead, "intra-action" is Barad's neologism that "signifies the mutual-constitution of entangled agencies" in contrast with interaction that assumes individuals that precede activity (33). Her account of the onto-ethico-epistemology of "agential realism" demonstrates how slices of reality "are iteratively configured through each intra-action" (ix). Nevertheless, there remain some meaningful parallels between Barad's account and the Jain fusion of existence, knowing, and action.

First, the co-arising of existence, knowledge, and action is not a human-only phenomena and, in fact, can be difficult to perceive and respond to adequately by humans. The agency described by Barad is of quantum entanglements, as difficult to perceive as the universal sentiency described in *Ācārāṅga I* which requires diverse modes of knowing such as testimony from Jinas or teachers, analogy, or memory. From her account of material physics, Barad argues that "What may seem evident to some is not simply a result of how things are, independently of specific practices of seeing and other bodily engagements with the world" (2007, 155). She continues, "[T]he seemingly self-evidentiary nature of bodily boundaries, including their seeming visual self-evidencse, is a result of the repetition of (culturally and historically) specific bodily performance" (155). Basically, habitual and repetitive modes of perception, representation, and response easily lead to overlooking, misperception, and misinterpretations of the agential intra-active becoming of quantum reality.

Second, the emergence of material reality is produced as much by what is excluded from intra-action as by what is included. While Barad presents the becoming of quanta as the positive, if difficult to control, intra-action of being, knowing, doing, this intra-action also involves what she calls the "agential cut," by which a quantum—or any entity—becomes separate from others in an "ontological condition of exteriority between observer and observed" (2007, 140). While the Jain cosmology presumes agential separateness in the multiplicity of infinite individual *jīva*s, the concept of exclusion, or in the Jain sense violence (*hiṃsā*), is explanatory of every *jīva*'s material existence. To be an embodied being in the Jain view means to participate in four instincts (*saṃjñā*) of craving for food, which leads to fear, desire for reproduction, and accumulation of things for future use, each of which requires the trespass, or exclusion of other beings (Jaini 2010, 284). Every harm—whether "'I did it;' 'I shall cause another to do it.' 'I shall allow another to do it.'" (ĀS 1.1.1.5)—has a material karmic effect whereby violent exclusion shapes the present and future toward, ideally, a transformation of view, knowledge, and less exclusionary conduct in a state of karmically-free liberation. As Barad states, "ethico-onto-epistemological questions have to do with responsibility and accountability for the entanglements 'we' help enact and what kinds of commitments 'we' are willing to take on, including commitments to 'ourselves' and who 'we' may become" (2007, 382). The way we become is, in no small part, what we are. In the Jain view, the aim is to practice restraints that enable one to become through less, and ultimately in the most advanced states of practice, no violent exclusion of others as the fullest expression of self.

Third, careful action is a form of knowing. The onto- and epistemic- commitments of *Ācārāṅga I* are both tested and reach their fulfillment in the restraint of action, often described as *carefulness*. "Look at the exterior (world from analogy with thy own) self; [then] thou wilt neither kill nor destroy (living beings); viz. out of reciprocal regard [well examining] he does no sinful act" (ĀS 1.3.3.1). Barad's "onto-ethico-epistemology" does not offer a prescriptive aim, though the implications for action are clear: "Making knowledge is not simply about making facts," she writes, "but about making worlds, or rather, it is about making specific worldly configurations—not in the sense of making them up *ex nihilo*, or out of language, beliefs, or ideas, but in the sense of materially engaging as part of the world in giving it specific material form" (2006, 91). Each agential cut is a specific intra-action of being, knowing, and doing that creates the material world of what is, and is to come. "We are not merely situated in the world" as pre-existent actors, Barad asserts, but "'each of us' is part of the intra-active ongoing articulation of the world in its differential mattering" (381) such that every action *is* a form of being and knowing.

Carefulness as a Mode of Ethical Knowing

Given the ubiquity and multiplicity of life in all of its six classes of beings, *Ācārāṅga I* repeatedly exhorts seekers to act, or make the world, with assiduous carefulness. Carefulness, here, is not just a translational choice, but a specific term (Pkt: *apammate*; Skt: *apramāda*) found in Jain texts that will eventually become a technical concept (Donaldson and Bajželj 2021, 53–70). At this point in the earliest canon, however, carefulness describes (1) general restraints of mind, speech, and body that follow from the recognition of beings' universal sentience and vulnerability to injury, causing one to accrue karma, (2) a pedagogical training space for mendicants who are not yet disciplined enough for nonaction, but who can remain in the community if they aspire toward carefulness. I'll say a word on each here.

Carelessness is understood as the most damaging kind of action, which results in karma that keeps one cycling in *saṃsāra*. In as much as one knows any of these modes of carelessness, they can know all the extreme emotions and sensory desires that lead to physical harm:

> [One] who knows wrath, knows pride; [one] who knows pride, knows deceit; [one] who knows deceit, knows greed; [one] who knows greed, knows love; [one] who knows love, knows hate; [one] who knows hate, knows delusion; [one] who knows delusion, knows conception, knows birth; [one] who knows birth, knows death; [one] who knows death, knows hell; [one] who knows hell, knows animal existence [one] who knows animal existence, knows pain. Therefore, a wise [one] should avoid wrath, pride, deceit, greed, love, hate, delusion, conception, birth, death, hell, animal existence, and pain.
>
> (ĀS 1.3.4.4)

And likewise, anyone who interrupts any of these habits through practicing restraints, can interrupt all, since, "if careful, thou wilt always conquer" (ĀS 1.4.1.3). The restraints are many and varied in *Ācārāṅga I*, a piecemeal assortment of bearing with external discomforts, sickness, and aging (ĀS 1.5.3.5), controlling oneself from killing and injuring when desires or aversions surge (ĀS 1.5.3.3), restraining the passions arising from sex (ĀS 1.3.2.3), restraining anger, imitating the equanimity of a learned teacher (ĀS 1.4.2.5). The text also recognizes that this path is very difficult, and even dedicated mendicants can "slide back" (ĀS 1.5.3.2) or even leave the path (ĀS 1.6.2.1).[8] While the text begins to make some concessions for monks in training, for instance, that "those who are afflicted and careless (*pamattā*) will be instructed" and presumably not kicked off the path (ĀS 1.4.2.2), it also remains bold in its claims that one must test the teachings against rival views as well as through practice to ultimately decide for themselves who they are and how they will strive to know complex reality (ĀS 1.2.6.5).

Conclusion

Ultimately, *Ācārāṅga I*, traced back nearly three millennia, begins with a realist diagnosis that every action has a cost within a universe of sentient, perceptive, and responsive life that fears death and desires to live. The text presents numerous forms of multispecies knowing that precede and, in some cases, are much more capacious than later formulations essential to modern animal ethics, ecological politics, and concepts of agential world-making far beyond the human. Knowing universal sentiency prefigures Bentham's utility calculations of pleasure and pain essential to animal rights but extends them to every entity in existence, which, in the early Jain view, can meaningfully be known and tested through analogy with the self. The Jain account of interchangeability of life forms through innumerable rebirths offers an ancient cosmological parallel to Rawls' theoretical "veil of ignorance" by which reciprocal suffering—imagined or remembered—is a transformative way of knowing life and shaping justice-producing action. The practice of careful world-making, notably expressed in Jain restraints of harmful action, arises simultaneously with right comprehension and knowing of a living multiplicity expressed later in the "three jewels," much as Barad illuminates the often imperceptible "intra-action" of "onto-ethico-epistemology" as they comprise the "agential cuts" of quantum world-making. If, as *Ācārāṅga I* asserts, "[One] who knows one thing, knows all things" (ĀS 1.3.4.1), then myriad forms of knowing—no less relevant to antiquity as they are in the present—are essential to, and enlarged by, careful actions of nonharm toward all living beings.

Notes

1 See Donaldson 2016.
2 All *Ācārāṅga-sūtra* I translations in this chapter are from Hermann Jacobi. Most Prakrit transliterations come from Jacobi, though I've transliterated certain terms using Bothara's translation (Bothara 1999).
3 Numerical references such as ĀS I.1.1.4 refer to *Ācārāṅga-sūtra*, book I, lecture 1, lesson 1, verse 4 in Jacobi's translation. Alternate translations or commentaries will be noted.
4 The terms are here in Sanskrit. The Prakrit (Pkt) is: *se āyā-vāī logā-vāī kammā-vāī kiriyā-vāī yā* (Schubring 2004/1926, 179).
5 Colors such as blackness or "variety of color" refer to karmically-induced states of one's *jīva*—a kind of "soul color" or psychic-spiritual coloring—rather than any pigment or racial connotations. This theory of color, known as *leśyā*, would characterize later aspects of Jain art and philosophy to express how the *jīva*'s qualities become obscured or darkened due to the intake of material karma and, conversely, inspire the shedding, or lightening, or one's karmic burden through ethical restraints. See Malde n.d.
6 See similar phrasing at ĀS: 1.1.3.4, 1.1.4.4, 1.1.5.4, 1.1.6.3
7 Prior to his transformation into the Buddha, Prince Siddhartha was said to experience "four sights" of illness, aging, death, and mendicant restraint which convinced him to renounce his material path toward enlightenment for the sake of all living beings.
8 For more on Bajželj's analysis of how early mendicant communities expanded to include novice monks, see Donaldson and Bajželj 2021, 46–51.

Bibliography

Primary Sources

ĀS [*Ācārāṅga-sūtra*]. 1884/2002. *The Ākārāṅga Sūtra* in Sacred Books of the East, vol. 22 (Jaina Sūtras: Part I). Translated by Hermann Jacobi. Oxford: Clarendon Press.
TS [Umāsvāti, *Tattvārthasūtra*]. 2011. *Tattvārtha Sūtra: That Which Is*. Translated by Nathmal Tatia. New Haven: Yale University Press.
US [*Uttarādhyayana-sūtra*]. 1895/2004. *The Uttarâdhyayana Sûtra* in Sacred Books of the East, vol. 45 (Jaina Sūtras: Part II). Translated by Hermann Jacobi. Oxford: Clarendon Press.

Secondary Sources

Appleton, Naomi. 2014. *Narrating Karma and Rebirth: Buddhist and Jain Multi-Life Stories*. Cambridge: Cambridge University Press.

Balcerowicz, Piotr. 2017. "Jainism: Disambiguate the Ambiguous." In *Indian Epistemology and Metaphysics*, edited by Joerg Tuske, 75–101. New York: Bloomsbury.

Balcerowicz, Piotr. 2021. "Beginnings of Jaina Ontology and Its Models." *Journal of Indian Philosophy* 49: 657–97.

Baldwin, Guy. 2023. "Rawls and Animal Moral Personality" *Animals* 13, no. 7: 1238.

Barad, Karen. 2007. *Meeting the Universe Halfway: Quantum Physics and the Entanglement of Matter and Meaning*. Durham: Duke University Press.

Bentham, Jeremy. 1843. *The Works of Jeremy Bentham, Vol. 1*. Edited by John Bowring. Edinburgh: William Tait. https://files.libertyfund.org/files/2009/Bentham_0872-01_EBk_v6.0.pdf

———. 2000/1781. *An Introduction to the Principles of Morals and Legislation*. Kitchener, Ontario: Batoche Books. https://historyofeconomicthought.mcmaster.ca/bentham/morals.pdf

Bothara, Surendra, trans. 1999. *Illustrated Acharanga Sutra, Book 1*. Edited by Amar Muni and Srichand Surana. Delhi: Padma Prakashan.

Bronkhorst, Johannes. 2011. *Karma*. Honolulu: University of Hawaii Press.

Derrida, Jacques. 1991. "'Eating Well,' or the Calculation of the Subject." In *Who Comes After the Subject?*, edited by Eduardo Cadava, Peter Connor, Jean-Luc Nancy, 96–119. New York: Routledge.

Donaldson, Brianne. 2016 "From Ancient Vegetarianism to Contemporary Advocacy: When Religious Folks Decide that Animals Are No Longer Edible." *Religious Studies and Theology* 35, no. 2: 37–57.

Donaldson, Brianne and Ana Bajželj. 2021. *Insistent Life: Principles for Bioethics in the Jain Tradition*. Oakland: University of California Press.

Garner, Robert. 2013. *A Theory of Justice for Animals: Animal Rights in a Nonideal World*. Oxford: Oxford University Press.

Haraway, Donna. 2016. *Staying with the Trouble: Making Kin in the Chthulucene*. Durham: Duke University Press.

Jaini, Padmanabh S. 2001/1979. *The Jaina Path of Purification*. New Delhi: Motilal Banarsidass.

———. 2010. "Fear of Food: Jaina Attitudes on Eating." In *Collected Papers on Jaina Studies* Edited by Padmanabh S. Jaini. 281–96. New Delhi: Motilal Banarsidass.

Long, Jeffery. 2009. *Jainism: An Introduction*. New York: I.B. Tauris.

Malde, Shruti. n. d. "Leśyā." *Jainpedia*. Accessed October 12, 2023. https://jainpedia.org/themes/principles/jain-beliefs/soul/lesya/

Miller, Christopher Jain and Jonathan Dickstein. 2021. "Jain Veganism: Ancient Wisdom, New Opportunities." *Religions* 12, no. 7: 512.

Ohira, Suzuko. 1994. *A Study of the Bhagavatī-sūtra: A Chronological Analysis*. Ahmedabad: Prakrit Text Society.

Rawls, John. 1999/1971. *A Theory of Justice, Revised Edition*. Cambridge: Belknap Press.

Riepe, Dale. 1961. *The Naturalistic Tradition in Indian Thought*. Seattle: University of Washington Press.

Schubring, Walther. 2004/1926. *Mahāvīra's Words*. Translated by William Bolleé and Jayandra Soni. Ahmedabad: L. D. Institute of Indology.
Vallely, Anne. 2014. "Being Sentiently with Others: The Shared Existential Trajectory among Humans and Nonhumans in Jainism." In *Asian Perspectives on Animal Ethics: Rethinking the Nonhuman*, edited by Neil Dalal and Chlöe Taylor, 38–55. New York. Routledge.

9 Multispecies Kinship and Kindness

Knowing Life Through Animist Etiquette and Ethics

Graham Harvey

We are surrounded by kin, most of whom are other-than-human, and immersed in a sensorium of multiple conversations, not all of which are sonic. Although Bruno Latour rightly says, "We have never been modern" (1993), too many of us have been seduced by the Modern idea that we humans are—or ought to be—somehow unique and separate from the larger-than-human community. This separation has been managed by asking the wrong questions, that is, questions that assume human exceptionalism, such as "how can we avoid projecting human competencies such as intentionality onto [other] animals?" and not asking the right questions, that is, questions that seek mutual understanding between species, such as "what are rats interested in during experiments" (Despret 2016, 89–96). Even those of us who want to resist the project of Modernity often contribute to policing its (fictitious) border between humans and a world that is supposed to be inanimate and to act mechanically or instinctively rather than intentionally or intelligently. The policing happens when, for example, we speak of "nature" as if there really is an inert realm distinct from "culture." It happens when we speak of "projection," "attribution," or "extension" of putatively human characteristics to other existences. As Isabelle Stengers argues, Modernity is a hegemonic "world-destroying machine" at war with other "worlds" (lived-in communities intersecting with others, each able to engage diplomatically with others), "destroying both politics and ontologies" (2018, 86). In John Law's phrase, it is a "One-world world" (2015) that literally de-animates the "world in which many worlds fit," which the Zapatistas evoke and struggle for (Marcos 1996).

What we should call the as-yet incomplete results of Modernity's project of human separatism is frequently debated. Rosi Braidotti summarizes several terms:

> the Anthropocene has already become another "Anthropomeme" (Macfarlane 2016), spawning an array of derivative terms, such as "Capitalocene" (Moore 2015), "Anthrop-obscene" (Parikka 2015), but also: "Plasticene", "Plantationocene" (Tsing 2015), "Mis-anthropocene" and "Chthulucene" (Haraway 2015).
>
> (2018, 32)

DOI: 10.4324/9781032659985-12

Perhaps it is enough to name the dominant power in this phase of planetary life and death, "Western Modernity" or "Euro-Modernity." This is not to acquiesce to the notion that this "Modernity" is a pure export of a natural, inevitable, or desirable developmental process from Europe and its settler colonies. Rather, Modernity is an interactive but unequal project of global "connected histories" (Bhambra 2007, 11, 32–33, citing Subrahmanyam 1997). Nonetheless, its deadly extractivist ambition is most clearly distilled and pursued in its "Western" forms—including in what several Indigenous friends have identified as the current era of "Most-Colonialism," referring to in the continuing (and sometimes escalating and even genocidal) assaults on Indigenous rights, lands, and lives. However, more is at stake here than names.

Still, maybe naming is one of the "Arts of living on a damaged planet" (Tsing, Swanson, Gan and Bubant 2017), part of the "getting to know you" etiquette that enables the recognition of monsters, and/or the making of allies and the renewal of relations. Maybe. Anyway, the naming I propose here is not the labeling of this or any other era. Rather, I seek to contribute to the project of "knowing life" by using names such as "animism" and "Indigenous" to evoke larger, longer commitments, obligations, and "arts of living." In particular, these names emphasize interspecies conversations in which humans already participate more fully than is encouraged, expected, or celebrated by Moderns. These are conversations that evidence Eduardo Viveiros de Castro's influential argument that in some cultural worlds, there are plural "natures"—a "multinaturalism" to rival "multiculturalism" (1998, 2007; see also Kopenawa and Albert 2013 and Skafish 2016, 395).

Much of what follows invites a narrowing of our focus from "knowing [all] life" to the etiquettes and ethics of "getting to know you" interspecies interactions. The particularity of what bodies sense and how they engage with others entails specific perspectives. All existences relate, communicate, and otherwise interact with the world(s) culturally, meaning that they have homes and families, taboos and ceremonies, culinary and other arts, and so on. Degrees of intimacy and immediacy lead to inclinations or indifferences towards members of other species or communities (Naveh and Bird-David 2013). Perhaps this means that there are diverse human and other-than-human cultures because, at least in part, communities live with(in) different "natures." They literally body their lived-in worlds. In this chapter, I engage with select animist and Indigenous ways of being and socializing in the larger-than-human world. Going beyond description alone, I provide examples of transformative scholarship that not only seeks more just and sustainable relations with which to contest the destructive machinery that threatens planetary life, but also offers a kind of *etiquette* enabling the recognition of monsters, the making of allies, and the renewal of relations.

Are These Stones Alive?: Beyond Correct Categorization to Perceiving Responsive Kinship

Seen in relation to the world-knowing and world-making (or worlding) of other worlds or cultures, the Modern "nature" is one culture's effort to relate to larger-than-human society. Yet, by imagining and engineering an un-animate and non-social "nature" lacking the ability to reciprocate human attention with any degree of agential mutuality, Modernity's cultural effort reveals the dangerous antagonism of its mode of relating. This simplified, limited and limiting naming of "nature" is one aspect of an effort to differentiate and distance humans from living, lived-in world(s). Such a distancing can only result in destruction as the "world-destroying machine" fails to recognize the embeddedness of humanity within this already dangerously damaged planet.

There are other worldings, however, and other "arts of living" among kin of other species. A conversation between the American anthropologist Irving Hallowell and the Anishinaabe ritualist Kiiwiich (Alec Keeper) is eloquent here. The conversation is quoted and considered in almost every recent discussion of animism, ontology, or the new materialism, and it deserves celebration and consideration. When asked by Hallowell, "Are *all* the stones we see about us here alive?", Kiiwiich "reflected a long while" and then replied, "No! but *some* are" (Hallowell 1960, 20). In case this conversation is not already familiar, what Hallowell was trying to understand (and then convey to his readers) was whether the grammar of Kiiwiich's native-tongue, Anishinaabemowin, had philosophical, ontological, behavioral, and perspectival implications. In that language, *asiniig* (stones) is an animate plural noun. Hallowell wanted to know if this grammatical animism expressed a general and generative idea of animation or liveliness. So the obvious question for a Modern (and an English-speaker) might be, "are stones alive or dead, animate or inanimate, personal or impersonal?" To put it another way, such a question is obvious to someone brought up to assume that stones are inanimate and impersonal. For such people, stones deserve the English-language inanimate pronoun "it." The use of that pronoun contributes to making the world Modern and to separating humans from the larger-than-human community. Distinguishing the knowledge of Moderns, such as disciplinary splits that demarcate biology from geology in the question of a stone's liveliness, from the cultural beliefs of other-than-Moderns contributes further to the Modernizing and colonizing project.

Robin Wall Kimmerer's (2015) proposal to use a new animist pronoun "ki" (singular) and "kin" (plural) may yet enable conversations that celebrate and (re-) assume the lively relationality of the world. In the meantime, Kiiwiich's response to Hallowell's question already alters everything. What emerges is that Hallowell's question needed significant adjustment. It began well, "Are *all* the stones we see about us here…" but ended less well with "alive?". At first, Hallowell and Kiiwiich are placed in relation to nearby stones. Specific stones and specific humans are located together. Even if Hallowell wanted to perform the then common anthropological task of constructing a systematic idea of the whole of Anishinaabe culture, he began with the specificity of relations rather than with a hierarchy of categories. But "alive" collapses the intercultural dialogue by seeking the kind of material that Edward Tylor might have included in his *Primitive Culture* (1871)—that is, information about misguided (if understandable) beliefs about spirits, souls, and metaphysics, and those categorical distinctions between life and death, culture, and nature. Kiiwiich's seemingly enigmatic and perhaps somewhat amused answer expertly headed the conversation elsewhere. He conveys a sense of relationality "all the way around." That is, what was of most importance to Kiiwiich was not a theory of existence, or a culture structured on oppositional or hierarchical differences, but an understanding of the obligation to behave well. '*Some* are" does not mean that some existences are not worthy of respect (let alone "not alive") but that relationships are more-or-less intimate or distant, more-or-less immediate and present, more-or-less hedged by etiquettes and expectations. Those stones "about us here" are potentially ready to give and receive gifts, or to join in other modes of interspecies communication. Some stones will interact with some humans. The right question leads to learning how to interact, how to act relationally, how to be kin.

Hallowell said that this conversation had a great impact on him. Rightly so. It taught him to continue conversations in a thoroughly relational world full of persons, most of whom are other-than-human, a multispecies co-evolving community of more-or-less present kin. It shaped Hallowell's discussion of the behaviors that arise from the pervasive Anishinaabe animist anticipation that interpersonal interspecies interactions might happen. As the scholarly project now named "the new animism" developed, the transformative issue has become increasingly clear. It is not about people's ideas ("beliefs") about whether things are alive, or how they know what is alive. The lively issue is how animists seek to live well with and among others, and how those ways are inculcated, adopted, and adapted. The short-ish animist answer is: assume that everything you encounter is a person (a being worthy of respect, of careful and constructive approach and attention), a relation, kith and kin, and seek to act respectfully by, for example, listening, and giving and receiving appropriate gifts. In Anishinaabemowin this is named *bimaadiziwin*, referring to good life/good living.

Animism as Obligations and Etiquettes

Kimmerer's advice about "Learning the Grammar of Animacy," which is the heart of her *Braiding Sweetgrass* (2013), advances our education about animate world words. She celebrates the impact of discovering that her ancestral language, Bodewadmimwin (a close relative of Kiiwiich's Anishinaabemowin), has a word for "the force which causes mushrooms to push up from the earth overnight," *puhpowee* (Kimmerer 2013, 49). Someone who learns to use that word is likely to have gained fresh appreciation for the way the world works. But it is the word *wiikwegamaa* that most powerfully teaches her and us about animist grammar and, therefore, about skillful animists living in an animate world (a world in which many worlds fit). *Wiikwegamaa* can be translated into English as "to be a bay." It does not name a defined area of geography, "a bay," distinct from other kinds of places. It is not a noun labeling a category. It is a verb. Someone encountering the activity that requires the word *wiikwegamaa* experiences the performing of bay-ing, the doing of bay. Such an action involves interaction with other actions—Kimmerer mentions the verbs "to be a hill, to be a sandy beach, to be a Saturday" (55). Then, most importantly, she points to the more-than-grammatical implications of these verbs. Water that acts as a bay (as English grammar requires me to write) has chosen not to act as "a stream or an ocean or a waterfall" for now. Water cannot bay alone. In collaboration with actions we may call "hill-ing" and "beach-ing," a community of relations enacts "bay-ing." Learning to speak this way involves learning to expect interactions and transformations, and to relate well, to act as kin, to respect the choices of others.

To summarize this animism more succinctly: animism grows from the knowledge that all existences are continuously interacting with others and are encouraged to do so respectfully, carefully, and constructively. Animism is *not* a theory that there is life in everything (as was proposed in earlier academic approaches). Rather, rooted in an understanding that every existence moves through, among, and *as* life (Ingold 2006, 14), animism is a commitment to the implications and obligations of kinship and pervasive relatedness. The scholarly project labeled "the new animism" is new only as a fresh academic approach to tried and tested relational cosmologies, ontologies, ethics and etiquettes of people who do not necessarily name any of this "animism." Linda Hogan says "we call it tradition" (2013). Others affirm and enact relationality in refrains such as "all our relations," which structures sweat lodge ceremonies.

The next paragraphs are not quite vignettes but more celebratory citations of some ways in which animists meet trees, deities, mountains, "spirits," food-persons, ancestors, and other kith and kin. They illustrate a few of the myriad ways of learning and performing animist relationships. In one way or another they are about "knowing life" on the assumption that "knowing = doing," and that this "doing" *know-how* shapes "good living"—in which "good" can (with practice) mean successful, proper, and long lasting.

Narratives of "Knowing Life" Through Animist Doing

Larry Gross' *Anishinaabe Ways of Knowing and Being* (2014) walks readers through a vibrant ecology of ways in which *bimaadiziwin* (good life/living) is inculcated, incorporated, and enacted in language, humor, storytelling, harvesting, and in dealing with postapocalyptic trauma. The "traditional" Anishinaabe world was nearly destroyed in the colonial genocide-and-ecocide; those alive now are recovering resilience (see also Vizenor 1994, Whyte 2018). Gross provides a resonant example of learning to be (more) animistic or (more) traditional—in which children accompany adults to collect maple syrup. They are told to be silent in the woods because "this is the deer's house and we are just visitors… . Just be careful with the [maple] sap. Remember, we are quiet in the woods"; then "the children broke the quiet rule as they run from tree to tree, laughing, wanting to be first to empty a full jug" (Gross 2014, 61, citing Northrup 2001). Far from being a failure, breaking the quiet rule is celebrated as an educative success. Children are allowed to be children and to take time to learn by (eventually) imitating their elders—and, above all, they *enjoy* the gifts given by the trees. Contrary to many assertions about the childishness of animism, elders become more animistic as they age and learn.

In part, elders do so by expressing gratitude more often, more explicitly, and more ceremonially. They also learn that the *gift* of sap is a sacrificial one on the part of the trees and that in giving the equivalent of their lifeblood the trees permit the violent act of being cut and bled (Gross 2014, 161). Grateful enjoyment is the right response.

Such intimate violence is central to the Māori scholar Te Pakaka Tawhai's assertion that "the purpose of religious activities here is to…do violence with impunity" (1988, 101). He illustrates this powerful and provocative phrase by referring to the felling and carving of trees and to the tending and harvesting of *kūmara*, sweet potatoes. Committing violence against such relations is required because it is necessary to shelter and feed guests. Tawhai's article says little more about these core elements of Māori sociality and does not fully set out the cosmic role played by trees in separating MotherEarth from FatherSky or (once cut, milled, and carved) separating floors from ceilings. Nor does he note that *kūmara* migrated to Aotearoa/New Zealand with Māori—establishing not just a horticultural ecology but an intimate kinship of co-habitation. Certainly, however, Tawhai's choice of examples was far from random: trees and *kūmara* are culturally, ecologically, and cosmically important relations. A nineteenth-century *tohunga* (elder/expert), Pita Kāpiti, provided an account of relevant rituals, including chants and actions (*karakia*) involved in *kūmara* tending, which express this relationality and the danger of acting violently towards them. If, for instance, a *kūmara* was damaged during the weeding of fields, the following chant was enacted, "Held aloft by me is the blood of this *kūmara*. Who will avenge, who will seek vengeance for your death? Tū [an *atua*/deity] will avenge, he will exact punishment for your death…" (Reedy 1997, 126).

Kāpiti then outlined procedures for healing the tubers. At harvest time, the unearthing of the *kūmara* was accompanied by a chant including the phrases: "You rise up bare from your resting place…you rise up bare from where you were put to rest…you rise up bare!" (Reedy 1997, 128). That is, the rested and "rising" *kūmara* is addressed as a celebrated personal being with a powerful guardian or avenger. Therefore, ceremonies—or "religious activities" as Tawhai says—are required to avert any danger resulting from either inadvertent or necessary violence (as a side note, the Modern distinction between technical skills and ritual, and between the powers of humans and "spirits," is powerfully challenged in Marshall Sahlins' *The New Science of the Enchanted Universe*, 2022).

Such traditional Māori ceremonies are among the many ways in which animists deal with the dilemma of life among relations in a world where every being consumes others. That dilemma was well expressed by the Iglulik shaman Aua, who told the Greenlandic-Danish anthropologist Knud Rasmussen that:

> The greatest peril of life lies in the fact that human food consists entirely of souls. All the creatures that we have to kill to eat…have souls, like we have, souls that do not perish with the body and which must therefore be propitiated lest they should avenge themselves on us for taking away their bodies.
>
> (1929, 55–56)

In her anthropological study *Why We Play*, Roberte Hamayon sums up the function of Buryat shamans as "to insure [sic] the group's survival and perpetuation, along with the renewal of its natural environment by sound management of their relationship" (2016, 81). Reindeer, wildfowl, and other potential food-persons must be encouraged to give up their lives and flesh. Even before hunters or herders seek prey, shamans and those who participate in their ceremonies must play their roles. For the shamans, this can involve gestures and vocalizations which evoke and please the desired food-animals and related "spirits." It can also involve enacting combative behaviors, like those of a powerful moose, to gain power over any beings who might resist efforts to sustain human communities. Just as shamans shamanize, youths play vigorous games (e.g., wrestling), and hunters tell amusing stories—and all conspire to entertain animals and "spirits" so that the group will survive in a nourishing, generous, and respected environment. "Knowing life" here means knowing the etiquettes of respect that will keep the multispecies community renewing itself.

The Etiquette of Learning and Living Well

It would be possible to proliferate paragraphs about animist interactions with(in) the wider world. The expanding scholarly and popular literature (on- and off-line) concerned with animist ideas and activities could be reviewed to make further points about multispecies kinship. Each of the previous paragraphs could be more insistent that some human relationships with stones, maples, vegetables, and "food-persons" are more than a kinder sort of culture or a resource for improving the one-world world.

However, such expansions and improvements could too easily become another catalog of the strange beliefs of putatively non-Modern or not-Modern-enough people—the kind of ideas that can be tolerated until they are replaced. Something more vital (and vitalizing) is possible. Indeed, climate disaster, the damaged world, mass extinction, global injustices, and other machinations of Modernity, demand that we re-view the worlds and our role(s) within them. ("We/our" here being scholars—a particular category of humans, disciplined to limit our focus and ambitions while claiming objective expertise). Perhaps it would help if we thought differently about what happens when scholars engage with others. The transformation of anthropologist Franz Boas provides some pointers.

Isaiah Wilner (2018) encourages us to rediscover how Kwakwa̱ka̱'wakw philosophers, especially the Kwakwa̱ka̱'wakw intellectual George Hunt, transformed Boas into a post-imperialist anthropologist. Wilner tells us that Boas was initially the kind of anthropologist who measured skulls to judge differences between "races" and cataloged the "primitive" art of "vanishing peoples." Encountering Kwakwa̱ka̱'wakw masks and other ceremonial objects in a Berlin collection, Boas was initially captivated by the "flight of imagination" they showed. Greater familiarity inspired his recognition of the "wealth of thought" in their designs and Boas began cataloging the meanings of artistic signs in order to construct a representation of Kwakwa̱ka̱'wakw culture. But such "meanings," "signs," and acts of "representation" insert distance between relations; they "perform division" (Law and Benschop 1997, 158) and "implement the coloniality of the modern constitution" (De la Cadena 2015, 99) by translating the many possible worlds into terms acceptable to the one-world world. Intervening against this colonization, Hunt—described as "an Indigenous intellectual from Vancouver Island in British Columbia" (Wilner 2018, 4)—wrote to tell Boas about a mask that had belonged to Lucy Homiskanis, Hunt's deceased wife. Rather than this and all the other masks being examples of "typical and tribal" work, they were "personal and particular" (Wilner 2018, 6). They were owned, worn, and danced by specific people to communicate messages, knowledges, and relations which built communities. Wilner sums up Hunt's influence and Boas's transformation:

> As a result of his work with Hunt at the [Chicago 1893] fair, Boas began to see anthropology not as a form of categorization but as a form of communication. Ideas would be like gifts, circulating between communities and, in the process, transforming them. Anthropology could become a kind of potlatch—a circulation of ideas as gifts—affording people from different places and traditions a chance to assert their identities, exchange views, and link lives. Boas's vision of mental exchange transcended the dehumanization of [racist] science to imagine the possibility of a ground-up global philosophy: an interactive process of communication that would draw on the local knowledge of all the peoples of the world to globalize the Western mind.
>
> (Wilner 2018, 26–27)

In short, Boas became an expert by becoming a learner. The version of anthropology that evolved through his relationship with Hunt and others has continued to evolve. Modernity has turned out to be one "local knowledge" among many—none of them ever entirely separate from others. Indigenous people (among others) turn out not to be distant "primitives" but contemporaries of Moderns (and of recovering Moderns). Local "informants" turn out to be experts in educating and transforming scholars and scholarship. Gifts of ideas and knowledges arising from more-or-less intimate relationships are offered and/or exchanged. Attention is drawn to the world-making ("worlding") entangled or embraided in the exercise of performing cultures—"ours" and "theirs"—and, therefore, to responsibilities to treat such gifts respectfully. This is not to say that scholars have always welcomed, or even recognized, the gift or the authoritative knowledge of their hosts. Anthropology and other ethnographic disciplines are not entirely decolonized. Clearly, we are not surveying the achievement of global justice or celebrating the end of the assault on the world(s). The human separatist movement continues, even thrives, at the expense of other kin. Nonetheless, the Modern pretension to possess a universal understanding—the one-world world—is revealed as a mask that limits the vision of its wearers. Other perspectives are available. Boas's transformation was a beginning of an ongoing and unfinished decolonizing process.

I am not sure if Boas ever said, "I cannot understand" something that Hunt and other Indigenous hosts told or showed him. Plenty of scholars assert their own seemingly secure and expert knowledge of reality and that other people "believe" what Moderns know cannot be true. It is, therefore, immensely refreshing that Marisol de la Cadena notes that "With my usual epistemic tools, I could not *know* Ausangate" (2015, xxv). Frequently throughout *Earth Beings* (2015), she identifies Ausangate as a Peruvian mountain and as an Earth Being. These are not the same things—mountains are areas of nature and geography, while Earth Beings (*tirakuna*) are dominant members of multispecies communities. But neither identification is wrong. So, like Hallowell asking a question based on one cultural assumption, I have misspoken. While "mountains" and "Earth Beings" are not the same things, neither are they different things. Ausangate is both a mountain and an Earth Being, but not all the time and/or not to everyone. Indeed, as a mountain (to some people or sometimes) and an Earth Being (to some people or sometimes), Ausangate is also potentially or actually other things: a resource, tourist destination, an owner who can make demands on others— just as human landowners and politicians can and do. Ausangate is always more than one of these "but never many" (De la Cadena 2015: xxvii). Like animist stones, *asiniig*, and water while "bay-ing," these are relationship terms rather than categories— or, at least, they work better that way to express what people know. Thus, De la Cadena's relationship with the larger-than-human community including and involving Ausangate enables her to write powerfully about what she has come to know. Introduced with appropriate etiquette (such as ceremonial offerings) and sometimes impelled by enigmatic conversations with a local Quechua *yachaq*/"knower" she learns and writes about the relationships which compose community and about what she cannot know with the "usual epistemic tools" of other-than-Indigenous/other-than-local people. Although she is not related in the same way as the local "knowers," she participates in the constant translation and composition processes relevant and necessary to the practices of worlds that overlap and exceed each other (5).

De la Cadena's rich political history and ethnography is about more than Ausangate and those related to the Earth Being/mountain in diverse ways. I have focused on what she says about Ausangate because it provides a perfect example of an animist pluriversal "political ontology" (see also De la Cadena and Blaser 2018)—especially as Ausangate is "more than one but never many." Ausangate's exemplary perfection arises because in grasping what can be known and what cannot be known in relational worlds, the question of how to speak about (and to, and with) Ausangate simultaneously improves our understanding of "Indigeneity" as a relational naming of the pluriversal constitution of a world of many worlds. That is, there are multiple Indigenous knowledge systems, communities, and relations with the larger world (perhaps "cultures" is the shorthand for this plurality), but they collaborate in a global strategy of confrontation with the ongoing (most-)colonialism of the Modernity project. (More than one Indigenous colleague has used the phrase "most-colonialism" to contest the notion that we are "post-" or beyond colonialism—and, indeed, that colonialism is increasing in many domains).

Kyle Whyte encapsulates the notion of Indigeneity, concluding that it is typically "used to express intergenerational systems of responsibilities that connect humans, non-human animals and plants, sacred entities, and systems" (2016, 145). Undoubtedly there are other systems that accept responsibilities for connecting humans with(in) the larger-than-human world(s). It is the multi-generational commitment to relational obligations that establish Indigenous contrasts with Modernity's fantasy of human separatism, exceptionalism, and universalism. This is what Quechua relations with Earth Beings like Ausangate demonstrate. And this is what De la Cadena's learning among *runakuna* (Quechua people) results in: further journeys in the decolonization of global knowledges, now rooted in respect for the relational constitution of all existences together, even when they tend towards opposition.

In this world in which many worlds are possible (but not all are just or sustainable—and in which some fit together better than others), there are many ways of "knowing life." In the project of Modernity, knowing life has been co-opted as an atomistic practice. It has, for instance, involved categorizing some existences as "biology" and others as "geology," with built-in assumptions about (in)animation, (in)agency, differential value, and more. It has been allowed to structure universities by dividing the "natural sciences" from the "social sciences." Despite Darwin's recognition of the co-evolving social kinship of all living beings (sketched as a tree of life in his notebook), Moderns have largely resisted the implications of relationality. In many Indigeneities, "knowing life" is not so much categorical as relational. Based on an assumption of pervasive relationality, the key "art of living" among animists is knowing how to live well with and among others—of whatever species. Such practices are inculcated in every domain of life as beings interact with others. Much of this is expressed in the near ubiquity of the term "respect" in contemporary Indigenous discourse and is one feature of the global cosmopolitical and decolonizing effort to celebrate and enhance the common ground that composes Indigeneity.

It is, again, important to stress with Latour that "we have never been Modern" (1993) and, indeed, that the ambition of Modernity is impossible to fully achieve. Humans cannot be separated from the larger community. Culture and Nature do not label realities but, at best, enable Moderns to focus on specific aspects of an always integrally interrelated world of many social worlds. Happily, perhaps, scholars in diverse academic disciplines are emphasizing and evolving more relational and participative approaches in their work. As feminist physicist Karen Barad writes,

> Practices of knowing and being are not isolable; they are mutually implicated. We don't obtain knowledge by standing outside the world; we know because we are *of* the world. We are part of the world in its differential becoming. The separation of epistemology from ontology is a reverberation of a metaphysics that assumes an inherent difference between human and nonhuman, subject and object, mind and body, matter and discourse.
>
> (2007, 185)

Nonetheless, anthropologist Marshall Sahlins demands more change, and critiques:

> a lot of received ethnography for a misleading conceptual apparatus composed of nearly equal parts of transcendentalist equivocation and colonialist condescension. The effect is an anthropology that disfigures both the discipline [anthropology] and the culture so described by maligning the people's mentality as a mistaken sense of reality.
>
> (2022, 11)

Knowing life more relationally, more justly and sustainably, and as a practice of improving kinship skills, may be achieved by learning from Aboriginal educators to "walk the land":

> not merely in a literal sense but in a paradigm-shifting epistemological sense as well. Rather than stepping back from the land, as the observer and theoretician do, we have actively to enter it, address it and engage it as a collaborator that can and will join forces with us in some vital venture…[and] in walking the land, not merely as an observer but as a collaborator in ventures of mutual concern to the land and oneself, one starts to experience the land's responsiveness to one's presence, where that cannot leave one other than profoundly moved. One has perhaps begun indeed to experience the kind of feeling—a kind of vital, participative awareness—that Senior Law Men such as Neidjie and Mowaljarlai have been trying to explain to non-Indigenous countrymen for many years. Once we have discovered this way of knowing, there will be no question of remaining ethically indifferent to the living world, marooned inside a plastic bubble of anthropocentrism.
>
> (Mathews 2021, xxvii)

Put succinctly, "The problem with ontology is not knowledge or representation, but engagement with and for a world" (Stengers 2018, 85). This chapter considers such engagement among animists and scholars. First, animism has been invoked as a decolonizing practice, contesting the destructive "machine" of Modernity. Examples of Indigenous animist relations (e.g. with *some* stones, trees, kumara, and mountains/Earth Beings) have generated thoughts about the implications for scholarly approaches to living among other-than-human kith and kin. More has been said, here, about languages and etiquettes of relationality than about ethics—but *bimaadiziwin*, "good living," fuses knowledge about relations with know-how to live respectfully. It entails obligations towards kindness within the all-embracing kinship of all existences. Secondly, and vitally, the chapter proposes that scholarly engagement should not be all about others (the "non" or other-than-Moderns, Indigenes and animists) but also about "us" (people who might learn from and among Indigenous people and other animists). Ethnography and other disciplines should accept the challenge that other-than-Modern worlds offer to the commonplace Modern ontology of, for example, culture/nature division and human exceptionalism. Because animism research and debate (and related fields) reach beyond the description of Indigenous and other animist knowledges and worldings, the chapter has touched on some transformative scholarship. Animism research entails efforts to do more than know *about* lives and life-worlds. It encourages further scholarly efforts towards "developing new languages for our changing world" and "stepping into the unknown" emerging endangered world by taking "a stand for life" (Gibson, Rose and Fincher 2015, iii).

Bibliography

Barad, Karen. 2007. *Meeting the Universe Halfway: Quantum Physics and the Entanglement of Matter and Meaning*. Durham: Duke University Press.

Bhambra, Gurminder K. 2007. *Rethinking Modernity: Postcolonialism and the Sociological Imagination*. London: Palgrave Macmillan.

Braidotti, Rosi. 2018. "A Theoretical Framework for the Critical Posthumanities". *Theory, Culture & Society* 36, no. 6: 1–31.

De la Cadena, Marisol. 2015. *Earth Beings: Ecologies of Practice Across Andean Worlds*. Durham: Duke University Press.

De la Cadena, Marisol and Mario Blaser (eds). 2018. *A World of Many Worlds*. Durham: Duke University Press.

Despret, Vinciane. 2016. *What Would Animals Say If We Asked the Right Questions?* Minneapolis: University of Minnesota Press.

Gibson, Katherine, Deborah Bird and Ruth Fincher, ed. 2015. *Manifesto for Living in the Anthropocene*. Brooklyn: Punctum Books.

Gross, Lawrence. 2014. *Anishinaabe Ways of Knowing and Being*. London: Routledge.

Hallowell, A. Irving. 1960. "Ojibwa Ontology, Behavior, and World View." In *Culture in History*, edited by Stanley Diamond, 19–52. New York: Columbia University Press.

Hamayon, Roberte. 2016. *Why We Play: An Anthropological Study*. Chicago: Hau.

Haraway, Donna. 2015. "Anthropocene, Capitalocene, Plantationocene, Chthulucene: Making Kin." *Environmental Humanities* 6: 159–65.

Hogan, Linda. 2013. "We Call It *Tradition*." In *The Handbook of Contemporary Animism*, edited by Graham Harvey, 17–26. New York, Routledge.

Ingold, Tim. 2006. "Rethinking the Animate, Re-animating Thought." *Ethnos* 71, no. 1: 9–20.

Kimmerer, Robin W. 2013. *Braiding Sweetgrass: Indigenous Wisdom, Scientific Knowledge, and the Teachings of Plants*. Minneapolis: Milkweed.

———. 2015. "Nature Needs a New Pronoun: To Stop the Age of Extinction, Let's Start By Ditching 'It'." *Yes Magazine*, March 30, 2015. www.yesmagazine.org/issue/together-earth/2015/03/30/alternative-grammar-a-new-language-of-kinship

Kopenawa, Davi and Bruce Albert. 2013. *The Falling Sky: Words of a Yanomami Shaman*. Cambridge: Belknap Press.

Kopenawa Yanomami, Davi. 2000. "Sonhos das Origens." In *Povos Indígenas No Brasil (1996–2000)*, edited by Carlos A. Ricardo, 18–23. São Paulo: ISA.

Latour, Bruno. 1993. *We Have Never Been Modern*. New York: Harvester Wheatsheaf.

Law, John. 2015. "What's Wrong with a One-World World?" *Distinktion: Scandinavian Journal of Social Theory* 16, no. 1: 126–29.

Law, John and Ruth Benschop. 1997. "Resisting Pictures: Representation, Distribution and Ontological Politics." In *Ideas of Difference: Social Spaces and the Labour of Division*, edited by Kevin Heatherington and Rolland Munro, 158–82. Oxford: Blackwell.

Macfarlane, Robert. 2016. "Generation Anthropocene: How Humans Have Altered the Planet Forever." *The Guardian*, April 1, 2016. www.theguardian.com/books/2016/apr/01/generation-anthropocene-altered-planet-for-ever

Marcos (Subcomandante) [Raphael S.G. Vicente]. 1996. *EZLN Fourth Declaration of the Lacandon Jungle*. III. Accessed January 26, 2023. https://schoolsforchiapas.org/wp-content/uploads/2014/03/Fourth-Declaration-of-the-Lacandona-Jungle-.pdf

Mathews, Freya. 2021. *The Ecological Self*. London: Routledge.

Moore, Jason. 2015. *Capitalism in the Web of Life: Ecology and the Accumulation of Capital*. London: Verso.

Naveh, Danny and Nurit Bird-David. 2013. "Animism, Conservation and Immediacy." *The Handbook of Contemporary Animism*, edited by Graham Harvey, 27–37. New York: Routledge.

Northrup, Jim. 2001. "Fond du Lac Follis." *The Circle* (Minneapolis), May 18, 2001.

Parikka, Jussi. 2015. *The Anthrobscene*. Minneapolis: University of Minnesota Press.

Piercy, Marge. 1976. *Woman on the Edge of Time*. New York: Random House.

Pflug, Melissa. 1992. "'Breaking Bread': Metaphor and Ritual in Odawa Religious Practice." *Religion* 22: 247–58.

Rasmussen, Knud. 1929. *Intellectual Culture of the Iglulik Eskimos*. Copenhagen: Cyldendalske Boghandel, Nordisk Forlag.

Reedy, Anaru. 1997. *Ngā Kōrero a Pita Kāpiti / The Teachings of Pita Kāpiti*. Christchurch: Canterbury University Press.

Sahlins, Marshall. 2022. *The New Science of the Enchanted Universe: An Anthropology of Most of Humanity*. Princeton: Princeton University Press.

Skafish, Peter. 2016. "The Metaphysics of Extra-Moderns: On the Decolonization of Thought. A Conversation with Eduardo Viveiros de Castro." *Common Knowledge* 22, no. 3: 393–414.

Stengers, Isabelle. 2018. "The Challenge of Ontological Politics." In *A World of Many Worlds*, edited by Marisol de la Cadena and Mario Blaser, 83–111. Durham: Duke University Press.

Subrahmanyam, Sanjay. 1997. "'Connected Histories': Notes towards a Reconfiguration of Early Modern Eurasia." *Modern Asians Studies* 31, no. 3: 735–62.

Tawhai, Te Pakaka. 1988. "Maori Religion." In *The Study of Religion, Traditional Religion and New Religion*, edited by Stewart Sutherland and Peter Clarke, 96–105. London: Routledge.

Tsing, Anna. 2015. *The Mushroom at the End of the World: On the Possibility of Life in Capitalist Ruins*. Princeton: Princeton University Press.

Tsing, Anna, Heather Swanson, Elaine Gan, and Nils Bubant. 2017. *Arts of Living on a Damaged Planet*. Minneapolis: University of Minnesota Press.

Tylor, Edward. 1871. *Primitive Culture*. London: Murray.

Viveiros de Castro, Eduardo. 1998. "Cosmological Dexis and Amerindian Perspectivism." *Journal of the Royal Anthropological Institute* 4, no. 3: 469–88.

———. 2007. "The Crystal Forest: Notes on the Ontology of Amazonian Spirits." *Inner Asia* 9: 13.

Vizenor, Gerald. 1994. *Manifest Manners: Postindian Warriors of Survivance*. Middleton: Wesleyan University Press.

Whyte, Kyle Powys. 2016. "Indigeneity." In *Keywords for Environmental Studies*, edited by Joni Adamson, William Gleason and David Naguib Pellow, 143–46. New York: NYU Press.

Whyte, Kyle. 2018. “Indigenous Science (Fiction) for the Anthropocene: Ancestral Dystopias and Fantasies of Climate Change Crises.” *Environment and Planning E: Nature and Space* 1, no. 1–2: 224–42.

Wilner, Isaiah L. 2018. “Transformation Masks: Recollecting the Indigenous Origins of Global Consciousness.” In *Indigenous Visions: Rediscovering the World of Franz Boas*, edited by Ned Blackhawk and Isaiah L. Wilner, 3–41. New Haven: Yale University Press.

Section III

Freeing Subjectivities

PRAXIS C

KNOWING ANIMALS IN THE ANTHROPOCENE THROUGH ANIMAL PHOTOJOURNALISM

Jo-Anne McArthur

As the founder of We Animals and a photojournalist documenting the plight of animals on all continents for over two decades, Jo-Anne McArthur explores animal photojournalism (APJ) and its place in the world of photography and media. Animal photojournalism is an emergent genre of photography that captures, memorializes, and exposes the experiences of animals who live amongst us, but who we fail to see.

Animal photojournalism is what I have been doing for over two decades, though I did not call it so until recently. In over 60 countries, I have crawled under cages to access mink farms, or scaled fences to visit egg production facilities. I have purchased tickets to document the experiences of bulls at *corridas*, and horses at rodeos. With my camera, I have bore witness to ranges of human callousness and kindness. Throughout my career, people have surmised that I must be a wildlife or pet photographer. And how could I be a photojournalist if I document animals? This specific work that I and many of my colleagues do needed a name.

In careful consideration with my team at We Animals, whose mission is to advocate for animals through photojournalism, we defined the genre of animal photojournalism because it gives clarity to both audiences and practitioners Additionally, we seek to build an understanding of the genre so as to integrate and normalize animal photojournalism across all parts of society and the media.

Wildlife and companion animal photography are exclusive to species. So is photojournalism, which is about the human condition. Animal photojournalism includes all animals. It includes the oft-overlooked animals who live crammed into cages by the billions, whose lives are cut short by trawlers and nets, who languish in labs. It includes the animals whose lives have been woven into the tapestry of industrialization.

DOI: 10.4324/9781032659985-13

The ideas presented in this inceptive definition of the genre of animal photojournalism (APJ) are initial guiding principles. We know that over time, they will be challenged and honed, and we welcome these iterations and evolutions.

What is Animal Photojournalism?

Animal Photojournalism (APJ) is an emergent genre of photography that captures, exposes, and memorializes the experiences of animals who live amongst us, but who we fail to see. At its core, the images in this pioneering field document the broader human-animal conflict and its resultant ecosystems of suffering. As global societies collectively awaken to the realities of our unjust exploitation of animals, APJ is of increasing interest.

From public and environmental health crises to zoonotic viruses, animals are inextricably linked to many areas of current global concern, and rightfully so. Our existence is intertwined, and the ethics of how we treat the other sentient beings with whom we share this planet are being called into question. Animal photojournalism aims to encourage swift and necessary change on behalf of the beings in the frame.

Animal photojournalism is, in part, defined by what it is not. Most often, we look at animals through the familiar lens of wildlife and portrait photography. In recent years, conservation photography has emerged as a prominent field, documenting wild, endangered, and threatened species impacted by humanity. APJ emphasizes the inclusion of all animals, particularly those historically underrepresented but with whom we, in fact, have very close contact: the animals we eat and wear, the animals used in research, and the animals we use for entertainment, work, and in religious practice. Animal photojournalism acknowledges their beinghood and brings their stories to light.

Like photojournalism and conflict photography, animal photojournalism often takes the form of an in-depth reportage or photo essay. It is relevant to current news and it shapes conversations about its subject matter. Additionally, APJ is used to further political, philosophical, and scientific pedagogy, as well as campaigning. Many APJs are invested in contributing to current worldwide campaigns to lessen, and end, animal exploitation.

Animal photojournalism descends from conflict photography in that the photographer is exposing a story that is kept hidden from the public through political and economic agendas. Thus, their images are necessary catalysts for change.

As with conflict photographers, APJs put themselves at physical and psychological risk in order to document a practice or an event. In common with humanitarian and conflict photographers, animal photojournalists can also suffer long-term psychological effects such as post-traumatic stress disorder as a result of bearing witness to violence and injustice towards others.

Because animals used by humans are often caged and concealed, APJs may need to surreptitiously gain access to a place of animal exploitation. Examples are industrial farms, fur farms, and laboratories in which animal experimentation takes place.

In recent years and in response to APJs, activists, and whistleblowers exposing animal industries, powerful lobbying efforts on behalf of large corporations have ushered in what are known as "ag-gag" laws, designed to dissuade and criminalize the documentation of animal use.

APJ is groundbreaking for two reasons. First, images in this genre demand radical empathy and self-awareness. Viewers must de-center themselves and consider the world through the eyes of a different species while holding the truth of humanity's undeniable role in the story. Additionally, it poses a fundamental threat to deeply embedded societal systems that continue, largely unchallenged. The act of seeking out these visual stories is itself an act of resistance.

For more information, check out We Animals, "What is Animal Photojournalism": https://weanimals.org/about/what-is-animal-photojournalism/.

***

Ron's records state that he was born on September 1, 1976. No one seems to have any idea whether he was born in the wild or in captivity, or whether he was a pet or a performer. What is recorded is that, in 1987, he was at the Laboratory for Experimental Medicine and Surgery in Primates (LEMSIP) in New York. Although his records are incomplete, we know that at LEMSIP, Ron was anesthetized with ketamine 105 times.

In 1990, Ron was sent to the Coulston Foundation, where in 1997 he was anesthetized 16 times in a five-week period. The following year, he was part of a six-month spinal dynamics study where a disk in his neck was removed. The charts state that following the invasive surgery on his spine, Ron was not given pain medication for eight days. When pain medication came, it was two Tylenol.

Mercifully, Ron was eventually rescued by Save the Chimps, and relocated to their 200-acre sanctuary near Fort Pierce, Florida.

During his decades being used for invasive research, Ron often lived in a 5x5x7 foot cage, suspended above the ground. On occasions, Ron was provided with "enrichment," the term used by zoos and laboratories to describe items like blankets, car tires, or foraging balls that animals are given to decrease their boredom and to provide us with the illusion that we're doing something meaningful for their welfare. Although at Save the Chimps, Ron had several acres of sanctuary space within which to roam with friends, he usually chose to stay indoors. He'd carefully arrange his blankets in a circle to form a nest. And it was there that he died in October 2011, peacefully but prematurely—a common fate for rescued chimpanzees used in research.

Save The Chimps was established in 1997 by primatologist Dr. Carole Noon, after the U.S. Air Force announced they would no longer conduct research using chimpanzees. Based on the treatment of the animals and the conditions the animals had endured, Dr. Noon sued and gained legal custody of their 21 chimpanzees. Since 2008, the sanctuary has rescued and provided a permanent home for over 330 individuals, including 266 chimpanzees from the bankrupted biomedical research lab, the Coulston Foundation. See Figure C.1.

Figure C.1 Ron the chimpanzee at Save the Chimps, Fort Pierce, Florida, USA, 2008.

Photo credit: Jo-Anne McArthur / Save the Chimps.

***

At just 25 years of age, Erin Wing became an undercover investigator and spent two years working at chicken, dairy, and salmon farms, documenting the institutionalized abuse of the animals in these industries. Animals were Erin's companions throughout her childhood, a time when she experienced household violence. It was because of this personal history and her deep connection with animals that Erin was convinced she was right for undercover roles. But even she was not prepared for all that would follow.

Now the Deputy Director of Investigations at the animal advocacy non-profit Animal Outlook, Erin relinquished her anonymity to speak out about what she saw on several American farms, and her testimony is confronting. Foremost in Erin's stories is the constant presence of violence. "Those environments are meant to take away all the better parts of yourself, all the parts that feel compassion, that feel happiness, that feel kindness, because you can't really feel much of anything," she says.

Figure C.2 Erin Wing and Ginger at Wildwood Farm Sanctuary & Preserve, Newburg, Oregon, USA, 2021.

Photo by Jo-Anne McArthur / We Animals.

Workers must become desensitized to survive while performing their jobs. Erin believes that members of the public who see her undercover footage will be surprised not only by the abuse shown but also by what farmed animals are actually like.

Erin doesn't think she can ever go back to a life outside of animal advocacy. Despite the horrors she has witnessed, she still has hope: "I see the future as being very bright."

Almost ten years ago, Ginger arrived at Wildwood Farm Sanctuary and Preserve after she and several other goats were seized from a neglect case in Oregon. Ginger was feeble and sick, and upon examination, found to be pregnant. Veterinarians were unsure if she or her unborn baby would survive. Despite the guarded prognosis and with attentive care, Ginger's health began to improve. With much anticipation and many ultrasounds later, the sanctuary welcomed her healthy baby, Rose, to the world. See Figure C.2.

***

A lone Baltic gray seal lay semi-immersed in a small pool while surrounded by hundreds of people listening to a presentation over a loudspeaker. I noticed that few people were looking at the seal. As the presentation wound to an end, people began to leave the enclosure to see other parts of the zoo. The seal was left alone again, surrounded by the paraphernalia of parties and, ironically, a lifesaving device. See Figure C.3.

***

Mink are solitary, long-ranging, and semi-aquatic mammals who busy their days with hunting small animals and improving their nests with grass, leaves, and the furs of their hunt. Kits (young mink) are quick to leave their homes once they have weaned.

Mink in fur farms live in tiny cages, with flooring that is uncomfortable to their sensitive webbed feet. Mothers can give birth to litters ranging from two to ten in

Figure C.3 A Baltic grey seal at a zoo in Vilnius, Lithuania, 2016.

Photo by Jo-Anne McArthur / The Born Free Foundation.

size. The kits grow quickly, and the animals soon find themselves in a cramped and stressful environment. In these conditions, mink and kits often fight, causing injuries that can lead to missing ears, scruffs, scalps, and paws. Cannibalism is also frequently seen at mink fur farms, but that animal loss is part of the cost of doing business.

In this photograph, mink kits with injuries react to my presence as I take a photograph. The remains of one kit—a skull—lies in the foreground. See Figure C.4.

***

On the left, a steer hangs from one leg, bleeding out after his neck has been slit, while another steer is led in to be stunned and slaughtered.

This is the first slaughterhouse in the Republic of Türkiye to install a pneumatic captive bolt gun, and practice stunning before slaughter. On this day, the captive bolt gun had been installed, and slaughterhouse staff were being trained to position and stun the animals. The bulls are led into a "knock box" and stunned with the pneumatic bolt gun, which shoots a six-inch bolt into their skulls, and then they are hung up by their hind leg. Dangling, a worker then slits their throat. In some instances, I witnessed the steer being severely injured by the imprecise placement of the bolt gun and then hung upside down while conscious. Hanging, the animals are left for a few minutes to bleed out while another animal is brought into the knock box just a few feet away. After bleeding out, the bulls are skinned, gutted, and processed.

Figure C.4 Four white mink kits and the remains of a cagemate. Sweden, 2010.

Photo by Jo-Anne McArthur / Djurrattsalliansen.

Figure C.5 Next for slaughter. The Republic of Türkiye, 2018.

Photo by Jo-Anne McArthur / Eyes on Animals / We Animals.

I spent time with some of the cattle before they were brought into the slaughterhouse. They licked our hands, accepted pats, and followed their farm hands down the chutes.

The name of the slaughterhouse is intentionally withheld.

See Figure C.5.

***

At a stockyard in Bath, New York, Gene Baur, Farm Sanctuary's co-founder and president, found Opie while doing an investigation. He was a small Holstein calf, immobile on the ground on a freezing cold morning in 1990. He was deemed to be mortally ill with hypothermia and dehydration. He was not even a day old and was still wet from birth.

Gene picked Opie up and put him in his van. With intensive veterinary care, Opie was nursed back to health. Gene and his staff bottle-fed Opie until he was weaned. Opie lived a happy and healthy life at Farm Sanctuary, growing to over six feet tall at the shoulder. There, he lived for twenty years before passing away. Gene and Opie remained friends to the end.

See Figure C.6.

***

Kiska was wild-caught off the coast of Iceland in 1979 with three other orcas when she was about four years old. Since her capture, she lived in Marineland where

Figure C.6 Gene Baur and Opie at Farm Sanctuary, Watkins Glen, New York USA, 2006.

Photo by Jo-Anne McArthur / Farm Sanctuary.

she gave birth to five babies, all of whom died. Although Kiska was retired from performing in 2011, she continued to live alone at Marineland in the ironically named tank, Friendship Cove. It took her less than one minute to circumnavigate her tank.

Kiska died in 2023. She was almost fifty years of age.

Orcas, also known as killer whales, have a highly complex social structure and spend much of their day traveling and socializing, usually in matrilineal packs. Orcas hunt for fish as well as larger mammals, such as sea lions and walruses. No one fully knows the extent of their range in the wild, but pods have been seen to travel as far as 160 kilometers (100 miles) in a single day. Orcas are not generally believed to be a threat to humans, although some in captivity have injured or killed their trainers. Female orcas can live fifty years in their natural habitat, with some reaching eighty or ninety years of age. In captivity, they rarely live beyond twenty-five years.

Tourists stand in front of a transparent acrylic wall, with Kiska on the other side. We point at her, but our fingers seem to be gesturing beyond—to the space around her, its barrenness and the unbridgeable expanse we've made between us and them. If we bridge that space, we must confront the aspect of ourselves that we've forgotten but secretly long for: to be part of an ecosystem instead of just its proprietors. See Figure C.7.

Figure C.7 Kiska at Marineland, Ontario, Canada, 2011.

Photo by Jo-Anne McArthur.

***

The Ballarat Saleyards, which is also known as the Central Victoria Livestock Exchange, constitute one of the biggest saleyards in Australia. According to its website, it offers "extensive open yards and pens, with cobbled and concrete flooring for holding sheep, cattle, horses and pigs."

Figure C.8 Sheep auction, Ballarat Saleyards, Victoria, Australia, 2013.

Photo by Jo-Anne McArthur.

When I took this photograph, an average of 32,000 sheep were brought in the morning from the farms, auctioned, and then either transported to slaughter or to another farm, or to ships for live export. During the auction, the buyers look at the animals' teeth and squeeze their rumps to see if they're healthy and meaty.

Sale yards are noisy. Humans yell out, asking for the best price; dogs chase and climb onto the sheep; transport trucks roll in and drive off. The sheep are afraid of the electric prods that descend upon them by human hands, and they try to flee from the noise of the hand-rattles that are shaken to keep them moving. As they run from one pen to the next, they often slip on the slick feces- and urine-covered cobblestones. Their stress levels are high, and some break their legs and necks while in motion. See Figure C.8.

***

In a field of horses and kangaroos, Animals Australia's Louise Bonomi kneels next to a sedated joey. Horses come near to see this unusual sight. Dr. Elaine Ong and Dr. Chris Barton, co-founders of Vets for Compassion, darted the injured joey so that they could inspect the animal's extensive burns, which were caused by the bushfires that raged in the area.

Now fully sedated, Louise covers the joey's face while the veterinarians determine that the young animal is beyond help and must be euthanized.

An estimated three billion animals were killed or displaced by fires that season. See Figure C.9.

Figure C.9 Louise Bonomi with a sedated joey in Mallacoota, Victoria, Australia, 2020.

Photo by Jo-Anne McArthur / Vets for Compassion / Animals Australia / We Animals.

Bibliography

Farm Sanctuary. Accessed January 13, 2024. www.farmsanctuary.org/

McArthur, Jo-Anne. 2023. *We Animals. 3rd ed.* New York: Lantern Books.

Save the Chimps. Accessed January 12, 2024. https://savethechimps.org/

Unbound Project. *We Animals Media.* Accessed January 12, 2024. https://unboundproject.org/

We Animals. Accessed January 1, 2024. https://weanimals.org/

Wildwood Farm Sanctuary. *Instagram.* Accessed January 13, 2024. www.instagram.com/wildwoodfarmsanctuary/

10 Gendered Assumptions in Narratives and Practices of Animal Welfare

Jacob Bull

According to the Humane League, seven billion day-old chicks are killed annually, "killed—or 'culled'—before their lives even begin" (Wills 2021). These are the males deemed a byproduct of the egg industry. Such males are not suitable for the meat sector as their sisters have been bred for their egg-laying qualities rather than weight gain. Companies such as the European *Respeggt* join agriscience and information technology to offer "solutions" to this issue by sexing eggs through an "endocrine gender identification" test so that eggs that would produce male chicks never hatch ("Solutions" 2023). This example points to the ways that the bodies enrolled in agriculture are gendered. However, what I want to draw attention to are the final words of the Humane League quotation with which I started this chapter: "before their lives even begin." Implicit in the statement is that, but for the culling, there was, what the United Kingdom's Animal Welfare Council might call, a "life-worth-living" (2009, ii). It is here, in the meaning of a "life-worth-living,"—in what it means to have a "good life" or to "fare well"—that I focus my analysis to show that, in addition to animal bodies being sexed, they are also structured and known by systems of gender. While animal welfare discourse and practice may appear to be detached from issues of gender, and that gender may seem as a uniquely human phenomenon, it plays a significant role in both the discussions of and practices of "good" animal caretaking. As such I illustrate that sex/gender operates in agriculture both in terms of which bodies are enrolled in agriculture as well as in the discourses and practices of care. Gender structures both how we know animals and sets limits and opportunities for animals.

Through this chapter, I argue that gender, with its sexed and heteronormative stereotypes, is embedded in the description of a "good life" in animal welfare policy and practices of animal care. I will employ three animal examples—a giraffe, a drift of pigs, and a bee population—to demonstrate the undisclosed gendered premises at work in shaping these discourses and praxes. From this analysis, it will become clear that gender is a highly relevant category of analysis, not only for thinking about human bodies, but also because of the implications it has for nonhuman bodies and the environment. While these examples are from specific contexts, the ways that gender structures how humans come to know animals and how we care for animals is not limited to this setting. Insights regarding the

DOI: 10.4324/9781032659985-14

relevance of gender for nonhuman bodies and the environment has the potential to lead to new, critical assessments of how animals might be provided with "good" care, since gender, according to this argument, is part of the condition of emergence for a variety of animals.

Approaching Gender and Animal Studies

In reviewing the development of the discipline of Animal Studies, Simmons and Armstrong observe that "many of the concepts, dispositions and sensibilities that comprise 'human nature' rely upon perceived differences and similarities between ourselves and other animals" (2007, 1). They argue that these epistemological assumptions, foundationally based on dualisms such as nature/culture, reason/instinct, etc., have significant implications for how understanding both the (auto) biographical project of 'the Human' and the lives of animals.[1] I suggest that the dualism of sex/gender operates similarly, with animals having "sex" while "gender" is seen as social or cultural and therefore beyond the realm of nature.[2] However, these views of sex and gender not only structure knowledge about animals and shape the spaces and discourses of care—particularly what constitutes a "good life"—in animal welfare, but also conditions how animals emerge within worlds structured by gender, which limits their possibilities of life and how they can respond to human-generated norms of sexed bodies.

Thinking about human-animal relationships through a gendered lens is not new. Considerable work has identified the ways in which power relations between the binaries human/animal and man/women overlap (Adams 1999; Donovan 2006; Donovan and Adams 2003). Such work illuminates how comparisons can be drawn between the violence perpetrated against women and the violence enacted on animals. Equally critical work has moved beyond direct comparisons between bodies and examined how gender, as an organizing structure, manifests itself in social relations and structures for example through the development of reproductive technologies (Franklin 2007; Gillespie 2014; Grasseni 2005). Other work has critically assessed the role of gender in dominant narratives of competition in the processes of evolution (Grosz 2011; Hird 2009), or investigated the gendered aspects of conservation discourse and practice (Haraway 2003, 2007; Heise 2016). Others have identified the roles that animals play in the production and performance of gendered identities and sexualities (Hayward 2010; Lönngren 2020; Weaver 2013). In contrast to previously named research, I examine the relatively unexplored area of animal welfare.[3] Furthermore, and to complement work that foregrounds the human-animal relationship, I foreground the animal *as a body* (like a human body) that is affected by the organizing structure of the sex/gender dualism.

Gender as a Condition for the Emergence of Animals

For this analysis I begin from a definition of gender as "the structure of social relations that centers on the reproductive arena, and the set of practices that bring reproductive distinctions between bodies into social processes" (Connell 2008, 11).

This definition is a useful definition in this instance because it combines the social, structural, and the material/corporeal aspects of gender. But more than this, it does not specify human bodies. As such it has an open potential to apply the foundations of gender studies to nonhuman bodies. This analytical move to include animal bodies in the structures and practices of gender is a step based on an understanding that gender is not equivalent to identity. This is in line with broader understanding of gender where "It would be wrong to think that the discussion of 'identity' ought to proceed prior to a discussion of gender identity for the simple reason that 'persons' only become intelligible through becoming gendered in conformity with recognizable standards of gender intelligibility" (Butler 1999, 22). Gender, thus, precedes and is part of the conditions for emergence of an individual's identity. How then, can we think similarly for animals?

This extension of gender beyond the human could be considered anthropomorphic in that it appears to incorporate nonhuman animals within rigidly human categories. But my intention is not to equate human and animal experiences of sex/gender. Clearly there are differences between humans and other animals, just as there are profound differences in how humans experience sex/gender. However, there are also significant overlaps. For instance, there are significant examples of how the sexed bodies of animals are treated differently within a binary framework of sex/gender. Indeed, as Kathryn Gillespie argues, the life experiences of male and female cattle are fundamentally different (2014, 1325–28). Female calves are often enrolled into the dairy industry's structural violence of repeated pregnancy and milking while male calves, on the other hand, are either slaughtered directly or become part of the veal industry. Alternatively, while most are killed, some male individuals are selected as part of breeding programs in which they achieve close to subject status (Bull 2014, 55). Sex and gender, therefore, operates as a structure that affects animal lives and it is within this structure that animal lives are defined, lived, and ended.

Gendered and sexed narratives of care also inform the definition of farm animal welfare. Pettit and Bull (2018) demonstrate that race, nationality, and gender are instrumental in the definition of good welfare in the sale of Swedish chicken. That is, in contrast to meat that makes little or no welfare claims, the advertising material for "higher welfare" chicken includes images of families, children, and is set against "Swedish" landscapes. As such, care, it is argued, is "placed" within the coordinates of white, heterosexual, Swedish families. Additional perspectives on animal welfare and gender have considered the ways that gendered ideas affect attitudes toward animal welfare in the slaughter industry (Wigham et al. 2020), among veterinary students (Pironne et al. 2019), and farmers (Kiliç and Bozkurt 2013), in addition to creating broader social awareness around the meaning of animal welfare (Mazas and Fernadez-Manzanal 2019), and how it matters who represents environmental futures (Dahl 2016). Gendered concepts have a clear impact on animal caretaking practices and the lives of animals.

In addition to understanding how sex/gender shapes the lives of specific categories of animals and how the gender of caretakers affects welfare practice, I submit that it is important to address how sex and gender structures human-animal relations in welfare practice beyond the effects on specific categories of animals or by specific categories of people. To do this, I turn to queer theory and Animal Studies to address animal sex, reproduction, and sexuality within the spaces and practices of animal welfare.

Queer Theory and Animal Welfare

Queer theory foregrounds sexuality as a structuring force within society and challenges the implicit and explicit assumption of "heteronormativity," meaning the assumed norm or default of binary heterosexuality (Butler 1999; Rich 1980). Queer theory argues that sexualities are qualitatively different, not a deviation or copy of, heterosexuality. Indeed, queer theory argues that sex (including much hetero-sex) is always already queer in that it is never fully and completely maps onto the heteronormative binary couplings of discrete individuals (Rubin 1984, 153–54). While queer theory has predominantly focused on human sex, kinship, and reproduction, significant steps have been made at the intersection of Gender Studies, Science and Technology Studies (STS), and Animal Studies that interrogate the intertwined ways in which narratives of nature, biology, and technology interact, particularly in relation to understandings of reproduction. Animals are increasingly understood through the lens of queer theory (Chen 2012; Giffney and Hird 2008; Rudy 2011, 2012) as well as trans* theory (Keir 2010; Hayward 2008 and 2010; Hayward and Weinstein 2015). Such work, often indebted to Donna Haraway's concept of "becoming with" (2007, 37), emphasizes that human sexuality is never an autobiographical achievement, but rather a sympoetic process enrolling a diversity of bodies, human and more-than-human, over a variety of timescales.

While there has been important work using feminist, gender studies, and queer and trans* theory to develop "more-than-human" ways to examine human sexuality and the intimacies and relations therein, there has been less attention to the implication of sex/gender and sexuality for animals, and even less within agriculture and animal welfare. I, thus, build on this work to argue that narratives of sex/gender and mammalian heteronormativity operate in defining what is considered a "good life" for animals since knowledges and practices of welfare and animal breeding technologies are structured by gendered norms and (hetero)sexuality. This follows theorists such as Hird (2009), Roughgarden (2004), and Ah-King (2009), who suggest that animal sex and reproduction exceeds the binary heteronormative frameworks that humans predominantly use to understand nature, and responds to the call for what Carolyn Dinshaw describes as the "inquiry into the ways that 'natural' has been produced by particular discursive matrixes of heteronormativity" (1995, 77).

Susan McHugh considers the implications of heteronormativity and the potentialities of attending to animal sexualities by proposing a move "away from the language of subjectivity (i.e. sexual identity) by situating it amid an even more queer spectrum of interspecies intimacies" (2011, 117). Employing the notion of companion animals and their "promiscuous sexualities," in which pair bonding is not typical, she claims that:

> the sentimentalizing of intersubjectivity across these particular species reflects and informs a distinctly "unnatural" pairing, that is, of the ideal monogamous heteronormative couplings among humans with the surgical mutilations of companion animals, who (like so-called aberrant or degenerate human) are thus demonized for their sexuality even as they are spared the ultimate punishment administered to their unwanted brethren.
>
> (McHugh 2011, 117)

This shows how the pervasive norm of heteronormative coupling has consequences for companion animals by explaining the perceived need to control animals' "excessive sexuality" so that it coheres with the norms of domestic life. McHugh's focus on companion animals and heteronormativity is likely to play out in other human-animal relationships in diverse manners. However, by attending to the sexualities of animals, the implications of heteronormative human norms and ideals for animals can be demonstrated. How humans respond to and understand animal sexualities, thus, becomes important as it has implications for both animals and human-animal relationships. In the next section, I offer two stories of animal caretaking and a description of bee breeding to illustrate how animal bodies are affected by similar assumptions about sex/gender and (hetero)sexuality.

Sex/Gender and Sexuality in Animal Welfare Policy and Practice

Sex, Gender, and the Avoidance of Incest in the Copenhagen Zoo

In 2014 the Copenhagen Zoo killed a healthy two-year-old giraffe called Marius. The incident made news headlines around the world and was widely criticized, as it emerged that Marius was shot, his body publicly dissected in front of school children, and then fed to the lions. In response to the criticism, the zoo management was forced to justify its decision. The argument followed genetic lines—that Marius's genes were already over-represented in the gene pool at the zoo and within the Northern European Zoo giraffe population. Both male and female giraffes that are not destined to be a part of the breeding program need to be removed from the zoo population to prevent inbreeding. Marius had no future as a breeding male and, therefore, he had no future.

Here we could argue that sex/gender legibility both enabled Marius to become recognizable as a subject and ultimately lead to his early demise. The way that assumptions about genes, reproduction, and difference are evoked within conservation/breeding projects has been amply discussed previously (Franklin 2007; Haraway 2003, 2007; Heise 2016). I do not want to focus here on the fact that Marius's genes did not have a future and, thereby, provided a justification for his demise. The justifications offered by the zoo were grounded in genetic science and current guidelines and were, thus, logical within their scope. Instead, I want to focus on the fact that Marius had a past. If his genes were over-represented in the captive population, why was he produced in the first place? Further, why was he permitted to live to the age of two years before he was killed? The zoo answered these questions in a public statement, stating that it was considered important for his mother to experience the process of pregnancy, birth, nursing, and raising a young giraffe:

> At the Copenhagen Zoo we let the animals breed naturally. By "naturally" we mean that they will get young within the same intervals as they would in the wild. That means that the animals get to carry out their natural behaviours. Parental care is a big part of an animal's behaviour. It is a 24-hour job in longer periods of their lives and we believe that they should still be able to carry out this type of behaviour also in captivity. Contraceptives have a number of unwanted side effects on the internal organs and we would therefore apply a poorer animal welfare if we did not euthanize.
>
> (Købehavn Zoo)

Marius's life and eventual death were, in part at least, a consequence of animal welfare decisions on the part of the animal caretakers *for his mother*. This story is not unique; similar narratives exist in pet keeping (McHugh 2011). However, what this points to is a heteronormative temporality within the construction of what it means to fare well for animals. Breeding, birthing, being a parent is considered *natural* for an animal in captivity; it is better welfare than using contraception and, further, it alleviates boredom since it is after all, according to the zoo's representatives, a 24-hour job. This narrative is in tension with the narrative of inbreeding and genetic management initially outlined; if copulation, pregnancy, birthing, and parenthood are important aspects of animal welfare, why does the genetic relatedness matter in this context? The reproductive capacities of the giraffes are always likely to exceed the spatial and financial limits of the zoo to provide homes to the offspring, so it is for those animals enrolled in the breeding programs that genetic management is most important.

Again, I underline that the zoo's motivations were justified according to existing guidelines and current theories in genetic science; however, what is interesting is that to "fare well" requires a conflation of two motivations that are not equivalent. On the one hand, zookeepers strive to manage the excessiveness of animal sexuality, which does not heed the human incest taboo and thereby risks "inbreeding." On the other hand, there is a perceived need for "natural" reproduction. Given that animals do not self-police the incest/inbreeding taboo, "natural" reproduction is not permitted. To resolve this tension, a heterosexual binary for reproductive[4] sex is the form of "nature" in this narrative of welfare. As humans increasingly exert control over (domestic) animal sexuality, there is a need to "straighten" the excessiveness of animal sexuality and naturalize heterosexual, binary, reproductive sex. A "straight" life temporality is enacted as part of this process, where animal welfare becomes the mechanism of its delivery.

Rare Breed Pig Sex

When I first started keeping rare breed pigs, I did a significant amount of research and sought the advice of others. A recurring theme was the need for good fences, particularly to keep the male boar away from the sows and gilts[5] when they were "in heat." Wooden fences with electric wire on the inside were built to keep the boar in his pen and out of the paddock where the females lived. This worked according to plan. The sows and gilts came into heat, the boar showed interest and attempted to enter the females' paddock, but was deterred after being shocked by the electric fence. As instructed, I had designed and built a fence that was successful in keeping the boar away from the sows. I had, however, failed to construct a fence that kept the gilts out of the boar's pen. A young gilt dug a trench under the fence, and thus shielded from the electrical current, she succeeded in making her way into the boar's pen. Mating ensued and, after 12 hours, she returned to her paddock the same way.

If preventing uncontrolled breeding was the goal, my fence construction was clearly a mistake and the fence construction has since been changed. However, upon reflection, this notion of how to construct the fence was grounded in advice that demonstrates the effect that gendered assumptions have on animal lives. I did not spot the gender bias in the advice I was given or read, as I built the fence to prevent an "active male" from accessing "passive females" or that a female might actively pursue her own sexual agenda. This active/passive binary maps onto what Malin Ah-King (2013) has identified in scientific accounts of sexual selection and ties to Hillevi Ganetz' (2004) discussion of a cultural boomerang between human ideas of gender relations and representations of animals. The boar, the male in the herd, is described in active, powerful, and dangerous terms. Structures are built according to these representations to control his "instinctual," active drive and to "protect" the (assumed to be) "passive" females. These representations clearly do not map onto the reality, not least because the boar in my drift is not very active and the gilt was able to pursue her own, active, sexual agenda. Both human narratives of masculinity and femininity embedded in pig-keeping discourse, as well as the fact that I unreflexively consumed this discourse contributed to this event.

In the aftermath of the "incident," a practical aesthetics of reproductive fitness was immediately deployed both to assess the welfare implications of this gilt carrying the offspring to term and to compare her against an idealized image of a good (pig) mother. The active gilt was one of approximately 600 of the Linderöd pig breed that currently exist; however, she was considered too young to breed and did not meet breeding standards, as she was small, narrow of hip, and the offspring of the same boar. Before this incident, she was destined for slaughter. Though, afterwards, there was considerable conversation and advice sought and offered about what should be done. The first aspect discussed was that her actions should make no difference to her destiny, or that slaughter should be hastened so that she was slaughtered before the embryos developed. In this regard, sexuality alone did not give her subject status—she did not automatically emerge as a subject because she had acted on her own sexual agenda. Further questions were also raised such as that if she was allowed to go to term, would she produce healthy offspring? This was in part an echo of the incest taboo discussed in the previous example, but there were further elements: welfare and breed standard issues.

As stated above, she was not a large pig, she was the smallest of a previous litter. Her age, size, and shape raised concerns about her ability to nourish, carry, and deliver the piglets. This was both a question of health and well-being for the piglets and her; would she need assistance at farrowing? Would there be long-term health implications for her or the piglets? Would she/they survive? This welfare discussion involved comparing her to an idealized form of a "good pig mother." This ideal is, in part, informed by a practical experience of animal caretakers and veterinarians, codified in handbooks and guidance notes, and informally discussed, passed on and inherited, as well as being informed by the breed standards, including both formal and informal breed descriptions circulated by the breed association. Despite the fact that she did not compare favorably to this ideal, the birth went well, and she raised a healthy litter. This story of the sow who, after the incident, was known as Cindy, demonstrates how discussions of animal sexuality are often intertwined with animal welfare normative ideas of human sex/gender and idealized images of gendered bodies.

In reflecting on this case, I would suggest that the aforementioned definition of gender (Connell 2008, 11) operated in the spaces, practices, and discourses of pig caretaking in at least three ways. First, welfare is the discursive and practical context for considering the possibility of animal sexuality beyond the basic notions of reproduction. Under the guise of welfare, animal sex is expanded beyond the mere reproduction of "livestock units" in a production system, or generations in a breeding plan or pedigree (as in the case of Marius), to include a broader understanding of animal sexuality. Even though the process is inherently tied to a production system that is typified by human control of animal sexuality, the pairing of animals is a care-filled process that also includes consideration of the well-being of animal individual(s) and their offspring. The reproduction of an animal is not taken for granted but is considered in relation to norms and breed standards and includes concerns about the animal's condition, physical development, and capacity to undergo the stress of gestation and farrowing piglets. This is not to say that welfare is a completely unproblematic form of care, but as a form of care, it is a space where a "good life" for animals is defined and, significantly, where animal sexuality is understood in more than simple reproductive terms.

Second, the fact that in building the fence, I overlooked the possibility that a female pig could find ways to actively pursue her own sexual agenda shows that the gendering of animals is based on normative, cis-heterosexually gendered human stereotypes. This also demonstrates that sex/gender norms and faulty human stereotypes are literally built into the spaces of animal caretaking. Acknowledging this in welfare and care practices requires a geography that both manages animal sexuality and avoids or at least controls the "excessiveness of animal sexuality" that McHugh highlights in relation to pet keeping. In the spaces of agriculture, this management is less about surgical or chemical control, as McHugh discusses, and instead includes a landscape of spatial and temporal separation of animals that regulates the excessiveness of animal sexuality to fit into agricultural production systems and ensures that the animals' welfare is upheld.

Third, this case clearly illustrates that while animal sex alone does not automatically result in subject status for animals, sex/gender operates as part of the condition of emergence for individual animals. By acting on what I refer to as animal sexuality, this gilt differentiated herself from the drift and emerged as an individual and her life course changed from slaughter to survival. The gilt received an individual name, Cindy, and became differently recognizable as a sow. Using sex/gender as an analytical refrain to think through welfare demonstrates how sex/gender is the condition of emergence for the animal in that it structures how animals become knowable or intelligible: sexuality is part of how animals are differentiated and individualized, gendered norms and faulty human stereotypes are built into the landscapes of agriculture, and welfare is recognizable as both a space where animal sexuality is discussed in more than simple reproductive terms while simultaneously remaining highly normative in the definition of a "good life."

Bee Breeding and Faring Well in Changing Environmental Conditions

The challenges facing honeybees are well documented in both the popular and scientific press which describes bees as confronted by a medley of pressures such as climate change, evolving constellations of parasites and diseases, changing agricultural landscapes and ecologies, and pesticide use. These "under threat" narratives are also reflected in the accounts offered by beekeepers of their experiences of keeping bees. Developing the limited but significant work reviewing these threats from an STS perspective (e.g. Suryanarayanan and Kleinman 2017), the focus of this section is to review interviews with beekeepers and discuss the implications of how breeding strategies are employed in an attempt to mitigate changing environmental conditions and allow bees to "fare well."

This section is based on results derived from a set of 36 in-depth interviews I conducted with beekeepers mostly in Sweden, but also in Norway and Denmark, from June to November 2015.[6] The research adopted a qualitative approach; interviews were semi-structured and lasted between one and four hours. Where possible, interviews were conducted in situ and included a visit to one or more bee yards. The situated nature of the interviews aimed at providing ethnographic context to the interviews, departure points for discussion, and opportunity to confirm, revisit, or refute claims made during the interviews.

Breeding is viewed as a mechanism, tool, or technology for (re)positioning bees in changing landscapes. Indeed, an article in the online popular science magazine *Science.org* provides an example of how breeding functions as a strategy for adapting beekeeping to changing environmental conditions; specifically, in this case, the article describes breeding bees that are resistant to the parasite *Varroa destructor*, claiming that "A more sustainable approach is breeding honeybees to fight the mites. Some bees are genetically predisposed to groom themselves and other bees; they kill mites during grooming by biting off their legs" (Stokstad 2019). According to this view, selective breeding is more "sustainable" than the chemical treatment of colonies and foregrounds genes as an explanation of bee behavior and, consequentially, their survival.

Breeding bees, like the breeding of other animals, has a long history. The focus on breeding as way to address environmental issues is, therefore, not a new trend. However, what is significant is that "new" knowledges and practices are emerging in response to environments that are positioned as *changing* and bees that are described as *threatened*. What I suggest in this section is that these knowledges and practices are underpinned by assumptions about (human and mammalian) sex and reproduction that may or may not be relevant for understanding bee sex and reproduction.

Bee reproduction is considerably different from mammalian reproduction. A colony of bees is made up of a single fertile female (queen), hundreds or thousands of fertile males (drones), and tens of thousands of female worker bees. The queen leaves the colony only once to mate with up to 30 drones. She stores this sperm and uses it to fertilize the eggs that she will lay over her lifetime of potentially several years. She lays her eggs in the wax cells built by the workers where they are fed a special high protein food known as royal jelly for 2–3 days by the workers, followed by a mixture of pollen and nectar until they pupate. If the workers build a slightly larger cell, the queen will lay an unfertilized egg. This haploid egg hatches and is raised by the workers and will become a male drone. This is only done when the environmental conditions are favorable, and the colony has significant resources of nectar and pollen. Similarly, if the queen encounters even larger cells, called "queen cups," she will lay a fertilized egg and, if the wider environmental conditions (e.g. nectar flow) are good and the pheromone levels within the hive are appropriate, the workers will feed the resulting larvae a diet of exclusively royal jelly and a new queen will be produced. The pheromones are emitted from the queen, but also from the developing worker brood, and it is the balance between these that determines whether this moment of reproduction is only intergenerational (succession) or also of multiplication (swarming). The complexity of bee sex and reproduction have led some to argue that the unit of selection for breeding decisions should be the colony rather than the individual (Delaplane 2017); bee sex and reproduction is a collective affair! With this complexity and collectivity in mind, I turn to the changing breeding practices.

Few, if any, of the respondents involved in the study on which this section is based would talk about bee sexuality. However, they were aware of the workings of bee sex and reproduction and had an awareness of bee science around genetics and inheritance. The respondents were selected because they were actively breeding honeybees. They all viewed breeding as a way to improve their productivity as beekeepers, but also as a way of caring for bees. They wanted bees that could survive and thrive in contemporary landscapes. What follows here is a description of the ways they went about breeding bees. It is not a summary of what traits they considered important, but the practices and technologies that they implemented.

The respondents implemented a variety of strategies in their bee breeding programs that can be described by three categories or strategies. These are: (1) open pairing, (2) closed pairing, and (3) instrumental insemination. In brief, open pairing involves the selection of particular "good" colonies by the beekeeper. These colonies are encouraged to make new queens, and therefore colonies. These resulting queens are then allowed to pair with whatever drones are in the area and the beekeeper then selects for "better" colonies after pairing, and replaces the queen in colonies that demonstrate "undesirable" characteristics.

The second strategy, closed pairing, occurs in isolated areas (many are on islands or in the middle of a forest). In this scenario there are restrictions on the drones introduced such that queens pair with drones from few (often up to four) colonies. The third strategy of instrumental insemination involves semen being collected from one (or more) drones, after which an unpaired queen is sedated, and then inseminated.

The first point of note is that when moving from strategy one to three there is an increased attention to the role drone (male) genes play in the breeding program. Open pairing embraces, or is at least unconcerned about, multiple pairings and make little attempt to control the drone population. Closed pairing pays closer attention to the drone genetics, and in instrumental insemination the drone genetics are given equal importance in the pairing such that possibly a single drone may be used. As such, in the shift from strategies one to three, bee breeding is increasingly understood in binary terms. That is, in instrumental insemination, the breeding method in which the beekeeper is most directly involved, bee sex and reproduction most closely resembles the binary pairings characteristic of breeding other animals. Bees are required to fit into binary heteronormative frameworks.

Second, as we move from strategies one to three, there is an increasing importance of gene-based understandings of bee fitness. For example, in an effort to breed resistance to a particular parasite (*Varroa destructor*), one breeder, Respondent 16, was developing lines of bees that displayed what is called "varroa sensitive hygiene" (VSH). He had a very ambitious breeding project which included travel to the U.S., Australia, and the Netherlands to "collect" genetic material to bring back to Sweden and introduce into his stock. Although this respondent spoke about this activity as introducing behaviors and traits, he linked the traits to genetic material, and the way to introduce this material was through instrumental insemination. As the respondent stated, "…to isolate these characteristics and concentrate them, you have to use [instrumental] inseminations" (Respondent 16).

Other respondents using other breeding techniques are also explicit in linking of VSH to genes and alleles. Respondent 13 stated, "He said [referring to another breeder] that there are four alleles that one can look at for functioning VSH and one could find them, find them yes, in the genome." In this quotation and through the conversation with Respondent 16, it is clear that VSH bees come with a particular sequence of base pairs each as a validation/valuation of their difference and suitability for contemporary beekeeping contexts. VSH becomes disconnected from the bodies that perform it and the environments in which it is enacted and instead becomes a "thing" that can be moved around the world and introduced into populations in the hopes that the resulting bees will "fare well" in changing environmental conditions. However, and as Haraway reminds us, and maybe all geneticists would agree, "the gene is not a thing" in terms of an isolated contained unit (2000, 89). Haraway argues that a "gene" is a way of talking about inheritance that locates and substantializes inheritance. Substantializing inheritance in this way can be linked to the technologies of instrumental insemination. These technologies were developed with the bodies of cows and humans (Wilmot 2007) and therefore have a mammalian history. With these mammalian histories come a cis-heteronormative framework for understanding sex and reproduction that insufficiently explains mammalian sex and fails to grasp the complexity of what might be called bee sexuality. It ignores the complexity and collectivity that contribute to the production of sexed bee bodies. These mammalian understandings of reproduction hide the collective interactions between bees in a colony, the environment of the hive, and the wider environmental conditions (such as nectar flows) that produce the conditions for reproductive moments.

The shift to instrumental insemination and the fetishization of VSH genes is not uniform amongst beekeepers and breeders. One respondent commented, "we see it very clearly, right, we see it, Bond bees [a type of resistant bee] have been moved to Germany, and Keyfuss bees [another type of resistant bee] have been moved to Germany, right, and they have no extra tolerance at all when they are in those environments" (Respondent 20). This respondent argued for a local specificity of trait expression—that VSH was not sufficiently explained by genes but that environmental factors affects how successful bees with the VSH genes are at combating *Varroa*. Trait inheritance was seen as more complex.

Several respondents implementing strategies one (open pairing) and two (closed pairing) challenged the gene orientated explanation and associated technologies. These contestations of instrumental insemination and gene fetishism were not anti-genetic or anti-science. Rather they focused on the more collective aspects of bee reproduction and in this regard were in line with some trends in bee science (see, for example, Delaplane 2017). Such scientific studies increasingly suggest that the colony should be considered a superorganism and that it is at this superorganism level that selection decisions should be made. Furthermore, related studies indicate that genetic diversity from multiple pairings is crucial to the productivity and resilience of the colony (Delaplane et al. 2015).

Colony-level, or superorganism, breeding plans align with the complexity and collectivity of bee sexualities and some respondents even suggested that instrumental insemination techniques resulted in queens that were of lesser quality than free-paired queens. This "lesser" quality was measured in the queens being rejected by the colony, or being accepted and then replaced by the bees after a short time, or queens dying at a younger age. Again, there are scientific studies to support such claims (Harbo and Szabo 1984; Peso et al. 2013) but this science is also contested. Susan Cobey suggests that instrumental insemination and free-paired queens are comparable in performance and also asserts that the experience and skill of the inseminator, the equipment, and care practices are responsible for differences in queen performance (2007; see also Cobey et al. 2013). However, while better guidelines might be developed, the technology (equipment) cannot be divorced from the operator or the conditions in which the technology is implemented. As a consequence, and in line with the questioning of the qualities of the queen in the accounts of respondents, might this not be an indication of bees resisting and exceeding the mammalian cis-heternormative frameworks of sex and reproduction imposed on them?

Conclusions

Through these three examples I have demonstrated that human gender norms shape how we come to know animals, and how animals emerge and are regulated through welfare policies and practices. The first instance, Marius the giraffe and his mother, demonstrates how gender, in terms of heteronormative assumptions and the incest taboo, are woven into animal welfare. Gendered norms are enrolled in the definition of a "good life" and what it means to "fare well." Similarly, in the second example of disorderly pig breeding, it is evident how gendered norms operate within the discourses and practices of pig care, but also how they are built into the physical spaces and structures of animal care taking, such as fences. These discursive and material gendered structures are the conditions of emergence for the animal; they are the frameworks and spaces where animals become intelligible and recognizable. Through the final example of bee breeding, I further demonstrate that the fact that gender structures the ways in which animals become knowable also has material consequences. As part of an effort to produce bees that fare well in contemporary landscapes, honeybees are increasingly subjected to binary, mammalian heteronormative understandings of sex and gender. While breeding projects may be motivated by care for bees as well as production, the accounts of queens produced through instrumental insemination being rejected by the colony points to a need for greater consideration of what might be termed animal sexuality and how it compares to the gendered norms in welfare policy and practice.

Notes

1 See also Buller 2013.
2 See also Meynall and Lopez 2021; Roughgarden 2004 ; Tuan 19 84.
3 From an animal studies perspective, Calvert (2013) does address some implications for welfare.
4 Reproductive here means that it leads to offspring that could be a part of the zoo's breeding population.
5 A gilt is a young female pig; a sow is a female pig that has farrowed (birthed) at least one litter.
6 Funded by the Swedish Research Council FORMAS (Project number 2014-01554).

Bibliography

Adams, Carol J. 1999. *The Pornography of Meat.* New York: Continuum
Ah-King, Malin. 2009. "Queer Nature, towards a non-normative perspective on biological diversity" In *Body Claims. Crossroads of Knowledge*, edited by J. Bromseth, L. Folkmarson Käll and K. Mattsson, 212–233. Department of Gender Research, Uppsala University.

Ah-King, Malin. 2013. "Queering animal sexual behavior in biology textbooks. Confero: Essays on Education." *Philosophy and Politics* 1, no. 2: 46–89.
Bull, Jacob. 2015. "Images of Cows, Stories of Gender" In *Ill Disciplined Gender: Nature/ Culture and Transgressive Encounter*, edited by Jacob Bull, 45–61. Dordrecht: Springer.
Buller, Henry. 2013. "Animal Geographies I." In *Progress in Human Geography* 38, no. 2: 308–18.
Butler, Judith. 1999. *Gender Trouble: Feminism and the Subversion of Identity*. New York: Routledge.
Calvert, Scout. 2013. "Certified Angus, Certified Patriot: Breeding, Bodies, and Pedigree Practices." *Science as Culture* 22, no. 3: 291–313. doi:10.1080/09505431.2012.718755.
Chen, Mel Y. 2012. *Animacies: Biopolitics, Racial Mattering, and Queer Affects*. Durham: Duke University Press.
Cobey, Susan, W. 2007. "Comparison Studies of Instrumentally Inseminated and Naturally Mated Honey Bee Queens and Factors Affecting their Performance." *Apidologie* 38, no. 4: 390–410.
Cobey, Susan W., David R. Tarpy, and Jerzy Woyke. 2013. "Standard Methods for Instrumental Insemination of Apis Mellifera Queens." *Journal of Apicultural Research* 52, no. 4: 1–18.
Connell, Raewyn. 2008. *Gender in World Perspective. 2nd Edition*. Cambridge: Polity Press.
Dahl, Ulrika. 2016. "Becoming Fertile in the Land of Organic Milk: Lesbian and Queer Reproductions of Femininity and Motherhood in Sweden." *Sexualities* 21, no. 7: 1021–38.
Delaplane, Keith, S. 2017. "Emergent Properties in the Honey Bee Superorganism." *Bee World* 94, no. 1: 8–15.
Delaplane, Keith S., Stéphane Pietravalle, Mike A. Brown, and Giles E. Budge. 2015. "Honey Bee Colonies Headed by Hyperpolyandrous Queens Have Improved Brood Rearing Efficiency and Lower Infestation Rates of Parasitic Varroa Mites." *PLoS One* 10, no. 12.
Dinshaw, Carolyn. 1995. "Chaucer's Queer Touches/a Queer Touches Chaucer." *Examplaria* 7: 75–92.
Donovan, Josephine. 2006. "Feminism and the Treatment of Animals: From Care to Dialogue." *Signs* 31, no. 2: 305–29.
Donovan, Josephine, and Adams, Carol, J. ed. 2003. *The Feminist Care Tradition in Animal Ethics*. New York: Columbia University Press.
Franklin, Sarah. 2007. *Dolly Mixtures: The Remaking of Genealogy*. Durham: Duke University Press.
Ganetz, Hillevi. 2004. "Skogens konung och djurens konung i TV: Natur, kultur och genus i naturfilm." *Nordicom Information* 26, no 1: 3–16.
Giffney, Noreen, and Myra Hird, ed. 2008. *Queering the Non/Human*. Aldershot: Ashgate.
Grasseni, Cristina. 2005. "Designer Cows: The Practice of Cattle Breeding Between Skill and Standardization." *Society and Animals* 13, no. 1: 33–50.
Grosz, Elizabeth. 2011. *Becoming Undone: Darwinian Reflections on Life, Politics, and Art*. Durham: Duke University Press.
Haraway Donna, J. 2003. *The Companion Species Manifesto: Dogs, People, and Significant Otherness*. Chicago: Prickly Paradigm Press.
Haraway Donna, J. 2007. *When Species Meet*. Minneapolis: University of Minnesota Press.
Haraway, Donna J., and Thyrza Nichols Goodeve. 2000. *How Like a Leaf: An Interview with Thyrza Nichols Goodeve*. New York: Routledge.
Harbo, John R., and Tibor I. Szabo.1984. "A Comparison of Instrumentally Inseminated and Naturally Mated Queens." *Journal of Apicultural Research* 23, no. 1: 31–36.

Hayward, Eva. 2008. "Lessons from a Starfish." In *Queering the Non/Human*, edited by Myra J. Hird, 249–64. Aldershot: Ashgate.
Hayward, Eva. 2010. "Spider City Sex." *Women and Performance: A Journal of Feminist Theory* 20, no. 3: 225–51.
Hayward, Eva, and Jami Weinstein. 2015. "Introduction: Tranimalities in the Age of Trans* Life." *TSQ: Transgender Studies Quarterly* 2, no. 2: 195–206.
Heise, Ursula, K. 2016. *Imagining Extinction: The Cultural Meanings of Endangered Species* Chicago: University of Chicago Press.
Hird, Myra, J. 2009. *The Origins of Sociable Life: Evolution After Science Studies*. New York: Palgrave Macmillan.
Kathryn, Gillespie. 2014. "Sexualized Violence and the Gendered Commodification of the Animal Body in Pacific Northwest US Dairy Production." *Gender, Place and Culture* 21, no. 10: 1321–37.
Kier, Bailey. 2010. "Interdependent Ecological Transsex: Notes on Re/Production, 'Transgender' Fish, and the Management of Populations, Species, and Resources." *Women and Performance: A Journal of Feminist Theory* 20, no. 3: 299–319.
Kiliç, Ibrahim, and Zehra Bozkurt. 2013. "The Relationship between Farmers' Perceptions and Animal Welfare Standards in Sheep Farms." *Asian-Australasian Journal of Animal Science* 26, no. 9: 1329–38.
Købehavn Zoo. 2008. "Visit the Zoo." Accessed February 1, 2024. https://web.archive.org/web/20140216000424/http://uk.zoo.dk/VisitZoo.aspx
Lönngren, Ann-Sofie. 2020. "That's When He Comes Rushing into Her Life': Swedish Literary Depictions of Cross-species Sexual Encounters at the Turn of the Twenty-first Century." *Humanimalia* 12, no. 1: 167–87.
Mazas, Beatriz, and Rosario Fernández-Manzanal. 2019. "A More Fine-Grained Measure Towards Animal Welfare: A Study with Regards to Gender Differences in Spanish Students." *Environmental Science and Pollution Research International* 26, no. 1: 844–54.
McHugh, Susan. 2011. *Animal Stories: Narrating Across Species Lines*. Minneapolis: University of Minnesota Press.
Meynell, Letitia, and Andrew Lopez. 2021. "Gendering Animals." *Synthese* 199, no. 1–2: 4287–4311.
Peso, Marianne, Elina L. Niño, Christina Grozinger, and Andrew B. Barron. 2013. "Effect of Honey Bee Queen Mating Condition on Worker Ovary Activation." *Insectes Sociaux* 60: 123–33.
Petitt, Andrea, and Jacob Bull. 2018. "Imag(in)ing 'Good' Swedish Meat: Gender, Sexuality, Race and Nation in the Sale of Higher Welfare Chicken." *Gender, Place and Culture* 25: 1622–45.
Pirrone, Federica, Chiara Mariti, Angelo Gazzano Mariangela Albertini, Claudio Sighieri, and Silvana Diverio. 2019. "Attitudes Toward Animals and Their Welfare Among Italian Veterinary Students." *Veterinary Science* 6, no. 1 (Feb 20): 19.
Rich, Adrienne. 1980. "Compulsory Heterosexuality and Lesbian Existence." *Signs* 5, no. 4: 631–60.
Roughgarden, Joan. 2004 *Evolution's Rainbow: Diversity, Gender, and Sexuality in Nature and People*. Berkeley: University of California Press.
Rubin, Gale. 1984. "Thinking Sex: Notes for a Radical Theory of the Politics of Sexuality." In *Pleasure and Danger: Exploring Female Sexuality*, edited by Carol S. Vance, 143–78. London: Kegan Paul and Routledge.
Rudy, Kathy. 2011. *Loving Animals: A New Approach to Animal Advocacy*. Minneapolis: University of Minnesota Press.

Rudy, Kathy. 2012. "LGBTQ...Z?" *Hypatia* 27, no. 3: 601–15.
"Solutions ". 2024. *Respeggt*. Accessed September 13, 2023. www.respeggt.com/solutions/
Stokstad, Erik. 2019. "Breeders Toughen Up Bees to Resist Deadly Mites." *Science*, July 25. Accessed September 13, 2023. www.science.org/content/article/breeders-toughen-bees-resist-deadly-mites#:~:text=By%20breeding%20bees%20from%20colonies,that%20helps%20create%20vigorous%20hives.
Suryanarayanan, Sainath, and Daniel Lee Kleinman. 2017. *Vanishing Bees: Science, Politics, and Honeybee Health*. Rutgers University Press.
Tuan, Yi-Fu. 1984. *Dominance and Affection: The Making of Pets*. New Haven: Yale University Press.
Weaver, Harlan. 2013. "Becoming in Kind: Race, Gender, and Nation in Cultures of Dog Fighting and Dog Rescue." *American Quarterly* 65, no. 2: 689–709.
Wigham, Ellie, Andrew Grist, Siobhan Mullan, Stephen Wotton, and Andrew Butterworth. 2020. "Gender and Job Characteristics of Slaughter Industry Personnel Influence Their Attitudes to Animal Welfare."*Animal Welfare* 29, no. 3: 313–22.
Wills, Lauren. 2021. "What Happens to Male Chicks in the Egg Industry." *The Humane League*, July 29. Accessed September 13, 2023. https://thehumaneleague.org.uk/article/what-happens-to-male-chicks-in-the-egg-industry
Wilmot, Sarah. 2007. "From 'Public Service' to Artificial Insemination: Animal Breeding Science and Reproductive Research in Early Twentieth-century Britain." *Studies in History and Philosophy of Science Part C: Studies in History and Philosophy of Biological and Biomedical Sciences* 38, no. 2: 411–41.

11 Unknown Patterns of Intersubjectivity

Zhuangzi, the Seabird and the Happy Fish

Jana S. Rošker

Chinese "Relational Epistemology"

In current research, the debate on the epistemological dimensions of Chinese texts and their role in the context of Chinese thought has been developed increasingly successfully under the aegis of rediscovering and applying specific traditional Chinese methodological approaches and categories (Lenk and Paul 1993). Chinese epistemology deals with problems such as the possibility of attaining correct knowledge of a given object—and thus the possibility of attaining a complete understanding of the Way (*dao* 道); the relationship between this knowledge and wisdom, on one hand, and morality on the other; the possibility of language to function as a conveyor of knowledge; and the relationship between knowledge (*zhi* 知) and action (*xing* 行).

According to the prevailing traditional European epistemologies, knowledge has mainly been gained through observation and reasoning. However, in traditional Chinese thought, knowledge has been understood in a much broader sense, namely as something that also (or primarily) stems from moral contents and which cannot be separated from (social) practice. The method which determined most of the epistemological teachings found in the Chinese classics was based on a holistic worldview, and was directed towards a comprehension that could be achieved through education and learning. The basic contents of these teachings were rooted in the premises of pragmatic and utilitarian ethics. Chinese epistemology was relational (Rošker 2010), meaning that it understood the external world to be ordered structurally, while the human mind was also structured in accordance with its all-embracing but open, organic system (li 理). The relational correspondence between the cosmic and mental structures thus represents the basic precondition of human perception and comprehension.

The naturalistic epistemologies that prevailed in Western discourses were dealing with the external world (or objective reality), which was to a great extent independent from the subject of comprehension. Chinese approaches to knowledge can be called relational epistemologies because they refer to relations. This applies not only to radical holistic epistemologies, which denied the notion of substance, but also to a number of contemporary theories which advocate a strict division between the subject and object of comprehension (Xia 2000, 4).

DOI: 10.4324/9781032659985-15

Chinese relational epistemology was based upon viewing the world as a complex. Specific Chinese models for investigating questions related to knowledge were thus premised upon a structurally ordered external reality. Since natural (or cosmic) order is organic, it naturally follows the "flow" of structural patterns and operates in accordance with structural principles that regulate every existence and are manifested in the concept of li 理. In this worldview, the human mind is also structured in accordance with this all-embracing but open organic system. The axioms of our recognition and thought are therefore not coincidental or arbitrary, but follow this dynamic structure. In this view, the compatibility or correspondence of both the cosmic and mental structures is the basic precondition that enables human beings to perceive and recognize external reality (Rošker 2010, 79–82).

In most traditional discourses (with the exception of the Mohist school, The School of Names, and certain representatives of the Neo-Confucianism), the focus upon relations was linked to the unity of the subject and object of comprehension. If we posit that the relation represents the object of comprehension, we must also specify that this object is not automatically to be seen as a counter-pole to the subject of comprehension. Relational epistemologies are not based upon a strict division between these entities, nor upon a strict (or necessary) demarcation of what, with respect to the subject of comprehension, we are accustomed to see as the external or internal world. Therefore, the methods used by certain philosophers in various currents of the Chinese tradition are by no means decisive for defining the positions of the subject and the object of comprehension, or the nature of their mutual relations. The methods for exploring (external) reality (*gewu* 格物) and introspective recognition (*fansheng* 反省) were both important as perceptive tools that primarily served to understand relations. However, these relations can be either continuous or discontinuous. This means that the relation between A and B can be changed into a relation between A and C. In Chinese epistemologies, such decomposition and changing of positions is in the nature of comprehension (Zhang 2002, 78). In this context, knowledge (*zhi*知) was primarily understood as recognition (*shi* 識) of the structural principles (*li* 理) of the all-embracing Way (*dao* 道), which, among all other entities of being, was also expressed in linguistic terms.

The relation as a basis or a central object and goal of any recognition manifests itself on all levels of comprehension and transmission of being. Hence, the relational aspect as a core of comprehension was already apparent in the specific structure of Chinese cosmology, which was based upon the holistic unity of humanity and nature (*tian ren heyi* 天人合一). The complexity and integrity of relations in nature and society, therefore, represent a basic aspect of Chinese epistemology. This aspect was expressed in all classical debates, which were based upon the elementary traditional epistemological categories of name (*ming* 名) and actualities (*shi* 實). Relations also formed the basic postulate of traditional thought, which defined the nature of the central epistemological relation between knowledge (*zhi* 知) and action (*xing* 行). Primarily due to the impact of Buddhist thought, the ancient holistic approach to perception and understanding reality through substance (*ti* 體) and functions (*yong* 用) was later replaced by the subject (neng 能) and object (suo 所) of comprehension. This kind of categorical demarcation that derived from the Indian tradition of thought would subsequently, in the 19th and especially the 20th centuries, help Chinese philosophers gain a better understanding of theories of knowledge in Western philosophy, which were based on the ontology of dividing substance from phenomena.

Interpreting Knowledge in the Modern Era: Dialogues With Gadamer

This relational epistemology, which guided traditional Chinese approaches to perceiving and understanding the external world, has led to various specifically Chinese modes of interpretation. Since the turn of the 20th century, however, most modern and contemporary Chinese scholars who have worked with classical Chinese texts have begun to employ interpretive mechanisms derived from Western hermeneutic theories.

In the first three decades after the establishment of the People's Republic of China (PRCh), Chinese intellectuals were, in this respect, mostly limited to the studies of Hegelian and Marxist theories[1]; from the 1980s on, they have yet again gained access to most of the classical works of modern Euro-American hermeneutics, including Schleiermacher, Dilthey, Heidegger and Gadamer, whose theory of the fusion of horizons soon gain a widespread popularity among Chinese theoreticians.

Overall, however, it has been the influence of Hans-Georg Gadamer that has seemed to dominate, even though more informed discussions of a relatively wider range of European hermeneutic philosophical works were already manifest in publications available during the early years of the twenty-first century (Pfister 2006, 4). As is well-known, his concept of the fusion of horizons was based on an elaborated version of Schleiermacher's notion of the hermeneutic circle, that is, on the idea that one's understanding of a text as a whole is established by reference to the individual parts, and vice-versa: the understanding of each individual part is being established by reference to the whole. Neither the whole text nor any individual part can be understood without reference to one another, which can be illustrated by a circular model of comprehension. According to this view, the "fusion of horizons" takes place between the writer and the readers, the speaker and the listeners, or the artist and the observers in the dialectical process of transferring meanings. In this sense, the concept of horizon refers to the particular situation into which every individual is embedded, whereas the situation is not limited to the vision or perception of what is nearby. Hence, the horizon implies an openness of existence and a possibility to overcome one's own prejudices.

The fusion of horizons always creates new meanings, for in this dialectical process of mediating and perceiving, it incorporates both horizons and surpasses, on the other hand, their individual limitations. On this basis of free and unlimited horizons, Gu Ming-Dong has drawn his method of traditional Chinese hermeneutic openness, which is rooted in the so-called metaphysical-aesthetic tradition of interpretations (Gu 2005, 9).

However, while in China, the theories of the Gadamerian discourse (including Gadamer's predecessors and successors) might represent a precious contribution for a successive establishment of transcultural hermeneutics. This overall enthusiasm could also be dialed back by raising some questions on the nature of this circle and this horizon.

Of course, the circular character of understanding does not make it impossible to interpret a text; rather, it stresses that the meaning of a text must be found within its context. But the problem is precisely that the very concept of the context has never been sufficiently explained and defined, particularly in view of different layers of reality and different modes of perception and categorization. In my view, the very term of contextualization and its respective contents can, for instance, at least be divided into external and internal contextualization. The former refers to the cultural, historical, and literary context of the text, and the latter to its inherent conceptual dimensions and their semantic and philosophical implications and developments. Hence, the hermeneutical method is still lacking a binding inherent consistence.

Recent sinological research has still not managed to develop a method of unifying these two kinds of contextualization, and the situation in Chinese-speaking academia is not much different in this respect. Scholars in China and in Taiwan have developed brilliant theories that help us understand the historical significance of the time and space, of politically, ideologically, and culturally-conditioned factors upon the interpretation of any important classical text. Such theories of external contextualization are mainly written in the area of intellectual history. Contemporary Taiwanese scholar Huang Chun-chieh, for instance, has written several significant works in this field that have clarified many important elements of the historical development of the exegesis of Mencius (Huang 1997, 2015). He has also illuminated how and why certain classical works of Chinese tradition have been—for various cultural, political, and ideological reasons—de-contextualized and re-contextualized in the course of their incorporation by Korean and Japanese cultures.

Conversely, let us briefly return in this context to Cheng Chung-ying, and his aforementioned method of onto-hermeneutics (*benti quanshixue* 本體詮釋學), that was rooted in Gadamerian hermeneutics, but further explored and developed in terms of the Confucian worldview. Cheng believes that the traditional Chinese hermeneutic is ontological; his model is rooted in the presumption, according to which the "understanding of reality and truth is simultaneously the source of meaning and the driving force for seeking understanding" (Cheng 2003b, 290). According to Cheng, no understanding or interpretation can be made without such a reference.

Irrespective of the questionable nature of the notion of truth in the Chinese worldview, the onto-hermeneutics is doubtless an interesting and innovative approach to the investigation of the traditional Chinese mode of exegesis. However, the problem that has to be discussed in this context is linked to the question of different contextualities underlying all these theoretical approaches. Although Huang and Cheng both write about traditional Chinese hermeneutics, the two kinds of hermeneutics they respectively propose obviously refer to two very different things. As I have shown elsewhere (Rošker 2021, 113), the first one is based on the external, and the other is rooted in the internal contextualization.

The former refers to the cultural, historical, and literary context of the text, and the latter to its inherent conceptual dimensions and their semantic and philosophical implications and developments. Consequently, the hermeneutical method is still lacking a binding, inherent consistency, not only regarding its various particular methods, but also the multifarious differentiations that can be detected within and between traditional explanations of its crucial terminology.

However, this is by no means the only problem linked to this method. In my view, what is even more questionable is its premise, which presupposes the existence of a normative intelligible meaning. Proceeding from a positive reevaluation of the function of prejudice[2] in the sense of Heideggerian anticipatory structures, Gadamer highlights that understanding always involves what he terms the "anticipation of completeness" (1982). In other words, it always involves the verifiable assumption that "what is to be understood constitutes something that is understandable, that is, something that is constituted as a coherent, and therefore meaningful, whole" (Malpas 2018, 3). In Gadamer's own words, the concept of horizon and the fusion of horizon respectively could represent a means to grasp the "own meaning" of a text:

> The concept of "horizon" suggests itself because it expresses the superior breadth of vision that the person who is trying to understand must have… . We are always affected, in hope and fear, by what is nearest to us, and hence we approach the testimony of the past under its influence. Thus it is constantly necessary to guard against over hastily assimilating the past to our own expectations of meaning. Only then can we listen to tradition in a way that permits it to make its own meaning heard. (1989, 305)

Obviously, this is still a conceptual view of hermeneutic understanding.[3] As we have seen, it still presupposes an "own meaning" of a certain tradition (or discourse). The problematic nature of such suppositions can be illuminated by Derrida's famous statement that "*il n'y a pas de hors-texte*" (1997, 158). This has often been misinterpreted in the sense that "there is nothing outside the text," implying that there was nothing outside the words (Derrida 1988, 136). He emphasized that what he really meant by "there is no outside-text" is that "there is nothing outside of context." He writes, "The phrase which for some has become a sort of slogan, in general so badly understood, of deconstruction ('there is nothing outside the text'…means nothing else: there is nothing outside context. In this form, which says exactly the same thing, the formula would doubtless have been less shocking" (136).[4]

The contextual nature of meaning has, hence, prevailed in several contemporary Western discourses, including psychoanalytical philosophy. On such a level, the very idea of the hermeneutical circle also becomes problematic; therefore, it is not surprising that Lacan also highlighted that such a circle is without semantic support, since the meaning is a product of an infinite sliding of the referential surface.[5] On this basis, Žižek (1976, 75) rejects the hermeneutical circle, which implies the anteriority of the entirety of the semantic horizon in particular statements. In his view, "hermeneutics proceeds to the edge of interpretation, but just before reaching it, it covers its eyes for the realization of the fact that there is no original meaning, which could provide a basis for a different referential network for the transmission of the reference, because the meaning is always relational" (75).

I do agree that meaning is always relational. Hence, Gadamer's model of the hermeneutical circle, which is based upon a conceptual view of horizons, is indeed problematic. However, instead of mourning over this fact, we should rather search for an un-conceptual foundation of the semantic unification in the process of interpretation. In my view, Gadamer's paradigm of horizons (which is, as we have seen, still conceptual in essence) should be replaced through an un-conceptual paradigm, such as *jingjie* 境界 (sphere, atmosphere, aesthetic realm), which has a Buddhist origin and belongs to crucial notions in traditional Chinese metaphysical and literary writings. Hence, in hermeneutical interpretations of Chinese philosophy, we could replace the notion "fusion of horizons," with the term "fusion of *jingjie*s or aesthetic realms (境界融合)."

Beyond the Hermeneutic Circle: The Fusion of Aesthetic Realms?

This important and typically Chinese aesthetic notion is a hermeneutic tool that can help us understand artistic and intellectual creations through the lens of the various manifestations of the human living world. The *jingjie* sphere can only be experienced, but it cannot be fully described in concrete language, nor can it be reflected upon in purely conceptual thought.

In the beginning, *jingjie* was a term pertaining to geopolitical discourses. Expressing a certain realm limited by boundaries, it was used for mapping out the geopolitical world of ancient China. It evolved into a philosophic-religious discourse of a mental or psychological "territoriality" after it was employed to communicate the Buddhist ideas of spiritual reality and enlightenment in the sense of crossing to the other shore (Han 2014, 86). In other words, the notion of aesthetic realm primarily pertained to the "objective" features of external reality. The internalization of the psychologically transmitted formations of this basic level of *jingjie*s is linked to the Buddhist interpretation of its nature. The unification of external and internal elements takes place on the level of transforming these outward formations and images into a specific mental realm (Cheng 1995, 92). According to Christina Han (2014), it further gained many new implications and complex semantic dimensions in the Neo-Confucian discourses of the Song and Ming Dynasties.

In pre-modern Chinese philosophy, the notion *jingjie*s was central to Wang Guowei's[6] (1877–1927) aesthetics. Wang himself defined the concept in as follows: "The 'realm' does not only refer to a landscape or scene. The emotions of joy and sorrow, anger, and pleasure also constitute a sort of aesthetic realm in the human heart" (Wang 2013, 18).

What Wang is referring to in the above quotation is the objectification of psychological states in which the external realm[7] is fused with inner world through subjective sensuality. In this way, *jingjie* is a paradigm that appears within an aesthetic *noumenon* in order to manifest a certain (in)significance of human life and to convey a certain meaning (Li 2010, 210). This manifestation of meaning is, therefore, rooted in the experience of the *noumenal* and, hence, in the merging of immanent and transcendent sections of time and space within the present moment of here and now.

In spite of the fact that Wang's theoretical framework was clearly strongly influenced by Western philosophy (especially Schopenhauer and Nietzsche), it can still be defined as the "revelation of life through the relationship between feeling and scene, and the objectified realm of the artistic subject" (Li Zehou in Samei 2010, xvi). However, this aesthetic realm cannot be reduced to mere integration of feeling and scene, nor to the emotion, sensibility, or motivation of its author or creator. It also implies dissolving the difference between self and the others, and it transcends all utilitarian purposiveness without having to negate the will, desire, and life itself.[8] Therefore, it could also serve as a tool for creating subtle blends between traditional Chinese hermeneutic notions, techniques, and approaches on the one hand, and Western philosophies of hermeneutics on the other hand.

The aesthetic realm conveys meanings with diffusive, continuously dispersing edges, which cannot be compared to meanings confined into the narrow semantic spaces with fixed borders of conceptual definitions. Wang Guowei continues, "Hence, if a text captures in words a real scene or a real emotion, it can be said to convey an aesthetic realm" (Wang 2013, 18).

This is not only true for poetry, painting, or other works of art. Because *jingjie* possesses a noumenal dimension, it can also be discovered in numerous (but certainly not all) philosophical works. They also contain insights, which convey a philosophical idea not only through conceptual phrases, but rather through that, which is engraved between the lines, creating a certain atmosphere, consisting of images, associations, sensations and emotions, experienced and expressed by their author, and also perceived and re-experienced by the readers. No wonder that—precisely through the realm of inner experience—philosophy is often linked to literature or poetry. Li Zehou, for instance, has defined philosophy as "science, imbued with poetry" (Li 2016, 4). As science, it offers us a systematic way of exploring and comprehending reality; as poetry, it walks with us through the opaque jungle of our life, on a long and intimate journey that not only offers us beauty and pleasure, but also forces us to confront fear and melancholy. Philosophy can be a way of life that is rational and artistic at the same time; it not only urges us to search for answers to eternal questions of being, but also to unceasingly raise new ones. It does not remain limited to discovering the world, but also allows for its ongoing creative change. *Jingjie*, or aesthetic realm, might be one of the most typical hermeneutic paraphernalia of philosophy that was created in China. Because it seems that precisely here, this affinity and subtle closeness between the philosophical depths of rationality and sensitivity could come to life and find its untroubled home—even when it comes to the ultimate existential concerns:

> *Jingjie*, attained through literary appreciation, is the sudden realization and cognition of ultimate reality that embodies the principle of truth, goodness, and

> beauty simultaneously. The experience contained in *jingjie* is not only aesthetic, but religious and existentialistic as well. Our close reading of Wang's remarks on *jingjie* has revealed the rich spiritual meaning of this concept: the fusion of subject and object through intuition, the consummation of truth, goodness, and beauty, and a deep existential concern.
>
> (Wu 2002, 450)

To a certain extent or in certain aspects, *jingjie* can be compared to Heidegger's understanding of "moods" (or attunements),[9] which, for him, revealed the Being of *Dasein*, for according to him, *Dasein* is always in an attunement; the world is discovered in a mood (Heidegger 1996, 126ff, 313ff). It is a very basis on which we can establish our being-in-the-world. In other words, moods establish *how we find ourselves in the world.* Similar to Heidegger's moods, aesthetic realms or *jingjie*s represent a pre-subjective and pre-objective sense of being-in-the-world.

The *Jingjie* of Zhuangzi: Birds, Fish, and the Circular Nature of Intersubjectivity

Hence, it is completely clear that *jingjie* is not a conceptual paradigm, but one that can at the most be grasped through situational, contextual approaches. Hence, like Chinese language and Chinese philosophy, interpretations through aesthetic realms are always linked to contexts, as much as concrete experiences. I will try to illustrate such a holistic, contextual interpretation with an example of two essays from Zhuangzi.[10] For this purpose, let us first take a closer look at Zhuangzi's famous essay about the seabird:

> Have you not heard of this? Once upon a time, on the outskirts of the land of Lu, a bird of paradise appeared. The Emperor of the land of Lu received it with the highest honors and expressed his heartfelt welcome to it. He had it transferred to the highest temple, treated it to the most select wines, and offered it the most enchanting music to please it. He had a good number of cattle and sheep slaughtered and he organized a sumptuous banquet. But the bird remained gloomy and depressed, not taking a bite to eat or a sip to drink. After three days, it died, just like that. This is what happens if one feeds birds with food for humans instead of food for birds. Those who know how birds should receive food suitable for them, would have let the bird fly into deep woods, where it would rest on trees. They would have let it fly merrily above sandy ground and soar above rivers and lakes. They would have let it hunt for its own food and feast on the small fish that it likes. It would have followed its flock, and stopped and rested wherever it pleased. It would have complete freedom of movement. Birds are most bothered by human voices. If symphonies of the greatest human musicians were played somewhere in the wild, birds would immediately fly away, all wild beasts would flee and fish would hide in the deepest waters. However, when humans hear such music, they gather around it and enjoy it. Fish can only live in water, but people die in it. Since man and fish have different

properties, what they hate and love differs as well. The ancient wise men did not measure individual features of ability and behaviour with the same criteria. The name is created before the reality. Meaning is effected through appropriateness. To follow these principles brings happiness and satisfaction.

(*Zhuangzi* s.d., Zhi le, 5)

That which is not a bird, therefore, cannot simply judge on its own and conclude that what is best for them is best for birds. This is a Daoist critique of the Golden Rule, advocated by the Confucians, which finds its most famous expression in the advice "not to impose on others what you would not wish done to yourself" (*Lunyu* s.d., Yan Yuan, 2). We must be aware of the fact that we are parts of different worlds, living in different realities.

But let us return to the basic message of the story. Because I am not a bird, I cannot inherently know the likes and dislikes of birds. This, naturally, is merely an assumption of Zhuangzi's method of perception and communication. It is not, by any stretch, a system of logical systematization. In Zhuangzi's basic work, his friend Hui Shi always represents the basic type of argumentation, which attempts to derive logical, universally valid conclusions from assumptions. And in doing so, naturally, he often makes himself abundantly ridiculous. If humans cannot know birds, since we are not birds, shouldn't it also apply that we cannot know fish, since we are not fish? Or, translated into logical inferences:

P1: Humans cannot know birds.
P1: Zhuangzi is human.

C: Zhuangzi cannot know birds.
P1: Humans cannot know fish.
P1: Zhuangzi is human.

C: Zhuangzi cannot know fish.

However, Zhuangzi is not quite convinced that it is all that simple[11]. Let us listen to an anecdote from the *Autumn Water* section in Zhuangzi's *External Chapters*.

	Zhuangzi and Huizi are strolling on a bridge over the Hao river.
Zhuangzi says:	How easily the white fish swims to and fro—this is the joy of fish!
Huizi says:	But you are not fish, so where can you know what is the joy of fish?
Zhuangzi then says:	But you are not me, so how can you know that I do not know what is the joy of fish?
Huizi says:	I am not you, therefore I cannot understand you; but you are also not fish, and therefore you cannot understand fish. That is all.

Zhuangzi says: Well, then, let's go back to the beginning. You asked me: Where can you know what is the joy of fish? So at the time you asked me that, you must have known that I knew what is the joy of fish. Well, I knew this on this bridge over the Hao river.

(*Zhuangzi* s.d., Qiu shui, 13)

Hmm. Was Zhuangzi here playing with sophisms? He has obviously been playing with words, for the Chinese interrogative *an* 安 can refer to time, space, or manner. It can thus mean *what*, *how*, *when*, or *where*. If it was understood in the latter sense, Zhuangzi provided a proper answer. However, if one takes into account the socio-cultural context of traditional China at the time when this work was (or is supposed to have been) created, we will readily think that Zhuangzi truly wanted to say something more meaningful with this anecdote, and to impart a message.

Of course, the following is only my subjective interpretation of the two stories—one under many, many others.[12] But since we have already rejected the idea of an absolute text or an absolute meaning, we also have to question the notion of an absolute interpretation. However, the real reason for my adding more water to this flood is because, with these interpretations, I would like to demonstrate how a new meaning and understanding can be acquired through the method of unifying (fusing) the aesthetic realms (*jingjie*s) that can be experienced in these two separated anecdotes, without relying on their strictly conceptual connotations.

If we try to connect and understand both stories in this way, namely considering the broader essential context of which they are both a part, we can easily see that they are both dealing with human relations (or relations between living beings in general). In Lao Sze-kwan's view, the "fundamental source" of both stories would probably be linked to the question of the nature of intersubjectivity. The common ground in both debates is doubtless connected to this problem. It shows us, to speak with Ram Adhar Mall that, in our attempt to understand one another, we meet to differ and differ to meet (2000, 6).

The first essay emphasizes differences between different beings. If one desires the well-being of everything that exists, one must—according to Zhuangzi—first get used to the fact that we are all different. Only on the basis of knowing this fact, i.e. the fact that we all live in different worlds, can one create close mutual contacts. (This—implicitly and latently proposed—fusion of *jingjie*s could apply to contacts and relations between birds and people, as well as those between Sinologists and the Chinese, or between authors and readers).

The creation of such contacts and communication, in turn, proves again that we all live in a single, unified world, as the second essay shows. Zhuangzi's comprehension of the joyfulness of fishes resulted from the entire context in which the fishes were observed. Zhuangzi was joyfully strolling in friendly nature, accompanied by his best friend, and he enjoyed the whole situation, of which the fishes were also a part. Hence, his joyfulness could not be separated from the fishes and vice-versa. It was precisely this very unification in joy (a fusion of this joyful *jingjies*), which made his innate, complete, and comprehensive understanding of fishes possible.

The fusion of aesthetic realms, experienced in both stories, show us very clearly that, ultimately, it is human individual subjectivity that determines what should be regarded as a genuine relationship. In this sense, we are presented a new, and more complex, image of intersubjectivity.

This kind of fusion is not to be mixed with a melting together of two different entities. The story of the seabird shows that what makes any aesthetic fusion possible, is precisely the experience of difference and separation. The happy fish from the second essay show, through the very fact of their happiness, that these differences and separations are instrumental of any genuine, vital, and creative unity, precisely because a fusion of aesthetic realms is always conditioned by diversity.

Conclusion

Hence, intersubjective understanding is not conditioned by the criteria of objectivity (with agreed-upon names) but rather by the thing itself, i.e. by understanding and experiencing the aesthetic realms in which the subjects are embedded. The apparent objectivity and independence of the human rational mind has repeatedly been proven a false, illusory chimera, which only leads to self-deception.

The dynamics of being limited to the intimate world of an individual, on one side, and muddled, continuous merging of all individual worlds into a single one, on the other, permeates our existence and positions us in what we call "time and space." Finally, the fusion of individual aesthetic realms is precisely the starting point for constructing a tiny bridge of understanding, connecting Zhuangzi and his reader, Chinese and Western philosophy, you and me.

Notes

1 Here, it has to be pointed out that in the early years of the PRCh, Hegelian hermeneutic was not particularly welcome, because these years were generally marked by a strong anti-Hegelian tendency. This did not change until the mid-fifties: "Engels' anti-Hegelian interpretation of Marx quickly became canonical. His anti-Hegelian approach to Marx has long influenced Russian Marxism and continued to influence Chinese Marxism. In broad terms Chinese Marxism is dependent on two main sources: Russian Marxism and the Chinese intellectual tradition. The shared Russian and Chinese anti-Hegelian interpretation of Marx was called into question in the middle of the last century through the relatively late publication of crucial Marxian texts, that is texts prepared by Marx and that are crucial for understanding his position" (Rockmore 2019, 56).

2 Gadamer offers the clearest explanation of his notion of prejudices in his *Philosophical Hermeneutics*, where he describes them as the "biases of our openness to the world." Gadamer writes: "Prejudices are not necessarily unjustified and erroneous, so that they inevitably distort the truth. In fact, the historicity of our existence entails that prejudices, in the literal sense of the word, constitute the initial directedness of our whole ability to experience" (Gadamer 2008, 9). In this sense, prejudices are simply "conditions whereby we experience something—whereby what we encounter says something to us" (9).

3 In this predetermined (and hence, conceptualized) notion of meaning that is implied in such a positive conception of prejudice, Gadamer redeploys the notion of our prior hermeneutical situatedness as it was developed in Heidegger's *Being and Time* (first published in 1927) in terms of the 'fore-structures' of understanding, i.e., "in terms of the anticipatory structures that allow what is to be interpreted or understood to be grasped in a preliminary fashion" (Malpas 2018, 3.1). Since in this view, understanding functions by means of such anticipatory structures implies that it necessarily implicates what Gadamer denotes as the "anticipation of completeness." Such an understanding is necessarily preconditioned by revisable presupposition that "what is to be understood constitutes something that is understandable, that is, something that is constituted as a coherent, and therefore meaningful, whole (3.1).

4 Indeed, for Derrida, the word "text" implies all possible referents. Hence, even when claiming that "there is nothing outside the text" this does not mean that "all referents are suspended, denied or enclosed in a book, as people have claimed, or have been naïve enough to believe and to have accused me of believing. But it does mean that every referent and all reality has the structure of a differential trace [*d'une trace differentiale*], and that one cannot yield to this "real" except in an interpretative experience'. The latter neither yields meaning nor assumes it except in a movement of differential referring (Derrida 1988, 148).

5 This critique of a static conception of meanings can also be connected to Jacques Derrida's critique of the "metaphysics of presence" (Huang Kuan-min 2020, 23). Derrida reproached this discourse for its systematic tendency to prefer or privilege notions such as identity, unity, and entirety over marginality, otherness, and difference. Particularly damaging has been, in his view, the tendency to conceive of linguistic truth as the "presence" of what is expressed by its representation in words. Indeed, the ungrounded nature of meaning—the fact that meanings are not given by a direct relation with things in the world, but only by their mutual structural connections—confirms that what is expressed is never fully "present," but is instead infinitely mediated by an endless chain of meanings. The concept of truth as "presence" is therefore not viable. In Derrida's theory, the two concepts of "différance" (a neologism that implies both a [spatial] difference and a temporal deferral) and "dissemination" characterize the infinite nature of meaning and the futility of metaphysics' attempts to reach a point of finality or closure.

6 Besides being the first scholar to introduce to China the works of Nietzsche, Schopenhauer, and Kant, Wang Guowei initiated the comparative study of Western and Chinese aesthetics. Similar to Li Zehou's thoughts, the idea of "human life" was the basic foundation of all aesthetic studies also for Wang Guowei . On this basis, he created an aesthetics of life in the sense of modern humanitarianism and purposelessness by elaborating Zhuangzi's category of "the use through the useless (*Wu yong zhi yong* 無用之用)" in aesthetic activities (*Zhuangzi* s.d., Nei pian. Renjian shi, 9).

7 This "external realm" can refer to both, an objective state in present or past external reality in which a subject is embedded, or a subjective reflection that is either thought of or imagined in their mind.

8 Wang Guowei used the term *jingjie* or aesthetic realm interchangeably with the concept of *yijing* 意境 or artistic conception (Cheng 1995, 93). This latter notion is a kind of imaginary domain, which is—similar to the aesthetic realm—also based upon a fusion of emotion and scene or situation (*qingjing jiaorong* 情景交融, Cheng 1995, 95). However, what the artistic conception implies, is more centered upon the mindful awareness of the here and now, which is conveyed by artistic creation. In his writings about the artistic conception Li Zehou (see for instance Li 2010) hence always highlights that it is—precisely

because of this fusion or the unity it implies—completely useless and redundant to seek comprehension through any kind of conceptual medium between feeling and object.

9 *Stimmung*. The term has often been translated into English as "mood," however in my opinion, attunement is a more appropriate translation.

10 For an informative analytical background regarding the central concepts of Zhuangzi's philosophy, which constitute the core of his aesthetic thought, see Sernelj 2017.

11 Zhuangzi was also quite deliberately not interested in a systematic uniformity of his philosophy (Heubel 2021, 273). François Billeter writes the following on this subject: "An overly coherent discourse would have seemed suspicious to him, for he was interested above all in the aporias of thought, the paradoxes and discontinuities that we encounter in the course of our experience of self and world" (2010, 34). In this context, Heubel refers to the phrase "an overly coherent discourse." In his view, such a formulation proves that Billeter assumes that the author Zhuangzi certainly does have a certain desire for discursive coherence. However, despite this desire, his main interest was still in the "paradoxes and discontinuities" of self- and world-experience. In Heubel's view, this is "an important insight" (Heubel 2021, 273).

12 Throughout the history, there have been created hundreds of different interpretations of this charming story. The most recent ones can be enjoyed in the collection *Zhuangzi and the Happy Fish* (2015), edited by Robert T. Ames and Takahiro Nakajima. The editors wrote in the description of the anthology (also on the book's back cover) that they have brought together "essays from the broadest possible compass of scholarship, offering interpretations that range from formal logic to alternative epistemologies to transcendental mysticism. Many were commissioned by the editors and appear for the first time. Some of them have been available in other languages—Chinese, Japanese, German, Spanish—and were translated especially for this anthology. And several older essays were chosen for the quality and variety of their arguments, formulated over years of engagement by their authors. All, however, demonstrate that the Zhuangzi as a text and as a philosophy is never one thing; indeed, it has always been and continues to be, many different things to many different people."

Bibliography

Ames, Roger T. and Takahiro Nakajima. 2015. *Zhuangzi and the Happy Fish*. Honolulu: University of Hawai'i Press.

Billeter, Jean François. 2010. *Notes sur Tchouang-tseu et la philosophie*. Paris: Allia.

Bradley, Arthur. 2008. *Derrida's of Grammatology*. Edinburgh: Edinburgh University Press.

Cheng, Chung-Ying. 2003b. "Inquiring Into the Primary Model: Yi Jing and the OntoHermeneutical Tradition." *Journal of Chinese Philosophy* 30, no. 3–4: 289–312.

Cheng, Fuwang, ed. 1995. *Zhongguo Meixue Fanchou Cidian* [A Dictionary of Chinese Aesthetic categories]. Beijing: Zhongguo Renmin daxue chuban she.

Derrida, Jacques. 1988. *Limited Inc*. Evanston: Northwestern University Press.

———. 1997. *Of Grammatology*. Baltimore: Johns Hopkins University Press.

Gadamer, Hans-Georg. 1982. "On the Problematic Character of Aesthetic Consciousness." *Graduate Faculty Philosophy Journal* 9, no. 1: 31–40.

———. 1989. *Truth and Method*. 2nd ed. Translated by J. Weinsheimer and D. G. Marshall. New York: Crossroad.

———. 2008. *Philosophical Hermeneutics*. Translated and edited by David E. Linge. London: University of California Press.

Gu, Ming Dong. 2005. *Chinese Theories of Reading and Writing: A Route to Hermeneutics and Open Poetics*. New York: SUNY Press.

Han, Christina. 2014. "Envisioning the Territory of the Sage: The Neo-Confucian Discourse of Jingjie." *Journal of Confucian Philosophy and Culture* 22: 85–109.

Heidegger, Martin. 1996. *Being and Time: A Translation of Sein und Zeit*. Translated by Joan Stammbaugh. Albany: SUNY Press.

Heubel, Fabian. 2021. *Was ist Chinesische Philosophie? Kritische Perspektiven*. Hamburg: Meiner.

Huang, Chun-chieh. 1997. "Characteristics of Chinese Hermeneutics Exhibited in the History of Mencius in Exegesis." *Zhongguo wenzhe yanjiu jikan* 7: 281–302.

———. 2015. *East Asian Confucianisms Texts in Contexts*. Goettingen and Taipei: V&R unipress, National Taiwan University Press.

Huang, Kuan-min. 2020. "Dissemination and Reterritorialization: Tang Junyi, Mou Zongsan, and the Renovation of Contemporary Confucian Philosophy." *Asian Studies* 8, no. 3: 15–33.

King, Po-Chiu. 2016. *Thomé H. Fang, Tang Junyi, and Huayan Thought: A Confucian Appropriation of Buddhist Ideas in Response to Scientism in Twentieth-century China*. Leiden: Brill.

Kubin, Wolfgang. 2005. "Chinese 'Hermeneutics': A Chimera? Preliminary Remarks on Differences of Understanding." In *Interpretation and Intellectual Change*, edited by Tu Ching-I, 311–20. London: Transaction Publishers.

Lenk, Hans and Gregor Paul, ed. 1993. *Epistemological Issues in Classical Chinese Philosophy*. New York: SUNY Press.

Li, Zehou. 2010. *The Chinese Aesthetic Tradition*. Translated by Maija Bell Samei. Honolulu: University of Hawai'i Press.

———. 2016. "Li Zehou duitan lu [Recordings of Li Zehou's Conversations]." *Dai yue ting yu zhu jilu*. Accessed May 19, 2016.

Lunyu [The Analects]. s.d. *Chinese Text Project*. Accessed March 1, 2017. http://ctext.org/analects

Mall, Ram Adhar. 2000. *Intercultural Philosophy*. New York: Rowman and Littlefield Publishers, Inc.

Malpas, Jeff. 2018. "Hans-Georg Gadamer." In *The Stanford Encyclopedia of Philosophy*, edited by Edward N. Zalta. Accessed November 12, 2021. https://plato.stanford.edu/archives/fall2018/entries/gadamer/

Ng, On-Cho. 2005. "Affinity and Aporia: A Confucian Engagement with Gadamer's Hermeneutics." In *Interpretation and Intellectual Change*, edited by Tu Ching-I, 297–310. New Brunswick and London: Transaction Publishers.

———. 2013. "Chinese Philosophy, Hermeneutics, and Onto-Hermeneutics." *Journal of Chinese Philosophy* 30, no. 3–4: 373–85.

Pfister, Laureen. 2006. "Hermeneutics: Philosophical Understanding And Basic Orientations." *Journal of Chinese Philosophy* 33, no. 1: 3–23.

Rockmore, Tom. 2019. "Hegel and Chinese Marxism." *Asian Studies* 7, no. 1: 55–73.

Rošker, Jana S. 2010. "The Concept of Structure as a Basic Epistemological Paradigm of Traditional Chinese Thought." *Asian Philosophy* 20: 79–96.

———. 2021. *Interpreting Chinese Philosophy: A New Methodology*. London: Bloomsbury Academic.

Samei, Maija Bell. 2010. "Translator's Introduction." In: *Li Zehou: The Chinese Aesthetic Tradition*, The Chinese Aesthetic Tradition by Li Zehou, translated by Maija Bell Samei ix–xix. Honolulu: University of Hawai'i Press. https://uhpress.hawaii.edu/title/the-chinese-aesthetic-tradition/

Sernelj, Tea. 2017. "Methodological Problems of Xu Fuguan's Comparative Analysis of Zhuangzi's Aesthetics and Western Phenomenology." *Asian Studies* 5, no. 1: 271–88.

Tu, Ching-I, ed. 2000. *Classics and Interpretations: The Hermeneutic Traditions in Chinese Culture.* London: Transaction Publishers.

———, ed. 2005. *Interpretation and Intellectual Change.* London: Transaction Publishers.

Wang, Guowei. 2013. *Renjian cihua* [Poetic Remarks in the Human World]. Beijing: Fenghuang chuban she.

Wu, Jiang. 2002. "What Is Jingjie? Defining Confucian Spirituality in the Modern Chinese Intellectual Context." *Monumenta Serica* 50: 441–62.

Wu, Kuan-min. 2004. "Distinctive Features of Chinese Hermeneutics. A Review of Classics and Interpretation: The Hermeneutic Traditions in Chinese Culture, edited by Ching-I Tu." *Taiwan Journal of East Asian Studies* 2: 233–47.

Xia, Zhentao. 2000. "Shijie zhexue yanjiude zouxiang 世界哲学研究的走向 (The Tendencies of Global Philosophical Research)." *Zhongguo renmin daxue xuebao* 2: 4–6.

Zhang, Yaonan. 2002. "Lun 'zhishi jieshi shuo' zai Zhong xi zhexueshi, 论'知识解释说'在中西哲学史 (On the Role of the 'Explanatory Theory of Knowledge' in the History of Chinese and Western Thought)." *Beijing Xingzheng xueyuan xuebao* 2: 73–79.

Zhuangzi. n.d. Chinese Text Project. Accessed March 12, 2016. http://ctext.org/zhuangzi.

Žižek, Slavoj. 1976. "Hermenevtika in Psihoanaliza." *Anthropos* 3, no. 6: 69–98.

12 The Genre of the Animal

Beyond Racial Capitalist Consumption

Che Gossett

Sylvia Wynter's destabilization of the human and its philosophical precepts and predicates proceeds through an interrogation of the onto-epistemological edifice that sustains its seemingly sovereign status. As Wynter stated in an interview with David Scott, her aim is "to move completely outside our present conception of what it is to be human, and therefore outside the ground of the orthodox body of knowledge which institutes and reproduces such a conception" (Scott 2000, 136). If we move completely outside of our present conception of the human and its theoretical topography, where will we be, and end (up)? What might our coordinates be outside conception or without philosophical ground? "So, you see," Wynter reminisces, "I was on the fringe, I was on the outside. Or, in a sense, I was on the other side..." (Scott 2000, 148–49). In this sense, Wynter is fugitive to the racial and colonial capture, while also insurgent; after all, Black studies, she remarks, was a war. This war is epistemic, ontological, cosmological and more than merely a prosecuting a war, in an adversarial sense. Wynter goes further: she undermines and refuses the very terms of engagement.

In political philosophy and theory, both the figure of the human and its corollary, the animal, share a proximity with the slave. While this proximity is inaugurated in the annals of political philosophy through Aristotle's categorization of man as a speaking and political animal, and the slave, as well as the foreigner, as being without reasoned speech,[1] the status of the slave undergoes a historical torsion with the anti-black incarnation of the racial slave and Middle Passage modernity. The political grammar of the human as an anti-black dispositif has been brilliantly deconstructed and radically upended by Frank Wilderson in his Afropessimist inflected writing, especially *Red, White and Black* (2010) and in Saidiya Hartman's phenomenal *Scenes of Subjection* (1997) where the slave is disclosed as the ground for the modern political subject. These Black radical interventions cut into political theory—which at once penetrate and simultaneously remain in a fugitive relation to its philosophical edifice. Early Marx wrote about the ways in which labor animalizes the worker through alienation from humanity. There is a certain version of what Nicole Shukin (2009) terms "animal capitalism" presented in Marx's writing and analysis. This version, per Shukin, morphs with the growth

DOI: 10.4324/9781032659985-16

of the factory farm and agribusiness, and also shifts with contemporary carceral animal law and protectionism. Marx theorizes capitalism as an animalizing violence that dispossesses the worker of humanity, outside of the circle of the human. Yet, following Cedric Robinson's towering interventions, Black radical thought and cinema have also been sites of interrogation of the ways in which blackness, animality, and the (lumpen)proletariat are sutured and bound up, fated to premature death. Charles Burnett's short films *Killer of Sheep* and *The Horse* fixate on scenes of premature death: the factory farm, Watts under the racial regime of capitalism in the 60s–70s, and the slow wearing down of life through a sense of melancholic hopelessness that is borderline asphyxiating. In this chapter, I think alongside Burnett, in the wake of Robinson, and to conclude, I revisit Wynter, who's prodigious work and fugitive human(ism) inspires—in the sense of breathing life into—this essay.

Marx: Capitalism Turns the Human Into Animal

The figure of the animal cuts across the division of Marx into early and late epochs, emerging first in his Hegelian-inspired writing, especially the *1844 Manuscripts,* and extending into the *Grundrisse* and the volumes of *Capital*. It is impossible to delimit Marx's writings on animality from his larger materialist enterprise. Marx's writing on animality is co-axial with his writing on ecology and materialism, and this specific writing is a fold of that critical endeavor, which is entangled with Engel's dialectical naturalism. If on one side lies Marx's dissertation work on atomism, Lucretius, Epicurus, Democritus, and the philosophy of nature, and on the other side lies Engel's dialectical naturalism, both these interventions rely on dialectical and Hegelian-influenced conceptions of nature that emphasize nature as processual, aleatory, contingent, and fluctuating, in which the human is enfolded. Judith Butler's article in *Radical Philosophy* (2019) "The Inorganic Body in the Early Marx: A Limit-concept of Anthropocentrism," approaches the question of anthropocentrism in Marx by close reading his writings on nature. Marxist sociologists and science studies scholars—Johnathan Bellamy Foster in particular, across his oeuvre *Marx's Ecology and The Return of Nature* (2020) especially—have analyzed Marxist contributions to ecological thought, such as the import of metabolic rift, which extends Marx's writing on capitalist extraction of natural resources that splits the human from the natural milieu. This body of work is critical for the environmental humanities and the critique of green capitalism.

The animal for early Marx of the *1844 Manuscripts* is an end in-itself. As Marx states, "The animal is immediately identical with its life-activity. It does not distinguish itself from it. It is its life-activity" (2012/1844, 75). Marx's and Heidegger's theory of the human and the nonhuman animal come to a chiasmus, or crossing point, in their imaginaries of the animal as lacking the capacity to envision the world outside of their own recursive and self-reflexive quasi-subjectivity and sphere of life, expressed in Heidegger's claim that the animal is "poor in the world" (1995, 185). The irony of Heidegger's statement is that it completely takes for granted animals as companion species and the psychic life of that relationship (if we keep Heidegger's formula of internal and external to human as worlding), including the social and affective bonds created between human and the nonhuman animal. Heidegger, to sustain his schema—Man in world, animal as lacking in world, and stone (representative of the inanimate) as without world—must leave out of purview the question of animal as laborer. More damningly, in terms of disclosing the extent to which fascism, anti-blackness, and anti-Semitism and Naziism were central to, rather than sidebars of, Heidegger's philosophical enterprise, the publication of his *Black Notebooks* revealed that he placed Jewish people outside of world completely: worldless (Brody 2014). Heidegger is not Marx; they are incommensurable, to say the least. Tracing the presence of animality in philosophy, as Derrida does within limits in his seminars on the beast and the sovereign (2010–11), is a project that requires discernment not only as to which ends the figure of the animal is activated, but also the coordinates through which the figure of the animal is located, which can be in totally different historical universes. In Marx's theory of mind, the animal lacks capacity for sociality and dimensionality. The animal is what it does (its life-activity), and the human is what creates and makes life-activity with intentionality. Marx argues that there is a recursive relationship and relay between the human and nature. Marx figures nature as the "inorganic body" of man, a kind of prosthesis (Butler 2019). He further argues that capitalism turns the human into the animal. Labor evacuates humanity leaving only bodily, animal-like functions of sleeping, sex, eating, etc.

Animal Capitalism as Racial Capitalism

Yet Marx's political-as-libidinal economy is haunted by the specter of slavery, as is shown by Cedric Robinson's *Black Marxism*, in which Robinson critiques the bracketing of racial slavery into the rubric of primitive accumulation while simultaneously tracing the ways in which capitalism arrives as racial (intra-European ethnicization as otherization) and anti-black (the figure of Negro as slave). Marx's writings on animality and the human as worker, which is always already a white subject, might be read for their unthought racial implications, inflections, and underpinnings. This acknowledgment is important, not for a dismissal or disposal of Marxism, but rather to continue to expand this form of thought to show how "animal capitalism" (Shukin 2009) is always already racial capitalism; moreover, abolition is a means to address all grammars of capture as well as the entanglement of animal captivity and industrialized killing with/in a broader context of racial carceral capitalism.

Much of the recent literature on blackness and animality approaches this entanglement through an ontological and literary studies register, given the sharp meta-ontological turn Black studies has taken in recent years. While this is crucial and a necessary part of the theoretical intervention into critical animal studies, I am curious about the ways in which late-stage post-Fordist and racial state capitalism orchestrates and co-joins anti-black and anti-animal violence. Towards this end, I turn to the Black radical cinematics of Charles Burnett.

Charles Burnett's "Killer of Sheep"

In his short films, *The Horse* (1973) as well as *Killer of Sheep* (1978), Charles Burnett explores how racial capitalism and animal capitalism are grafted onto one another through the cinematic aesthetics of the Black radical tradition, which illuminate how the survival of the Black worker and the animal are entangled under the regime of racial capitalist violence. *Killer of Sheep* has an unconventional and even paradoxical career as both one of the first fifty films inducted into the Library of Congress national film registry, considered a classic in the vein of *Citizen Kane* (1941), while simultaneously being largely unrecognized by mainstream American audiences and Hollywood. One of the main reasons the film remained on lower frequencies was that Burnett was not able to purchase the rights to the music of the soundtrack. The music, however, is central to the affective force of the film, especially regarding the figure of the animal. This neorealist film follows a working-class Black family that exists in the interplay between the proletariat and the lumpenproletariat—perhaps more accurately described as the Black precariat. The film follows Stan, his partner/wife, and two children as they survive in the late seventies in Los Angeles. Only men are named in the film, while the partner and children remain unnamed, raising further questions of gender, Black masculinity, and representation even as the film refuses a certain version of the revolutionary hero.

Stan labors and languishes in a slaughterhouse. The film is an affective critique of racial capitalism: "I'm just workin' myself into my own hell," Stan says at one point as the viewer is privy to his despondency as it escalates into a loss of subjectivity. Here, Burnett might agree with Marx, but only to an extent: it isn't that the worker is reduced to the status of the animal, as the animal in the film is exterminated, but rather that both the worker and the animal are mediated by premature death under the regime of racial capitalism. The worker does experience a tragic alienation and dispossession that attends the violence of racial capitalism—in this case, the extended and longitudinal—and seemingly interminable and always unbearably slow temporality of that violence as it is visited in the film, especially onto the character of Stan, who vocalizes his suffering.

Suffering and death in *Killer of Sheep* are instrumentalized by the anti-black technologies of capitalism. The children also figure into the film in a meaningful and intense way. The poor Black Los Angeles neighborhood of Watts is a scene of survival and also a kind of social inertia and static temporality of everyday life with a post-apocalyptic feel. The children play in dilapidated housing, with old toys, creepy animal masks of dogs, and around old train tracks. Even as there are moments of joy for the children in the film, it is never outside of the vortex of racial capitalist destitution and the haunting and lingering sense of despair—the atmosphere that Burnett conveys. It is, however, not an abject portrayal, but a scene of life and death that resists the aesthetic and cinematic registers that would either lionize or demonize Black working-class subjects.

In an interview, Burnett recounted meeting a slaughterhouse worker on the bus (the form of public transport most frequented in Los Angeles by workers, including migrant workers, who cannot afford cars). Much contemporary work, such as Shukin's *Animal Capital* (2009) named above, addresses the centrality of animal suffering and extermination to capital, not only through agribusiness and the factory farm, but also how film, in its very technology of production, is linked to animal death, since the gelatin used to produce film is historically derived from animal remains. But Burnett's film *Killer of Sheep* goes further, presenting animal disposability as immanent to racial capitalism, as in how the Black, Brown, and undocumented workers—like those taking the bus to the slaughterhouse—are central to the reproduction of racial capital. The exceptionalization of the animal by animal rights activism and animal studies ignores the ways that capitalism is a necropolitical project—part of the afterlife of racial slavery and the plantation. Leftist critiques of agribusiness and industrialized killing of animals and factory farming that ignore the raciality of capitalism while also analogizing the animal to the figure of the slave repeat the ruse of analogy, failing to think about how extraction and racialized bodies produce capitalism. The slaughterhouse in the film is part of, and a technology of, racial capitalism, which is the condition of its possibility.

It is the Black working class that is at the core of Burnett's film. "The film was never designed to be shown theatrically, it was responding to how people were making films about the working class and usually oversimplifying the issues they were involved in" (Evry 2007). Burnett is interested in the texture of life of the Black working class. Texture implies roughness, layers, substance, and resistance beyond gloss and one-dimensional renderings.

The Black working class is historically a trope for the academic and filmic gaze, generally not appearing in films or, if so, as tropes without the interiority that Burnett allows for. Working class life is not dramatic in Burnett's film in the sense of the spectacular or in the sense of a crescendo of a tragic plight and downfall of a character—usually following cultural pathologizing and/or sociological tropes of drugs, gangs, and the constraints and dilemmas of inner-city strife. There is a cottage industry for this latter type of gritty film figuring working class criminality that has also been prominent on television, especially series like *The Wire* and *The Corner*. In the academic sphere, the trope of the Black working class has been most visible in their portrayal in anthropology and urban sociology, dating back to Robert Park and the Chicago School, and to the infamous culturally pathologizing 1965 Moynihan Report.

Burnett takes an alternate theoretical and aesthetic avenue through the gesture of making a film about the Black working class that refuses a reductive or redemptive depiction. Burnett's vision of the Black working class is irreducible to a narrative that would figure them as monolithic and revolutionary proletariat; he likewise refuses a romantic and redemptive narrative by which the Black working class becomes a unified object of theory. Rather, Burnett opts for nuance and the quotidian. Burnett offers an alternative to the secular messianic temporality of revolutionary vanguard. His temporality is that of endurance and he draws the spectator's attention towards what happens in the interstitial spaces, an approach that the logic of scale (the party form—such as the vanguard) cannot account for.

Instead of militant rage, Burnett opts for a different affectivity of blackness. Burnett's directorial choices reroute the viewer away from the filmic and heroic figuration of the Black radical tradition that still is the dominant representational mode for Black masculinity especially, including the trappings of a cinema of racial uplift, and instead moves the viewer towards micro-affective circuitry of Black sociality under the regime of racial capitalism. Such a move can be understood in light of Sianne Ngai's *Ugly Feelings* (2005), in which Ngai directs the reader towards the aesthetic and affective impasse. Burnett alerts us to feelings that are non-recuperative, un-redemptive, and evade the register of revolutionary agencement, what Ngai terms "minor" or non-cathartic feelings that index situations of what she powerfully diagnoses as "suspended agency" (2011). Yet, what if agency, as Saidiya Hartman has shown through her work, especially in *Scenes of Subjection* (1997, on which more below), is never available to be suspended in the first place? How might such minor affects challenge the libidinal-as-political economy of anti-black violence? Furthermore, how might such minor affects refuse the very grid of intelligibility through which a revolutionary figure becomes legible?

The film also employs the multitude—or better, the swarm—rather than the figure of the leader. The many children who accent nearly all the scenes were, in fact, part of the filming. Burnett engaged the children in the process of making the film itself, in an act of artistic co-production and aesthetic experimentation as they joined with him by not only appearing in but also shooting the film as an intergenerational and undercommons-like invitation into the creative fold of Black radical cinema (Kapsis 2011, 9). This type of historiographic and cinematographic choice flies in the face of Hollywood and the racial respectability of heroism. Cedric Robinson's work *The Terms of Order* (1980) and Erica Edward's *Charisma and the Fictions of Black Leadership* (2012) dissuade the reader of the symbolic power of the electrifying leader and illustrate how social movements are generated and sustained, as the engine and hydraulics, by the people in the indeterminate and extra-individuated sense, as opposed to being held together by the cohesive glue and affective enthrallment of the figure of the leader. The figure of the leader works to occlude the Black radicality indexed by the problematic of "the people." The radical immanence of the people exceeds and precedes the singularity of the leader. Perhaps former president Barack Obama would best represent the contemporary figure of the race leader, or "race man" in Hazel Carby's political grammar in our present neoliberal juncture (2000).

It is against this narrative of leadership (nearly always appearing as heteronormative)—whether it be realized in the figure of the president or in the outlaw figure of the Black radical and of the revolutionary event as a singular and historical rupture instead of a perennial process—that Burnett counter-poses and stages another scene. Burnett taps into an infrastructure of black feeling in his depiction of Black working-class sociality. This narrative cannot be heroic since none of the figures are redeemable; there's no neoliberal and utopian version of hope that is salvageable, neither in overcoming racial capitalist quotidian life and slow death through respectability, nor through revolutionary comportment. In a way, the film depicts an impasse—what happens when both upward mobility and revolutionary potentiality are exhausted? What happens in everyday racial capitalist life when people are just trying to get by? What are the intimacies that are forged and sustained in unlivable settings? The characters in the film all make life within conditions of racial capitalist destitution and despair. Rather than the affective urgency and charged electricity of revolutionary fervor, we are instead often met with resignation. "Don't nothing ever want to make you want to smile?" Stan's wife asks him at the kitchen table. Yet, the children in the film are also joyful and resilient and the family endures; the grief that is felt is expressed through the animal as well as the people in the film, dramatizing how capitalism itself is a slaughterhouse.

Burnett's "Going Home"

A distinct economy of black affect is staged through the visuals but also the music of *Killer of Sheep*. There is a remarkable scene in *Killer of Sheep* where the anthropomorphic cinematic gaze is troubled and interrupted. At the site of the most intense violence levied at the animal, in this case, the sheep being slaughtered, the human subject is subtracted and taken out of the picture. We see the scene of sheep, a mass of them, channeled and funneled and walled into a chamber. They flow like a current, to their collective death(s), which are not shown. Two goats, who are termed in animal husbandry parlance "Judas goats" lead the sheep to the slaughter. "Lots of folks gathered there—all the friends I knew"; Paul Robeson's melancholy and heart-wrenching version of "Going Home" plays as the spectator watches what becomes the funeral procession of the sheep, already captive, now being led to their demise. The song acts as an elegy. Premature death in the film is not solely the condition of the captive animals; rather the foreclosure of Black life under the regime of racial capitalism bonds worker and animals together in the film. To close this brief reflection on radicality and animality, I want to turn to another insurgent genre: that of anti-prison and animal liberationist activism.

Collective Liberation

The film *If a Tree Falls: A Story of the Earth Liberation Front*, follows Daniel McGowan and traces the downfall of the activist group Earth Liberation Front (ELF) due to infiltration, surveillance, and neutralization (Curry 2011). McGowan became politicized by virtue of his own incarceration following his participation in a series of insurgent acts of environmental solidarity that, in historical parallel to the Weather Underground and the Black Liberation Army, involved acts of dangerous sabotage resulting in property destruction and intensive prosecution. One of ELF's actions was in response to the government capturing wild horses on federal land in rural Oregon and sending them to slaughterhouses. One of the slaughterhouses, located in Redmond, Oregon, became so oversaturated with horse blood from the killing floor, that it leaked into the water treatment facility: overkill turned overspill (Curry 2011). McGowan raises a profound question in the film about political violence and the violence of the political: "Why are we so gentle in our activism when this is what's happening?"

As in the political logic of the riot, wherein tactics of civil disobedience toward sabotage are deployed, though at a high price, to yield gains that were not possible through legal channels, the ELF members were able to interrupt, however only fleetingly, the slaughterhouse as usual. As Marshall Curry explains, "in one night they had accomplished what years of letter writing and picketing had never been able to do" (Curry 2011). *If A Tree Falls*, in dramatic documentary style, traces the implosion of the activist cadre as McGown and ELF go on the run, while the group also fragments due to dissensus about the extra-legal action and whether it proved just, strategic, or futile. Fugitive, McGowan is eventually captured, along with most of the remaining members of the ELF. McGowan is placed in controlled management units, which are racialized social control cells—prisons within prisons—designed for psychic and social deprivation, and to sever activist communication. When released, McGowan spoke out about this type of treatment as being a response to his dissident status, and was re-incarcerated, in violation of his already curtailed freedom of expression and, by the state's logic, his parole. He was then released again, as it was proven that McGowan was within his rights to critique the conditions of the prison system. Yet, the conditions of his parole barred him from participating in animal liberation and ecological justice organizing, even of the cast of PETA (People for the Ethical Treatment of Animals) and other liberal organizations that disavow animal and eco-liberation tactics and ideologies, and who were also, despite this disassociation, monitored during the period of the so-called "Green Scare" of governmental prosecution of animal and environmental groups in the 2000s.

The case of the ELF raises unique questions about human rights as much as the status of animal rights which I explore more fully in the following section. Moreover, the case illuminates the stakes and ethics of political violence and resistance; such questions can be found in social and political theory and philosophy, but hardly appear in animal philosophy discourse, which seems largely preoccupied with the ontological status of animality. There is an uncanny moment in the film where a sibling of one of the ELF members in the film tearfully and mournfully recounts confronting his brother, asking him why he chose to privilege the life of an indifferent animal over his own brother and family, knowing that he would likely go to prison for his actions and yet still choosing them. The inquiry reveals an interesting moment of tarrying with the negative—both in the sense of negating the nuclear family and re-configuring kinship through seemingly unilateral attachments to animals, and in the sense of disturbing easily digestible relations with animals by offering attachments that unhouse and dislocate one's coordinates of belonging.

Legal Subjects as Racial Subjects

The modern legal subject is also a racial subject. As Denise Ferreira da Silva diagnoses: "the racial figure of the human being plays the same ethical role for global capital that the notion of the nation played for industrial-state/empire-capital for most of the twentieth century. It allows the demarcation of whom falls on either side of the law, namely the protective and the punitive…" (2017). Therefore, while there are definitive political possibilities for rights-based redress that accompanies and enables the incorporation of the figure of the animal into the category of legal personhood and its corollary protections, what might the limits to legal personhood be? To what extent is personhood itself haunted by the specter of the slavery and its carceral afterlife? Saidiya Hartman's interventions are critical for conceptualizing the long durée of abolition. Hartman's work in *Scenes of Subjection: Terror, Slavery, and Self-Making in Nineteenth-Century America* (1997) discloses how the political ontology of "freedom" is slavery. Hartman also critiques racial liberalism, tracing its emergence not with the contemporary multicultural discourse of inclusion, but rather to a project of slavery's extension. Hartman asks, in the preface and interlude to her series of chapters that reflect on "the subject of freedom," "How did emancipatory figurations of a rights-bearing individual aimed at abolishing the badges of slavery result in burdened individuality?" (1997, 122). Abolition then, requires the undoing of freedom as well as slavery, and also undoing a rights-based freedom for animals that is premised on inclusion within, rather than a challenge to, the paradigm of carceral democracy.

Homo Narrans as Telling New Stories Beyond Violent Exceptionalism

To conclude my reflections on the entanglement of the genre of the human and its corollary of the animal, I return to Sylvia Wynter's theoretical embrace—which is also a political embrace—of "bios and logos" as the defining traits of the human (McKittrick 2015, 1). I read Wynter as making a nuanced argument that diverges from the tenets of linguistic anthropology and genetic reductionism. Wynter argues that humans are *homo narrans*, storytelling animals. While this is now the dominant narrative about the human, she does not simply treat it as exceptional but, rather, opens up the potential inherent in aesthetics and the art of storytelling and narration as the capacity to tell *new stories* other than the received or accepted ones. In short, Wynter treats *homo narrans* as a verb rather than a noun. Here, the human can be reverse-engineered and the dominant story—the story that writes the racial and colonial genre of the human into a cultural and sociopolitical script—can be reworked, edited, even abolished. I first address the limits of linguistic anthropology and the privileging of certain features—speech especially—as key to the idea that humans are unique in their "storytelling" capacities, given that this singularity, if true, is used as an alibi for exceptionality. Wynter allows for the possibility for *homo narrans* to be re-thought such that it does not reproduce the same violence since it is about changing the narrative as a creative act, versus treating the human and the story that articulates this figure, as fixed.

Homo narrans is troubled by Charles Burnett's films, where the animal is also an agent in the aesthetic genre whose sphere is historically seen as exclusively human—that of fiction and more generally, art. Are humans the only storytelling animals? Who decides what the signature of the human is and how is this always already a political act? What if the singularity of language or storytelling doesn't require exceptionality? What is the role or voice of the animal in the films about them? This is the landscape of "zoological surrealism" raised by the concept of "nonhuman cinema" which I would revise and consider a cinema of entanglement, given that the human presence is definitely orchestrating/directing (Cahill 2019). How are these roles subject to constant negotiation, failures, and contingency? Where in the story does the narration by the human, or the animal, begin and end?

"Homo narrans" has been treated as a noun—a socio-scientific fact that substantiates human exceptionalism—versus a verb, in the way I see Wynter activating it, which allows for the story to change. The logocentrism of the dominant version of *homo narrans* forecloses consideration of nonhuman animals as well as other nonhuman actants and narrative forms. Wynter's turn towards theory of mind, neurophilosophy, evolutionary biology, and linguistics has potential for deconstructing the meaning of "bios and logos" for the human. Contrast Wynter's approach with biolinguistics, cultural anthropology, and cognitive psychology (represented by thinkers such as Noam Chomsky, Daniel Dennett, Steven Pinker, respectively) which place storytelling at the core of human subjectivity in a move that reinforces the binary distinction between the human and the nonhuman animal. Yet language itself arrives with colonial associations.

For Locke, Rousseau, and Herder and other early modern language theorists, language was bound up with sovereignty and civilizationism. The cutting edge of Black studies lies at the theoretical horizon of the abolition of the racial and colonial genre of the human, seen and discerned with different inflections in the writing of Fred Moten and Stefano Harney (2004), Saidiya Hartman (1997), and most explicitly Frank Wilderson (2010). Wynter helps us to ask, "how do we change the story of the storytelling creature?" and I wonder, along with her, how does this expansion of storytelling change our ethics and our intimacies with other entangled and accompanying forms of multispecies life?

After the End of the Racial and Colonial Genre of the Human

The notion of animal as lacking storytelling and language capabilities is one that undergirds biolinguistic theory about language. I think we can extend Wynter in her reformulation of the creative potential of bios and logos, to disrupt the logocentric logic of biolinguistic theory. Breaking with the existing line of such theories, to extend Wynter we need to think about how the genre of Man is entangled with the genre of the animal—as seen in the entanglements of carceral, animal, and racial capitalism that is the subject of the films—and how her theorization has the capacity to serves as a critique of human—that is, of racialized and biocentric—exceptionalism.

Note

1 For a rigorous discussion of the triangulation of slave, foreigner (figured as "barbarian") and citizen in ancient Greece, see Arendt 2013, 31.

Bibliography

Arendt, Hannah. 2013. *The Human Condition*. Chicago: University of Chicago Press.

Brody, Richard. 2014. "Why Does It Matter If Heidegger Was Anti-Semitic?" *New Yorker*, March 27. www.newyorker.com/culture/richard-brody/why-does-it-matter-if-heidegger-was-anti-semitic

Burnett, Charles, dir. 1973. *The Horse*. 16mm film.

______, dir. 1977. *Killer of Sheep*. 35mm film.

Butler, Judith. 2019. "The Inorganic Body in the Early Marx: A Limit-concept of Anthropocentrism." *Radical Philosophy* 206 (Winter): 3–17.

Cahill, James Leo. 2019. *Zoological Surrealism: The Nonhuman Cinema of Jean Painlevé*. Minneapolis: University of Minnesota Press.

Carby, Hazel V. 2000. *Race Men*. Cambridge: Harvard University Press.

Curry, Marshall. 2011. *If A Tree Falls: A Story of the Earth Liberation Front*. Oscilloscope Laboratories. Film.

Derrida, Jacques. 2010. *The Beast and the Sovereign, Volume I*. Chicago: University of Chicago Press.

———. 2011. *The Beast and the Sovereign, Volume II*. Chicago: University of Chicago Press.

Edwards, Erica Renee. 2012. *Charisma and the Fictions of Black Leadership*. Minneapolis: University of Minnesota Press.

Evry, Max. 2007. "Killer of Sheep: Interview with Charles Burnett." *Blackfilm.com*, April 9. www.blackfilm.com/20070406/features/charlesburnett.shtml

Ferreira da Silva, Denise. 2017. "The 'Refugee Crisis' and the Current Predicament of the Liberal State." *L'Internationale Online* (March 8). Accessed February 22, 2021.

Foster, John Bellamy. 2020. *The Return of Nature: Socialism and Ecology*. New York: Monthly Review Press.

Hartman, Saidiya. 1997. *Scenes of Subjection: Terror, Slavery, and Self-Making in Nineteenth-Century America*. New York: Oxford University Press.

Heidegger, Martin. 1995. *The Fundamental Concepts of Metaphysics: World, Finitude, Solitude*. Indianapolis: Indiana University Press.

Kapsis, Robert. E., ed. *Charles Burnett: Interviews*. Jackson: University Press of Mississippi.

Marx, Karl. 2012/1844. *Economic and Philosophic Manuscripts of 1844*. New York: Dover Press.

McKittrick, Katherine, ed. 2015. *Sylvia Wynter: On Being Human as Praxis*. Durham: Duke University Press.

Moten, Fred, and Stefano Harney. 2004. "The University and the Undercommons: Seven Theses." *Social Text* 22, no. 2: 101–15.

Ngai, Sianne. 2004. *Ugly Feelings*. Harvard University Press.

———. 2011. "Our Aesthetic Categories: An Interview with Sianne Ngai." *Cabinet Magazine* 43 (Fall). www.cabinetmagazine.org/issues/43/jasper_ngai.php

Robinson, Cedric J. 1980. *The Terms of Order*. New York: SUNY Press.

———. 2020. *Black Marxism: The Making of the Black Radical Tradition, Third Edition*. Chapel Hill: University of North Carolina Press.

Scott, David. 2000. "The Re-enchantment of Humanism: An Interview with Sylvia Wynter." *Small Axe* 8, no. 120: 119–207.

Shukin, Nicole. 2009. *Animal Capital: Rendering Life in Biopolitical Times*. Minneapolis: University of Minnesota Press.
Sorentino, Sara Maria. 2022. "Slave/Animal/Labor: Marxist Incapacity and The Direction Analogy Flows." *The Comparatist* 46, no. 1: 29–51.
Wilderson III, Frank B. 2010. *Red, White and Black: Cinema and the Structure of US Antagonisms*. Durham: Duke University Press.

13 The Forbidden Turn

Decentering Modern Subjectivity by Integrating Animal and Mystic Affects

Balbinder Singh Bhogal

In order to not think about affect but to consciously allow affect to reframe what thinking itself is and can be, this paper argues for a reversal: to move from cognitive abstraction back to the affective body's iconic registers of instinct and intuition. This foregrounding and *a priori* nature of the body's perceptual sensorium and intuitive imagination implies decentering the self—narrowed to "I think" by secular modernity. Such an orientation agrees with the Sikh Gurūs' view that unselfing begins with listening not only to one's body but all bodies as one's own, thus forming a reality where the world and Word are not two things—for every form communicates. Self-loss, however, constitutes a "forbidden turn" even beyond the affective and participatory turns of recent scholarship. These perceptual orders are re-framed under the metaphors of the "animal" (*pasū*) and "mystic" (*gurmukh*). Their I-less instincts and intuitions profoundly trouble representational epistemologies of the human and the humanities. The direct communication-as-communion of the *pre-epistemic semiosis* of the "animal" and the *post-epistemic soteria* of the "mystic," that unfurls beneath conscious egoic awareness, may then not only trouble normative ways of knowing but also offer a decolonized and pluralized ontology marking a new point of departure.

Beyond Knowledge or Why There Are No Animal Libraries

> Now what should I think? I have given up thinking.
>
> (Gurū Arjan, Gurū Granth Sāhib (GGS) 1209; trans. Bhogal[1])

> I have not tried to do anything through knowledge.
>
> (Gurū Arjan, GGS 894)

> That life which has no contact with the mystic/mystical [*sādhū*] is useless…[2]
>
> (Gurū Arjan, GGS 810)

I have argued elsewhere, and assume here, that the "human" is a liminal vector *in medias res* between the instinctive "animal" and the intuitive "mystic"—neither

DOI: 10.4324/9781032659985-17

of which order the world through abstract knowledge and therefore lack the need to store data or information; they live affectively and immanently (Bhogal 2012). Three academic lacunae structure this chapter. Firstly, the omission of animal communicators and contemporary mystics from what is considered the purview of knowledge. Secondly, Heidegger's elision of naming the mystic in his charting of the differences of the animal from human failing to notice the startling overlap between animal and mystic qualities. Thirdly, the fact that secular modernity, which privileges discourses of human exceptionalism and rationalism, leads to a double erasure of the *muted* animal and the *ghosted* mystic. Whilst recognizing the welcome corrective that the Affective Turn has proffered, I here argue that its focus on the animal's "pre-epistemic *semiosis*" reveals a reiterative secular bias at the expense of the mystic and their "post-epistemic *soteria*."

"Animal" (*pasū*) and "Mystic" (*sādhū*, *pīr*, *sant*, etc.) are here used as umbrella terms and metaphors signifying a range of diverse subjectivities describing broadly two orders of iconicity.[3] Both terms figure—the former unconsciously, the latter super-consciously—a connectedness and spontaneity that the *manmukh* or "ego-centered-One" misses, given his dependence on representational language. Unlike the human ego-idol, the animal and mystic icons lack origin and cannot be owned. Both enchant precisely because they precede and exceed any naming logic; they confound our desire to totalize or order sense-data into a known fixity, remaining firmly undecidable, incommensurable, and open to a livingness beyond knowledge. Both render the invisible visible (as icon) rather than reduce it to the visible (as idol). Together, their aporetic natures challenge the hegemony of the universalized "I think" as the ground of being. And yet, though much scholarly ink has been spilled on the absentee animal, much less has been enunciated about the forgotten mystic, and barely a jot on how the two orders of iconicity may be differing sides of the same affective coin.

With the focus on affective living and perceptual/intuitive orders, here aligned with Gurū Nanak's concept of truthful living (*sac-ācār*), I use this chapter to challenge the scholar's sacred cow of the *thinking self* as the sole ground of valid knowledge. By re-engaging pre- and post-epistemic (re)sources, I argue, we may be better able to dethrone the divisive, colonial, and secular authority of the racialized and gendered "I think" that disguises as neutral a Euro-American geography of knowledge production. Humbling egoic subjectivity through contact with animal and mystic subjectivities may aid us in relocating ethics beyond thought to feeling and to a participatory communion of liminal selves: *bhagati* (loving devotion), like many religious contexts, often dictates love rule over knowledge. The Western hierarchy of (symbolic) cognition over (semiotic/soteric) perception is not only overturned, but integrated and re-balanced more equitably where feelings and affects are restated as the foundation of thinking.[4] The point is not merely to undermine and humble our inherited epistemologies but to acknowledge pre- and post-epistemic affective sources and their attentional intensities as essential resources to inform thinking.

Namely, I suggest that before, within, and beyond human knowing exist the *pre-epistemic semiosis* of the "animal" and the *post-epistemic soteria* of the "mystic"—together forming iconic orders of unmediated communication. Both animal and mystic inhabit an ahuman/nonhuman temporality, a liminal Word/space/time beyond the thinking self's abstractions framed by secular time. The consequential unselfing of the mystic thereby illustrates the symbiosis between Word and world as an "ecology of participation" (Cabot 2018) such that an "ontological flooding" (Hunter 2015, 16) of multiple subjectivities and worlds may occur, displacing the ethnocentric and epistemic stranglehold of colonial modernity.

Rejecting the idea of "sacred" language(s), I am using the religious *Word* (synonymous with *Life* and *Intelligence*) as a universal that includes all languages and all forms of communication, verbal (cognitive) and non-verbal (affective), conscious and unconscious. This is in line with, and will be anchored through, the Sikh case of the Word (*sabad*). As noted in the epigraph, Gurū Arjan consciously stopped thinking, having integrated the human mind back into the animal body to reveal an "enlivenment" (Weber 2013) beyond European "Enlightenment" achieving the perfected state of the political mystic (*gurmukh*), exemplified by the Sikh Gurū as the true king (*saccā-pātisāh*).[5]

Knowing Life (Anthropomorphically)

The animal-mystic consciousness transcends the subject-object duality that forms the very basis of Cartesian knowledge. Unavoidably, then, there is always something fishy (or perhaps not fishy enough) about thought as it tries to "know life" and reduce it to anthropocentric meaning.[6] Rather, as Boaventura de Sousa Santos elucidates, "[a]t the core of ecologies of knowledge is the idea that different types of knowledge are incomplete in different ways and that raising the consciousness of such reciprocal incompleteness (rather than looking for completeness) will be a precondition for achieving cognitive justice" (2014, 212). Is it in fact, then, possible to *know* life when life outstrips knowledge at every turn? Does not the longevity of the affective arts (including poetry, music, and ritual) speak to this very limit of knowledge production? The relatively obvious step of transforming rather than adding to our limited epistemology in the presence of other limited epistemologies by questioning the sovereignty of the "I think," and by deepening our subjective perception via some praxis, is persistently missed if not unconsciously resisted.

Life may have a single source, but it is not other than the diversity that blooms within or without. This is recognized within the Sikh hymns themselves: *ek-anek*, "One is not One," "One is all," hereby offering a radically alternative epistemology in that knowledge itself cannot avoid contradiction. At the root of all being then exists not a simple or distinguishable Oneness but a radical interdependence that does not take away from the iconic incommensurability of each being's sovereign existence: *ek-anek* or "Oneness" as "One-many-ness" is ineluctably mysterious, aporetic.

This is not to say that communication between beings is an impossibility but rather that, in recognizing we are strangers both to ourselves and others, we have the capacity to resonate with others in multiple grammars of relation and exchange, making a plethora of epistemologies and ontologies not only possible, but natural, and perhaps necessary. In recognition of this incommensurable plurality in knowing life, one must shift from the knowledge-identity-meaning paradigm to a translational frame of ethical encounter, where a reciprocal grammar of listening and perceiving (iconicity) sideline unidirectional law and naïve classificatory correspondence (idolatry/ideology). For an ethics of translational listening, an *unselfing* (*āp-gavāi*) is required. That unselfing, however, constitutes for modern subjectivity a forbidden turn to the private as *also* authoritative.

The incommensurability of life discloses the nature of representational knowledge to be provisional and inherently myopic. Knowledge taxonomies can never be wholly severed from analogy and imagination, yet anthropomorphism may—despite its obvious limitations—offer limited but illuminating insight into animal/mystic being. As Daston puts it, "[u]nderneath the smooth word 'anthropomorphism' are hidden a multitude of *anthropoi*, of kinds of humanity—in this case, of kinds of mind—as well as the multitude of *morphoi*, of shapes of understanding other minds" (Daston and Mitman 2005, 51).

This reading of anthropomorphism beyond mere projection or incommensurability is clearly an important though striking one. For their continued absence in academic discourse are those who have long claimed to know "the lived experience of other life forms"—that is, the mystics, shamans, and animal communicators who not only imagine but psychically or telepathically intuit the worlds of biodiverse beings. This is the advantage and real enchantment of anthropomorphism beyond anthropocentrism: for the "animal" and "mystic" crypts are not singular entities or classes, but diverse congeries—the biodiversity of material beings matched by a biodiversity of psychic beings, moods, voices and images.[7] This communication (as communion) of the *Word as world* in both animal communicators and mystics, is a mode of engagement resistant to the sedative forces of knowledge (words *about* the world).

The following three sections—each noting the affective, participatory, and forbidden turns respectively—reflect classifications taken from the study of mysticism: knowledge *about* (objective thinking, descriptive, indirect), knowledge *through acquaintance* (subjective feeling, participatory, direct), and knowledge *by merger* (unselfed becoming, ever-fresh immediacy that enters into the Being of beings).[8] My argument is that both the affective and participatory turns, despite their corrective moves, nevertheless maintain the subject-object duality that sustains egoic consciousness, and it is only in a further turn—as yet proscribed by secular academia, namely of self-loss—that this duality can be fundamentally challenged.

That is to say, the egoic subjectivity of "I think" still remains the ground of knowledge production in affective and participatory studies; only in the Forbidden Turn is the ego dislodged to experience subjective liminality. This praxis opens the door to potential shifts, not only from thinking to feeling, but from knowledge to unmediated communication-as-communion to the point of "revelation" of the "inner-knower" (*anatarjāmī*) expressed in a variety of ways from intuitive awareness to transformative insight (*surati, samādhi, anabhau, giān, drisaṭi*).

Affective Turn and its Cognitive Limit

Frédéric Neyrat's *Atopic Philosophy* (2017) helpfully encourages readers to begin with "*difference starting from existences, and not identity starting from objects*" (7, emphasis original) but nonetheless this, for Steven Shaviro who writes the foreword, is a move towards "a new way of *thinking*" (xi, emphasis added). This "ability" for thought/philosophy "to dig holes, and open up gaps, in what is otherwise a suffocating (and even totalitarian) world of hyper-presence" (xi), makes *Atopic Philosophy*, Neyrat argues, "a work of ethics" (xii). I remain unconvinced by this claim, given that thinking itself is not radically decentered. I have previously argued against the periodic though undying erections of "new thinking"—deemed "different," "radical," or even "revolutionary"—that push towards thinking's de-ontologization and decolonization (Bhogal 2021). Below, I argue that even affect thinkers are still captured by *thinking about*—making the notion of an affective turn moot, because it is not existentially embodied. No matter how much distance Neyrat's atopic philosophy puts between its own disruptive nomadism and fixed idealism, thought is still the liberating tool par excellence. Surely, starting from existences and not identity demands a suspension of thought such that affective modes of opening, connecting, and listening may be engaged? Isn't *being* prior to *affect*, and *affect* prior to *thinking*? We are not born speaking but crying (before a sense of self is formed).

In a short section titled "On the Subject of Animals," Neyrat argues for the subjecthood of animals, countering the humanist and anthropocentric biases of Western philosophy:

> This theoretical proposition is necessary to counter a humanist and anthropocentric mode of civilization. It is, however, insufficient, even harmful, if it ultimately maintains the figure of Man behind the subject. Is it not from the concept of the subject itself that it is necessary to depart, insofar as this clings to the humanist schema?
>
> (2017, 47)

This important insight—that it is not enough to make subjecthood more inclusive but that it is subjectivity itself that needs to be troubled—is central to the religious/Sikh task of dismantling the egoic "I think" for is it not thought itself that "maintains the figure of Man behind the subject"? Neyrat, however, ultimately pulls back from the promise of this radical insight in returning to Wittgenstein:

> If a lion could speak, it would be time to find some alternative between an immediate pseudo-comprehension that annuls alterity by translating it into human language and an *a priori* deafness, stripping in advance the capacity for generating their own ways of subjectivation from all other animals.
>
> (2017, 50)

This precisely exemplifies the problem with thinking as a tool of analysis and not acknowledging or engaging affective and intuitive skills as modes of communication: a tracker is not caught by this intellectual quandary, not least for the fact that her comprehension is not considered a falsehood (pseudo). Neyrat's in-between remains an effect of thought itself, and not a reflection on what is actually (affectively) possible.

Why assume that by translating animal affects into human language that this would automatically "annul" their "alterity"? The communication of affects, though ever-changing and incomplete, carry practical meaning in their embodiment. What makes us believe we can make abstract sense out of this infinite sea of affects and represent it in thought without confabulation? Perhaps, it is not that thought gives one access to the outside (*dehors*) but rather *creates it*—and hence why the figure of the thinking subject feels so impossible to leave behind. This impossibility is tempered, though, once we turn to the Indic traditions in which these thought-worlds are recognized as illusory projections (*māiā*) to begin with, where truth is often delinked from "objective description" and tied to the subjective transformation of awakening.

The new materialists and the new animists have rightly complicated the simplistic reading of religion-as-belief, charting the complexity of forces affecting and effecting the vulnerable "I." However, they do not venture beyond the "I" and consequently restrict and reduce epistemology to what the "I" can think. This point is not simply a tangential one but targeted at the very foundations of this long-held academic prejudice against epistemologies beyond the "I think."

Jane Bennett's clarion call for an "aesthetic-affective openness" that "stretches received concepts of agency, action, and freedom sometimes to the breaking point" promises the decentering of the rational human (2010, x). She writes:

> If human culture is inextricably enmeshed with vibrant, nonhuman agencies, and if human intentionality can be agentic only if accompanied by a vast entourage of nonhumans, then it seems that the appropriate unit of analysis for democratic theory is neither the individual human nor an exclusively human collective but the (ontologically heterogeneous) "public" coalescing around a problem.
> (2010, 108)

The closeness of this "public" to Sikh *ek-anek* heterogeneity is encouraging; Gurū Gobind Siṅgh chants of "One Form perceived through many perspectives" (*ek mūrat anek darasan*).[9] However, in Bennett's language of bearing witness to the vital materialities around us, she is still positing a viewing subject, no matter how entangled this subject may be with other subjects (whether nonhuman animals, rocks, trees, microbes, or tools). Recognizing matter as lively or vital is crucial, as too the expanded understanding of personhood in new animism, but what mystic subjectivities disclose is not only recognition of multiplicity from the outside but absorption with them on the inside—a perpetual contact like a gushing river of affect beyond knowledge—in an ever-fresh immersion. Indeed, it is this very image that is used by Gurū Nānak: "You are the River (of Life): all are within You" (GGS 11, M4). Thought can't take us to emergent affects, no matter how sophisticated and inclusive our theorization.[10]

Bennett elaborates her notion of "public" via Karen Barad's important distinction between "reflection" and "diffraction." Reflection assumes an ontological/epistemological binary of knower/known that separates *interacting* "things" (forming a second-order reflection); diffraction assumes an onto-epistemological pluralism of *intra-acting* "phenomena" (of a first-order nature). Reflection attempts to find accurate representations through a gaze from afar, a discrete subject surveying a field of objects, while diffraction assumes mutual responsibility in recognition of an ethics of entanglement and an ontology of contingent co-emergence. It is diffraction that moves us away from the indirect seductions of representationalism closer to the mystic's third direct, immediate knowledge of merger, and to a public marked by affective intra-action of humans and nonhuman assemblages.

Though Bennett talks convincingly about the need for "a cultivated, patient, sensory attentiveness to nonhuman forces operating outside and inside the human body," this "attentiveness to things and their affects" (2010, xiv) stops short of an active cultivation of liminal modes of absorption or self-loss that would allow that direct contact beyond the meditated subject-object epistemology. Indeed, "in egoism one cannot remain awake and aware" (GGS 1346, M3). Rather, Bennett calls on others "to devise new procedures, technologies, and regimes of perception that enable us to consult nonhumans more closely, or to listen and respond more carefully to their outbreaks, objections, testimonies, and propositions" (2010, 108). But do we not already possess these "regimes of perception" in the previously existing figure of the mystic across a variety of traditions (Yogic, Bhaktic, Tantric, Siddha, Sūfī, Sahajīyā, Sikh, Jain, Buddhist, Śaivite, Vaiṣṇava, Śakta to name just Indic ones)?

Here, I am following Zayin Cabot's correction of monist materialism, i.e., "ontologically one, formally diverse," to a pluralist flooding, i.e., "ontologically diverse, formally one" (2018, 252). Resonating with the Sikh notion of the affective and processual Word as world, Cabot argues that reality is not independent of our perception; there is not one world, nor even many worlds, but a process of worlding, becoming, or concrescence: "I do not think reality is given at all. Rather, it is co-created. And I feel language is a key that unlocks diverse forms of participation. Language is code for the mysterious creativity whereby meaning, lived experience, and worlds come into existence" (Cabot 2018, 11, 131, 135, 168–70). Compare Gurū Nānak's "In egoism does the world emerge" (GGS 946, M1).

Importantly, Cabot levels a critique of recent scholarship that shifts, as I frame it, from *thinking-about* things only then to continue to *think-about* feelings: "The new materialists, new animists, various continental thinkers, and many affect theorists tend to assume an animist ethnocentrism" (2018, 147). He includes here works by Jane Bennett (2010), Mel Y. Chen (2012), Donovan Schaefer (2015), Manuel Vásquez (2011), and Viveiros de Castro (2014).

> Each of these authors means to defend the importance of diversity in our planetary setting, but not of people, insides, or subjectivities. They either take these for granted and/or assume a shared interior (Culture), and then defend diverse bodies (natures). This is interesting, in that it aligns with my ecology of

> animism, but it is also disturbing. They have clearly subsumed diverse cultures within the context of a singular Culture.
>
> (Cabot 2018, 149–50)

This animist ethnocentrism wrests "power away from the liberal self" but then limits the field of agency to the material by claiming "power is not found in angels but in animality" (Cabot 2018, 163). This privileging of the animal is also matched by a privileging of affect—the "diversity of emotional bodies," rather than a more extensive "diversity of selves" that may include non-affective bodies (164). By its animist ethnocentrism and monist materialism, theorists of affect and animal studies generally unveil their undisclosed secular positionality that ignores religions and their onto-epistemic worlds—the very pluralism Barad's diffraction arguably calls for. From the Sikh perspective of the world as Word, reflection (words) would need to integrate diffraction (Word—*sabad*); the cognitive knowing-about of the ego (*haumai*), would need to contact the affective sensorium and intelligence of life itself (*hukam, sabad, gurū*).

Participatory Turn: From "Landscape" to "Taskscape"

How did thought come to be viewed as *the* singular authority, such that merely *thinking about* an object was sufficient for "valid" *knowledge* of it? Why does thinking remain *the* authoritative architecture even by those who see cracks in its foundations and holes in its walls? Steven Shaviro has argued that this aggrandizement of thought precedes Descartes and Kant and is rather founded in George Berkeley's "correlationism"—that "things cannot exist without being thought or perceived" (2014, 108–9). The irony that Cabot's view seems to reflect Berkeley's is not missed here and shows how similar orientations can lead to diametrically opposed conclusions. Shaviro, a speculative realist, argues instead for "noncorrelational thought," which occurs below the cognitive-conceptual level by refusing the distinction between consciousness and its object. In elaborating noncorrelationism (similar to Barad's diffraction), Shaviro shifts to an affective register, which he reframes as "sentience" that is "nonintentional and noncognitive," and better understood as "feelings rather than articulated judgments" (2014, 131). So why then persist in calling it "thought" if thought is the wheelhouse and prison of thinkers?

Thinking further about sentience, Ralph Acampora's "pre-epistemic semiosis" speaks to the affective work that goes on before, beyond, beneath, and within thought. Shifting from an "objectivist theory of meaning to an embodied/experiential view," Acampora defines pre-epistemic semiosis as a "sensibility of common creaturehood," wherein "sentience itself can be shot through with semiosis," from the "perceptual, emotional" to the "imaginative," i.e., "a pre-linguistic 'bodily logos' that builds on the evolutionary basis of species-typical patterns of motion, gesture, sound, and posture."[11] Still, *thinking* about sentience is not to directly *engage* sentience, nor should the "embodied" be reduced to merely the animal, by forgetting the mystic's transformed consciousness. The limits of intellect (commonly framed by a subject-object duality) reveal themselves in the fundamental binaries that modern secular thought keeps returning to, whether or not it is dressed in new garb. Thought remains inextricably bound to individuated subjectivity, and hence, the move to affect appears to be a shuffled reiteration from the deck of "I think."

Tim Ingold, however, arguably provides us an advance. Having noted through his fieldwork how the subject-object binary is a "western projection and formation," he argues that perception and action reflect the inseparability of biology (body) and culture (mind), noting how "practical enskilment" is "conceived as the embodiment of capacities of awareness" (2000, 4–5). Moreover, the discipline of learning new perceptual skills and capabilities develops a "kind of sensitivity and responsiveness"—what Ingold calls "intuition" and "feelings"—forming a "sentient ecology" that is "both pre-objective and pre-ethical" (25). Ingold contrasts this dynamic "poetics of dwelling" with a Western view of subjectivity as a constructing force akin to "building":

> The contrast, I repeat, is not between alternative views of the world; it is rather between two ways of apprehending it, only one of which (the Western) may be characterised as the construction of a view, that is, as a process of mental representation. As for the other, apprehending the world is not a matter of construction but of engagement, not of building but of dwelling, not of making a view *of* the world but of taking up a view *in* it.
>
> (2000, 42; emphasis original)

The former tends towards a "landscape" in which one constructs meaning and truth before one acts in the world; the latter tends towards a "taskscape" in which meaning and truth are disclosed as one acts—living without a library of knowledge as animals and mystics demonstrate. Apart from neatly summarizing a key thematic difference between Abrahamic-Western *theoria* and Asian *praxis*, Ingold writes, "it is through such *attentive engagement*, entailed in the very process of dwelling, that the world is *progressively revealed* to the knowledge-seeker" (2000, 216; emphasis added).

Here we have the beginnings of a secular expression of revelation, or what I am calling post-epistemic *soteria*. Ingold notes how an indigenous hunter tracking an animal through an attentional intensity *felt* the overwhelming presence of the animal "as if his own being were somehow bound up or intermingled with" it, "a feeling tantamount to love" (2000, 25). I see a direct resonance between the animal-becoming of the hunter with the mystic's "becoming other"—though the latter is here considered a conscious harmonization, a *merging with* rather than a perceptual *emerging of*. This leads us to what I have introduced as the Forbidden Turn. Below, I build upon the insights of the affective and participatory turns, whilst also recognizing their limitations—i.e., the remainder of a secular bias that keeps alive the "I think" and a persistent exclusion of the mystic, whether by positing a monist materialism or an ethnocentric animism. This Forbidden Turn seeks to extend Acampora's notion of the pre-epistemic semiosis of the animal's "bodily logos" to the post-epistemic soteria of the mystic's "psychic logos" and thereby recognizes the implications of the ontological flooding of the Word as world.

The true import of the affective registers of the animal and mystic is the liminal delight and terror of unselfing. As the fifth Sikh Gurū Arjan sings, "The thinking self (*ahambudh*) and love of illusion (*māiā*) are the most severe of chronic diseases. Hari's [God's] Name is the panacea the Gurū has given me as the yoga (*jog*) to be practiced" (GGS 502, M5).[12] To slay "I think" (via *nām-jog* for example) is to turn away from the familiarity of egoic habits by looping representation back to encounter (i.e. within the sentient ecology of animal and mystic communications).[13] To better explore the Forbidden Turn, it is necessary to sketch out the Sikh view as an exemplary case depicting not only how the world is actually a constant communication of the Word, but also how its communication pivots on unselfing through an attention that re-members the animal and mystic grammars dwelling within the body beckoning enlivenment through resonance.

Contacting Oneness Through the Sikh Word (*Sabad*) by Unselfing as Listening

It may be helpful to differentiate *Sikhī* (a continuous tradition of dynamic learning and engagement) from *Sikhism* (a static reified reading of tradition). Within Sikhism *ikk oaṅkār* is often translated as monotheism but in Sikhī *oaṅkār* is better understood as the singular Being of beings—both revelatory Word (*sabad*) and revealed World—thus meaning not "One God" but the "One-Many Resonance." This Being of beings is unseen (*purak alakhai*). It is unfathomable, mysterious (*alakh abhev*); it is non-possessable incomprehensible, invisible, and inscrutable (*agam agochar alakh abhev*). Yet mystics can contact this Formless Being merging within their own formed being, and throughout this section, I will incorporate many descriptions of this contact recorded in the *Gurū Granth Sāhib* (GGS) by its different cross-traditional authors (Sikh Gurūs, Sūfī Pīrs, Bhagats and Sants). By merging with this Oneness, the Gurūs perceived, sensed, and intuited this imperceptible Being of beings: "Wherever I look, I see that One (*ekai*) alone. Deep within each and every heart, He Himself is contained…With my eyes, I perceive the Embodiment of Bliss to be everywhere" (GGS 387, M5). The Sikh One is always Many (*ek-anek*).

Approaching this contact with Oneness through the World's diversity is a humbling and educational experience and, as such, forms the central aim of Sikh praxis: to move from a prideful ego (*manmukh*) towards a selfless servant (*gurmukh, dās*). This transformative movement rests upon the recognition of there being One Awareness (*surati*) among all created beings and that this awareness is both inherent to creative Being and created beings, i.e., Word and world (GGS 24–25, M1). "Hari Himself is the Word; He Himself is the awareness, attuned to its music" (GGS 165, M4).[14] It is this very awareness that makes the affective contact of the mystic possible. Given that Oneness "pervades all places and interspaces" (GGS 269, M5), and that "whatever is seen is His form" (GGS 723, M5), both "within the body and outside it, [we may come to] know that One" (GGS 1343, M1). Hari, the Being of beings as Word is simultaneously the World.

The Being of beings, expressed in the GGS, is thus "all forms yet formless" (*ākār-niraṅkār*), "One yet All" (*ek-anek*) (GGS 238, 250, M5), "concealed yet revealed (*soī guputu soī ākāru*)," "with attributes yet without attributes (*saraguṇa niraguṇa thāpai nāu*)" (GGS 387, M5). The *affective* existential Oneness of Sikhī is composed of a nondual awareness—such that Form, the Formless, and the Word cohere in a single living unity of *surati* (awareness) (Bhai Nand Lal *Rahit Nāmā* 10 in Ganda Singh 2000, 231).

In Sikhī the human tragedy is that this nondual awareness is forgotten, its having been occluded through a process of individuation that hardens into "the veil of egoism, (*haumai*)". This ego-driven forgetfulness is compounded by the "fascination and attachment to illusion (*māiā*)" that puts all into a "hypnotic somnolence" (GGS 205, M5), sundering that single all-encompassing point of Oneness into a binary subject-object conception (*dubidhā*) where "I am me (not you!)." The awareness of Oneness, the Being of beings, is split and suppressed during the process of individuation leaving all to suffer.[15] The Sikh Gurūs explain it thus: "Wherever I look, I see the very same pain (of 'I')" (GGS 1153, M1). That is to say, when "afflicted by ego, duality arises" (GGS 893, M5). The one who turns away from ego desires, who understands their seductions as ultimately unfulfilling (*māiā*), may turn to the Word to uproot suffering that follows from attachment.[16] Duality is overcome via a sensorial affective and psychic merger (*liv lāi*) recalling that always-already-existing One Awareness (GGS 1185, M5). As the first teacher Gurū Nānak states, "Eradicate duality, then recognize the Word" (GGS 1343, M1), and "Destroying the ego, one finds the Gurū's Word" (GGS 228, M1). Word and Awareness are, therefore, inseparable in awakening:

> When one's light merges into the Light, and one's awareness (*surati*) is joined with the Awareness (*surati*), then one's violent instincts (*himsā*) and egoism (*haumai*) depart, and skepticism and sorrow disband. Hari (Being of beings) abides within the mind of the mystic (*gurmukh*), who merges through the Gurū.
>
> (GGS 21, M1)

Contrary to the modern *thinking* subject then, affective modes of non-thinking and spontaneity are valued to aid this recollection-as-merger. Indeed, one could chart a movement from the rational ego to the affective self; for within man awaits the woman (*nārī*), but within both awaits the innocence of the child (*bāl*, *bālak*). Yet within the child lies the animal (*pasu*), but even within the animal, the ever-fresh mystic consciousness resounds (*cautā-pad, nirbānī-pad, dasam-duār*, *joti sarūp*) (Bhogal 2012, 12–18). This is not a movement from outer form to inner formlessness, but to re-experience form from the depth of the formless, experience time from the vantage of timelessness (*akāl-mūrat*). Many other metaphors and analogies are used that break open the abstract delusions of "I think" that the Gurūs identify with: the fool (*mūrakh*), blockhead (*mugadh*), idiot, stupid, (*iānā*) beggar (*mangat, jācik*), dumb-struck, mute (*gūngā*), insane (*baurānā, divānā*), intoxicated (*mātā, rucit*).

Interestingly, the esoteric vocabulary and psychic powers of the jogis/yogis, (*siddhi, dasam-duār, cauthā-pad*, etc.) are not dismissed per se within the GGS but rather realigned to the childlike innocence and spontaneity of the bhagats (lovers of the divine) and their vernacular terms of *acint, acet*, and *sahaj*. The innocence of the child's mind (*bālak budh acet*) is extolled within the GGS as it is seen to reflect a more attuned level (GGS 75, M1). The child-mind is contrasted with the ego-burdened mind (*manmukh acet*). The child-mind or beginner's mind is carefree but spontaneously absorptive, resisting an over-identification with thought (GGS 214, M5). It is in light of this appraisal that Gurū Arjan's eulogizing of *acint* or non-thought as a natural spontaneity should be understood: "Spontaneously, I take His Name...am saved by the Word...my works are perfectly accomplished...my enemies have become my friends...brought my mind under control...I am rid of hatred...I behold God in each and every heart...my doubts are dispelled...all forms of discernment have welled up within me" (GGS 1157, M5).[17] Here *acint* (non-thought) rather than thought is where true (i.e. nondual) discernment arises, accompanied with great bliss (GGS 371, M5).

Like the spontaneous, carefree non-thinking of *acint,* the term *acet* is used positively as a childlike *sahaj* (the natural unburdened self-described above), but also negatively as delusionary ignorance when referencing the adult's ego-identified mind (*manmukh acet*). This latter unconscious mind must be awakened (GGS 1110, M1). It is this awakened mind that we have referred to as the mystic—less a mode of critical consciousness than *transformative* consciousness arising from the spontaneity of the intuitive mind that contacts *surati* (the undifferentiated awareness of oneness).

> The Gurū is the River, from which pure Water is obtained forever; it washes away the filth of dualistic thinking. Finding the True Gurū, the perfect cleansing bath is obtained, which transforms even beasts and ghosts into gods.
>
> (GGS 1329, M1)

> Attaining effortlessness [*sahaj*] in an instant, doubt is dispelled from within; meeting Him, light merges in Light.
>
> (GGS 205, M5)

This is how the unconscious become conscious, through the taskscape of Sikh affective dwelling, which enables one to contact this *self-rejuvenating* flow of the poetry of life (GGS 177, M5). The Sikh sentient ecology is formed by embodying *simaran* ("mindfulness"), *sevā* ("self-less service"), *sat-saṅgat* ("association with the true assembly"), and *kīrtan-kathā* ("singing scriptural hymns with exegesis"). Importantly, the mystic's transformative merger reveals a psycho-somatic liminal spatiality. As Gurū Arjan sings, "Twenty-four hours a day, the wind breathes Your Name...In the four sources of creation, and in all speech, You dwell" (GGS

130, M5).[18] Via the Word, the outside is inside and vice-versa: "The Creation is in the Creator, and the Creator is in the Creation" (GGS 1350, Kabīr). Yet, the main wall (*pāl*) that temporarily disguises this liminal nondual dynamism is the ego's subject-object consciousness; the baby develops the protective shield of the ego structure through the acquisition of desire and language to compete for resources and defend itself from others (Bhogal 2012, 14–17). As instrumentalists, all egos "make countless efforts for pleasure; no one strives for pain" (GGS 1428, M9). As a result, all egos suffer their own thinking for none can avoid pain nor secure pleasure indefinitely.[19]

> O Mind don't be so proud of yourself as if you know it all; the Gurmukh is humble. Within intellect resides ignorance and I, through the true Word their filfth is cleansed. Be humble, surrender before the True Gurū, don't perceive only yourself.
>
> The world is ablaze with ego and pride; see this, before you also lose yourself.
>
> (GGS 441, M3)

The only way out is within to contact this psycho-somatic liminal Word, beyond the ego structure, "God/Ram is in the soul and the soul is in God/Ram… recognized through the Word suffering is dispelled; 'I' is killed" (GGS 1153, M1).[20] This Being of beings lies "deep within each and every heart, always awake and aware" (GGS 868, M5), and open to all those who humble themselves. Wisdom (*giān*), not-I knowing, opens in humility; knowing closes in ego:

> Through the True Gurū's Word the Primal Being [of beings] is revealed;
> so attune your consciousness to His Word.
> Hearing the Gurū's Word my mind has been softened and saturated with it
> and returns to its own home deep within.
> The Unstruck Melody resounds there continuously,
> the stream of nectar trickles down constantly.
>
> (GGS 1308, M4)

The world comes out of the Word such that reality is a dynamic simultaneous process of "Wor(l)ding" revealing multiple epistemologies and ontologies.[21] Yet two primary modes of human "dwelling" (borrowing Ingold's term here) emerge: (1) the selfish mode of the ego-centered *manmukh*, and (2) the unselfing mode of the gurū-centered *gurmukh* who constantly returns to the Word/Name.[22] *Manmukh* and *gurmukh* are not two beings in the world, but rather two modes of being and apprehending the world (the former being dualistic, the latter nondual). Denying the truth of death in the World, the egoic *manmukh* vainly desires eternal life; the *gurmukh*, who relinquishes desire in yearning for the plural One, faces death squarely, and is gifted *eternal birth*—which yields an ever-fresh subjectivity in tune with *hukam*, or the natural, self-sustaining intelligence that informs all life.[23] It is this that constitutes awakening: "When I turned away [from Worldy attachments], and became dead-while-yet-alive, I was awakened [to the Word]"

(GGS 221, M1). This is the integrative ideal of *jīvat-marai*:[24] to be dead-to-the-ego-*while*-alive-in-the-Oneness, that is, the second-order reflection of the ego is now led by the first-order diffraction of animal-mystic affects, for it is then that one understands *hukam* (GGS 555, M3).[25] Having transcended the thinking of the burdened ego (*haumai*), the mystic (*gurmukh/sādhu*) flows with "intuitive ease" (*sahaj*), "spontaneously" (*acint*), by recovering the "ever-fresh child mind" (*bāl-budh*). This return to a natural harmony unveils the supra-ethics of *hukam* (organic intelligence contra the artificial intelligence of the ego).

The Sikh practice of remembering the Name in meditation and chanting (*nām-simaran*) is a first-order affective praxis rather than a second-order cognitive or doctrinal belief. *Nām* frames both Word and world: "everything is singing [His Name]"; every form in the world communicates the Word (GGS 5, M1). The emphasis here is on affective musicality: "The Word is the Gurū, and mindfulness of its music, is the disciple. Speaking the Unspoken-Speech, I remained unattached... contemplating the One Word, the Gurmukh extinguishes the fire of egoism" (GGS 942, M1). Therefore, the proper response is to listen, attune to, dwell within, and participate in the songs within life's relational field. "Listening and listening I have crossed over [to *nirvāṇa*]" (GGS 660, M1). *Nām-simaran* is then a practice of self-loss (*āp-gavāī*), in that the unconscious pre-epistemic *semiosis* and superconscious post-epistemic *soteria* actively displace *haumai*—to listen the ego has to be absent, like the child's or animal's listening. In this emphasis on affective and transformative practice, revealed speech may then be better understood as both a singing and an un-speaking, where words arise uncolonized by *haumai*: "As you cause me to speak, so do I speak, O Master, what other power do I have?" (GGS 507, M5). Hence, why Gurū Arjan does nothing through second-order knowledge.

The Sikh world as Word is backed by a logic-destroying "union" of form with the formless, allowing the One to be Many. What I have hoped to show here is that the Gurūs had long addressed—albeit in vernacular, devotional terms—a plurality of epistemologies and ontologies, ideas that are only now being entertained in the academy.

Forbidden Turn: Liminal Portals to Animal and Mystic Wor(l)ds

As we have covered above, modern subjectivity is normally habituated in the positionality of "I think"—which suppresses the "I feel" and "I sense"—as any therapist knows. It certainly displaces "feeling or sensing *without I*" where a meditative flow (*dhiān*), or spontaneity (*acint*, *acet*, *sahaj*, *bāl-budh*, etc.), mentioned above, is attained, rather, returned to. To elucidate Gurū Arjan's cryptic rejection of "I think," it is helpful to recall Bhagat Kabir's crucial reversal: "Hari (Being of beings) has taken away my intellect (*buddhi*) and replaced it with perfect perception/intuition (*siddhi*)" (GGS 339, Kabīr). This constitutes the mystic turn, what I have called so far the Forbidden Turn, educing a movement away from thought and abstracted knowledge to the direct perception of truth/reality. Such non-knowing (i.e., enlightenment as enlivened sensing awareness) may be found in other religious traditions—for example, Chuang Tzu's *ziran* ("naturalness," "spontaneity") and Dogen's *hi-shiryo* "*non*-thinking"—which lies beyond the egoic duality of thinking and *not*-thinking:

> The stage of non-thinking is beyond egocentric cognition. If you reach the state of non-thinking you will realize the true luminous nature of mind. Non-thinking must become the eye through which you view phenomena. The activity of every buddha is based on non-thinking.
>
> (Loori 2003)[26]

Of course, such unrestricted modes of consciousness are not only the domain of Asian spiritual masters. Beyond Western mystics, we can turn to "interspecies communicators" to see a similar turn to more spontaneous and *non*-thinking forms of engagement.

M.J. Barrett and Anna Breytenbach are "animal communicators" who expand the term "animal" to "interspecies." Barrett, featured in this collection, is also a scholar of "intuitive interspecies communications" (IIC) that promotes a theory of liminality and a variety of research methods to contact and communicate with nonhuman beings. As a professor in the School of Environment and Sustainability at the University of Saskatchewan, she works to bridge knowledges and multiple ways of knowing (including indigenous ones). Breytenbach is a South African interspecies communicator, animal activist, conservationist, and public speaker. She has been subject of a full-length documentary,[27] and a particularly poignant section of it, involving her non-verbal telepathic "dialogue" with a black leopard released on YouTube, went viral (Postma 2013).

Both Barrett and Breytenbach demonstrate a similar reversal to Kabir's in their claims to engage in two-way silent intuitive communications with animals, trees, plants, etc., through a subjective depth and subtlety, that similarly requires a loss of—or at least a radical loosening of—the ego-self and its addiction to second-order thinking as the sole epistemic ground of authoritative knowledge. That is to say, by decentering thought and thinking, animal communicators such as Barrett and Breytenbach enter a *liminal* world whose gross physicality reveals a subtle psychic form. In this we hear resonant echoes of the Sikh Gurūs' hymns.

In harmony with the Sikh Word/world, the material world is recognized by Breytenbach as a field of *affective* communications composed of a universal *intuitive* language of "energetic signatures," "mental images," "sounds," "emotions," "thoughts," "mini-films" as "open and clear two-way channels of communication" that can be dialed into like "radio frequencies" and "along which information passes freely," allowing one to "feel the fabric of we instead of I" (Thiyagarajan et al. 2012). When tracking an eland (antelope), for example, Breytenbach "sees" through the eland's eyes as it looks to the west towards the sunset, and "feels" the sensation of a low bush on her own back that the eland is brushing against (Thiyagarajan et al. 2012, 34–36 min.).[28] This is clearly reminiscent of Ingold's account of the hunter's expanded consciousness, and indeed Breytenbach similarly references the capabilities of indigenous trackers; both of them refer to this capacity as love. Indeed, such intuitive interspecies communication has an important relation if not overlap with the GGS's *bhagati* (loving-as-becoming-Oneness) and its "speaking the unspoken-speech."

The point here is that interspecies communicators, like the mystics, are not *thinking about* animals, but sensing and feeling *with* them. In his book "Spirit of the Sikh," the highly revered scientist, mystic, and poet, Pūran Singh (1881–1931), wrote "In this great sympathy lies the secret of creation. When this is found, all is found; when this is known, all is known. All true culture lies in this culture of Self-Realization. Feeling is all in all" (2000, 2). Barrett's intuitive interspecies communication is "non-verbal" and, like Breytenbach, draws on a "mutual exchange of visceral feelings, emotions, mental impressions and thoughts, embodied sensations of touch, smell, taste, sound, as well as visuals in the mind's eye" (Barrett et al. 2021, 151). Moreover, physical proximity to the animal is not necessary and *psychic* communication "can also occur over great distances and without the need for visual, auditory, olfactory, voice or other cues"; hence, the practice is often referred to as "telepathic" (151). The ability manifested by her students to communicate and even *merge* with the animal is often classed as *parapsychological* (156) and hence marginalized in the academy as spurious. It is at this point where the secular bias of Western modernity becomes apparent.[29] I argue that in identifying a continuity of experience across cultures and histories, we can better address the exclusionary politics of colonial modernity and its secular subjectivity. Asian spiritual teachers, indigenous hunters, contemporary interspecies communicators all speak of a shared, expansive consciousness that transcends thought and individual identity. Why not turn to centuries of religious practice to better understand this shared experience? If we continue to ignore these traditions for the simple fact that they are judged as "religious" (read untrustworthy), we risk missing vocabularies, grammars, and methods of great insight. Current interspecies communicators know that this is not a dead knowledge but a living one; knowledge might die, but not affective "knowing."

David Gordon White's excellent work on Indian yoga is a valuable exemplar of such scholarship. He locates the practice of non-local consciousness of yoga within a broader history and philosophy of super-sensible "rays" (*raśmīn*) from Vedic times onward in a chapter titled "The Science of Entering Another Body" (White 2009, Ch 4). He covers a variety of cases across Indic scriptures: from the ascension of great beings, yogic suicide, gods producing multiple forms, gods entering statues (*yoga-mūrtīs*) "to reveal themselves anthropomorphically before deserving humans" (2009, 141), gods entering human forms often to impregnate them with their rays, not dissimilar to "solar impregnation in the Christian doctrine of the incarnation of the Logos" (142). In Mahāyāna and Tantric Buddhist texts "rays become the prime media for the transmission of teachings and the transformative process of initiation and yoga" (147), as well as the accounts of the miraculous perception of the "Buddha-eye" and "wisdom/*dhamma*-eye" (157). Two specific examples will suffice to connect this to our discussion. White references the *Mokṣadharma Parvan* of the *Mahābhārata* which states, "That connoisseur of yoga entered (*praviveśa*) into the king, having conjoined (*saṃyojya*) his consciousness with [her] consciousness, [his] eyes with [her] eyes, and [his] rays (*raśmīn*) with [her] rays (*raśmibhiḥ*). With the bonds of yoga did she bind him" (117). Secondly, texts like the *Yoga Sūtras* and the *Vaiśeṣika Sūtra* and their commentaries, describe how the "vision of the siddhas" (*siddha-darśanam*) and super-sensible seers is said to involve "'flashes of insight' (*pratibhaṃ jñānam*), intuitive knowledge arising from the direct contact of self with mind, [that] permit them to see things as they truly are" (157). White concludes that the "yogic body was conceived as an open system, capable of transacting with every other body—inanimate, animate, human, divine, and celestial—in the universe" (166).[30] The implication of the Forbidden Turn is that we are all potential yogis in this regard and this has been known for millennia.

To take the numerous cross-cultural and transhistorical accounts seriously, i.e. to include such notions in our theorizations, would "open the flood gates of ontological possibilities" (Hunter 2015) and thereby risk the destabilization of the hierarchized binaries of Western modernity. If a Western ontological absolutism "protects and reinforces the mainstream 'consensus reality,' what we might call ontological flooding destabilizes it, and opens it up to questioning, exploration and expansion…" (2015, 16). In a much-needed addition to the field, Hunter, in his book with Diana Santo, expands the "matter-mattering" discourse to cover invisible matters focusing "On The Materiality of Unseen Things" (Santo and Hunter 2021, Introduction). Rather than assume a real world in contrast to an "enchanted One," Hunter, like Cabot, believes "that *worlds* are performed and enacted through different forms of relationality and thus *become* real"—reflecting here both the participatory and forbidden turns. That is to say, there is no decisive break dividing the sentient and nonsentient, the material and spiritual, the interior and exterior (2021, 3), just as there is no such break in the Word and world in Sikhī. Santo and Hunter open the door to precisely the (Western) mystics whose absence I've been lamenting, like Emanuel Swedenborg's "angelic visitations and visions of the otherworld" as well as Joseph Smith Jr.'s "All spirit is matter" (2021, 7).

Redressing the imbalance, Hunter analyzes communications with ghosts and spirits of the dead, just as M.J. Barrett studies multispecies communications with animals and trees. Both unveil an obvious *inseparability* at play along liminal forms of contact (whether this contact be described as sensory emergence, clairvoyance, or possession). Both actively open up to and engage with diverse states of consciousness; both are interested in communication that transcends the logos of representational, symbolic language and its parochialization of consciousness to the individuated self (and its brain organ); both are interested in a diversity of embodied practices involving sensory and psychic forms of communication that open up the animal-mystic, affective-intuitive continuum even to the "occult side." Earth itself is recognized as an intelligent organism, composed of inter- and intra-acting "beings" of diverse forms (from grass to humans; from material to the energetic) who are constantly communicating and intra-acting.

Against the parochialism of the "I think" epistemology (that paradoxically justified a politics of universal conquest), *pre-epistemic semiosis* and *post-epistemic soteria* of the body discloses its immersion in the livingness of an endless sensing environment composed of the atemporal non-locality of affect in/as nature. The crucial point here is that liminality is necessary to access animal-mystic affects, that guard the gates to their realms of spontaneous *gnosis/giān* with requirement of self-loss. It is no surprise then that Barrett forewarns her audiences, "scholars will need to be open and receptive to transformations that may, at times, involve their whole being and cannot be neatly contained and left at the 'research bench.' When engaging IIC, it matters who and 'how' we 'are' (our be-ing) as researchers and humans" (2021, 158).

Now, that we have charted how the limitations of current affect and participatory turns are tied to the assumption of a fixed secular subject being always in place to order knowledge through a variety of binaries (reflexive/diffractive; correlational/non-correlational, landscape/taskscape, conceptual/perceptual, cognitive/affective etc.), as well as note how the Forbidden Turn revolves around liminality and self-loss, the dimension and import of the animal-mystic resonance comes into view.

Rather than interpret the behavior of animals, or trust the words of IICs, why not include in our scholarly study contemporary mystics and simply ask them—what's happening? For, unlike the animal who subconsciously lives "the world as Word" through its *pre-epistemic semiotic* sensorium, the mystic does so consciously through a *post-epistemic soteria*. This Word-intelligence-energy-consciousness beyond intellect is tapped into by all beings at different levels of affective awareness, including the conscious practices of shamans, diviners, mystics, yogis etc. As the contemporary and hugely popular Indian mystic Sadhguru (2018) states, reiterating Kabir's reversal from intellect (*buddhi*) to perception and intuition (*siddhi*):

> Intellect is being used as the sole tool but it's only good for dissection: we cannot know another by dissection, only through inclusion, engagement. Using a scalpel to understand the world you will just leave it in tatters…All we are doing is studying the footprint of the tiger and gathering huge amount of evidence to feed our theories, but as soon as we encounter the tiger all that knowledge vanishes.

The Forbidden Turn engages that tiger through this reversal returning to the psychosomatic sensorium. Barrett's, Cabot's, and Hunter's "knowledges" are not theories of an outside world but live intra-actions that shift from "*representation of*" to "*contact with*" requiring new liminal methods. However, Barrett's "trans-species episteme" that reframes *animals* as "social peers, communicative partners, and co-researchers" is perhaps too narrow, without also including contemporary *mystics* as well. Excluding Cabot and Hunter, what does it mean that affect scholarship in general includes matter and the animal but neglects spirit and the mystic?

To redress this let's take the Lingayat Shaivite Gurū Nisargadatta Maharaj (1897–1981),[31] whose book *I am That* (2005/1973) became a well-known "spiritual classic"; in it he writes:

> I am neither object nor subject…Look, my thumb touches my forefinger. Both touch and are touched. When my attention is on the thumb, the thumb is the feeler and the forefinger—the self. Shift the focus of attention and the relationship is reversed. I find that somehow, by shifting the focus of attention, I become the very thing I look at, and experience the kind of consciousness it has; I become the inner witness of the thing. I call this capacity of entering other

> focal points of consciousness, love...Love says 'I am everything.' Wisdom says 'I am nothing.'
>
> (Maharaj 2005, 268–69)

Beyond Deleuze and Guattari's *conceptual* "becoming-animal," isn't Breytenbach's embodied becoming-eland also "becoming-mystic"? Deleuze and Guattari's secular bias, makes us ignore a plethora of rich religious texts, traditions, and mystics narrowing our epistemology to postulate only a "bodily logos" without connecting it to a "psychic logos" as we saw above—and keeps us addicted to only what can be thought and publicly claimed rather than privately intuited and felt. Unlike pre-epistemic perception this post-epistemic speech can reflect on perception in such a way to reveal differences between conscious humanity and unconscious animality.

The essential node of self-loss which separates the Forbidden Turn from previous turns is starkly communicated by Nisargadatta when elaborating his "clear and empty" state: "I mean free of all contents. To myself *I am neither perceivable nor conceivable*; *there is nothing I can point out and say*: *'this I am.'* You *identify* yourself with everything *so easily*; *I find it impossible*" (Maharaj 2005/1973, 268; emphasis added). Whose voice is this? Whose perception is Breytenbach's when seeing through eyes of the other? The modern subject equates "I" with memory/events, and if you lose your memory, you lose your sense of self (at least you become unrecognizable to significant others). But here there is a mode of being beyond memory, that suggests the fiction of equating of self with memory/karma. As we know, *pre-epistemic semiosis* and *post-epistemic soteria* are I-less too: pre-I and post-I. This is why the animal connects with the mystic as both reside in the body's unconscious and superconscious affective modes that are inherently immeasurable but radically *a priori*. The iconicity of animal perception resonates with the iconicity of mystic imagination as both represent the Word-side of the world that far transcends the symbolic, representational side of egoic speech and thinking. It seems clear that at some point affect studies overlaps with consciousness studies.

The shift to a pre- and post-epistemic *perceiving* and *intuiting* that animal communicators and mystics show, in turn, implies a de-universalization of the modern Western subject through contact with *ahuman* temporal affects that *re-ontologize* language in their multiple voices and direct experientiality. Then the door has to be opened beyond representational language to include a deictic, talismanic, mantric, yogic, darshanic, shamanistic, clairvoyant language. Affect, Animist, and Continental theory, limited to the realm of thinking through a medium of referential language without a practice of "deepening the senses" towards a "fusion of signifier with signified," may be "inspired" but rarely becomes "revelatory." But with the practices of the Forbidden Turn there is a chance to engage the affective Word directly. Cabot says we need to "try on different [linguistic] ontologies. We need to wear them, live them, think them" to rediscover these ignored subjective depths (2018, 8). Cabot provides a four-fold model of ontologies, each with its own *affective*, *perceptive*, and ignored figures: "Shaman," "Diviner," "Mystic," and "Agent" (2018, 35–37).

My point is that if we are able to recognize *the shaman as the human-animal* who, because still animal, can communicate with interspecies (diverse forms), then perhaps we may invite the mystic, no less a human-animal, as someone who can communicate intra-species (diverse ontologies). Is it not urgent then to perceive more keenly that colonial modernity is really constituted and repeated by the violent imposition and universalization of a singular, dualistic subjective register that claims a sovereign autonomy and sole legislative power over all others, but is gendered male, racialized white, and regionalized as European? Without the iconicity of material semiosis redolent in the interconnections of nature or the religious soteria of psychic culture, modern subjectivity is locked into a flatland of symbolic thinking alone.[32]

Conclusion

I argue that "multispecies epistemologies" not only need to look outward to other bodies of species (inter-actions), but inward to other psychic subjectivities (intra-states). If we are prepared to grant animals, plants, and rocks meaningful voices via new materialist and new animist theories, why are some affect theorists so quick to deny mystic speech in all its variety? Why only look *outside* to the animal for alternative epistemologies when we have a plurality of mystic resources *within*? If the animal communicator troubles the Western axiom of the impossibility of knowing another, so too the mystic troubles the supposed impossibility of being without thinking. There is a missed affective resonance across animals and mystics: both simply walk out of the built architecture of egoic thought. Mystic consciousness returns us to, but also distinguishes us from, the animal as pre-egoic instincts are different from post-egoic intuitions.[33]

To avoid the power of othering in hegemonic forms of knowledge, Cabot cites Santos: "It is not only useful to discern these diverse [subjective/affective] modes," he argues, "it is necessary" (2018, 38). This diversity of internal subjectivities demonstrates that "people *matter* the invisible in a variety of different ways" (Santo and Hunter 2021, 3). Modern subjectivity's carnophallogo-ocularcentric culture,[34] that erects itself through a violent domination of God, nature, and the animal, occurs through human words, thought, and language that territorialize the world mapped by the dominance of an "enlightened" colonial seeing.[35] Turning that gaze within —de-photologizing it, closing the eyes to see in darkness, contacting that affective inside—has always been secularism's unthought spawning various subcultures. Affective, participatory, and forbidden turns recover multispecies epistemologies within an ontological flooding of the world as Word to rediscover subjectivity as different to itself and in conversation with inestimable others as others.

Weber (2013) argues for the urgent need to upgrade the Enlightenment frame of domination, efficiency, and growth born out of individuals in competition, for nature is not separate from us, nor is it "dead matter" or "dumb" as Žižek puts forth "a materialist acceptance of the meaningless stupidity of nature" (2022, 6). The omission of nature's *subjective* dimension of evolution is its crucial flaw; we are not just genetic machines but *embodied subjectivities* that bring forth ourselves.

Subjectivity, as an intelligent inter-being, is not extraneous to evolution but central to any and all organismic growth—as we all share in an "open-source" intelligence before and beyond "I think." Our commons is not merely a geography but a psychogeography.

I have argued that in recovering the *pre-epistemic semiosis* of the animal, we are still overlooking the *post-epistemic soteria* of mystic others. It is easier to look at the animal than to gaze in the mystic mirror as thought, number, and algorithm impress us more than the affective Word. Is it not time for an epistemic endarkenment of the Enlightenment via animal-mystic affects as our enlivenment? Or is our "vessel" turned upside down still (being ego-centered)?

> One whose tongue naturally/intuitively tastes Hari's Nectar,
> by the Gurū's grace is absorbed in the True One.
> Merged with the Truth, one discerns the Gurū's Word (and)
> drinks the Ambrosial Nectar from the immaculate stream within.
> The Name is collected when the vessel (of the mind) is upright.
> Nothing is collected if the vessel is upside down.
> Through the Gurū's Word, the Name dwells in the mind.
> O Nanak, true is that vessel which thirsts for the Word.
>
> (GGS 158, M3)

Even in the humanities, we are not only caught with our pants down revealing our bestiality, but also with the cup of our mind's full to the brim with "I think." The Buddhist, Taoist, Zen, and Sikh cups only gain their *raison d'être* because they never forget their fundamental emptiness. Isn't this a civilizational advantage to be taken note of?

Notes

1 Gurū Granth Sāhib, hereafter GGS, which includes various authors: M1: Gurū Nānak (1469–1539); M2: Gurū Aṅgad (1504–52); M3: Gurū Amar Dās (1479–1574); M4: Gurū Rām Dās (1534–81); M5: Gurū Arjan (1563–1606); M9: Gurū Tegh Bahādur (1621–75), Bhagats Kabīr and Ravidās, along with other bhagats and court poets. All translations are mine.

2 "Sādhū" is only one of many terms used within the GGS meaning both ascetic, mystic, saint and/or the mystical, saintly.

3 Both clearly relate to many pairs in Sikh thought that elaborate the one/many (ek/anek) which is with and without qualities (saraguṇ/niraguṇ), with and without form (akār/ niraṅkār), seen and unseen (drisaṭ/adrisaṭ), among others (bāṇī/bāṇā; jogī/bhogī; sant/ sipāhī; nadar/karam).

4 I have critically adapted Donovan Schaefer's (2015, 58) combination of the two streams of Deleuzian and phenomenological affect theories in Bhogal (2021, 185) and assume a similar point of departure here.

5 Elsewhere I understand this state as an "animal sublime" (2012).

6 Compare the many resonances (that follow) with Michael Marder's Biblical contrast between the affective Tree of Life and the cognitive Tree of Knowledge of Good and Evil, in this volume.

7 Cabot (2018) for instance elaborates a four-fold model, see closing parts of this chapter below.

8 Cf. Aurobindo (2005/1914–19), Forman (1990), Katz (2013). Scholars in the field and mystics engage William James's and Bertrand Russell's (1910) philosophical distinction between "Knowledge by Description" and "Knowledge by Acquaintance," and initially classed mystical experience as part of the latter. Steven Katz notes, "mystics—have what philosophers call knowledge by acquaintance as compared with knowledge by description" (2013, 20; emphasis original). However, later scholars distinguished three categories. Robert Forman cites two mystics, "Franklin Merrell Wolff, for example, a twentieth-century mystic, describes his relation with the one as a 'knowledge through identity'," and "Maharishi Mahesh Yogi echoes this notion when he says that the mind knows the Absolute 'by being it'," which, unlike the other two, "involves at least a radical shift of self-identity" (1990, 222). But the mystic Aurobindo had, between 1914 and 1920, noted a four-fold structure: "knowledge by identity, a knowledge by intimate direct contact, a knowledge by separative direct contact, a wholly separative knowledge by indirect contact are the four cognitive methods of Nature" (2005/1914–19, 544). Knowledge by identity or being is clearly of a different nondual order such that it is more of a non-knowledge and an unknowing.

9 Jāp Sāhib, verse 81. Sri Dasam Granth Sahib (1999, 14).

10 This perceptual, non-conceptual, affective, intuitive, noetic contact, because inherent to the cosmos, occurs across cultures and time in, therefore, an ahuman temporality (Bhogal 2021); the ancient Greek Hermes Trismegistus/Egyptian Thoth writes, "Now be entirely present, as far as your mind and ability are capable. For the knowledge of God is to be attained by a godlike concentration of consciousness. Such knowledge comes like a rushing river tumbling in flux from above to the depths beneath. By its headlong rush it outruns any effort we make as hearers, or even as teachers" (Salaman 2007, 55). Both ancient insights are repeated by modern writers. William James discusses a mode of consciousness attained through his "anesthetic revelations" (with nitrous oxide), as "perceptual, not conceptual," and that those intuitive facts came "so fast" and "rapidly" that his "intellectual processes could not keep up the pace" (1987/1910, 1274–75). Psychiatrist Richard Bucke writes upon naturally and suddenly entering an altered state of consciousness that he calls "cosmic consciousness" (1905, 6), and speaks of himself in the third person, "he learned more within the few seconds during which the illumination lasted than in the previous months or even years of study, and that he learned much that no study could ever have taught" (1905, 8). These noetic "states of insight into depths of truth unplumbed by the discursive intellect" (James 2002/1902, 295) makes James wonder "Is . . . consciousness already there waiting to be uncovered?" (1987/1910, 1280), rather than "generated de novo [anew] in a vast number of places" already existing "behind the scenes, coeval with the world" (James 1898, 47). Akin to James' view, the Sikh Word, as mystical Consciousness, "always stood there to be known" (James 1987/1910, 1274).

11 Cf Acampora, this volume.

12 Ahambudhi connotes "egotistical intellect," "prideful intellect".

13 As I have argued elsewhere (Bhogal 2007, 26–27), liminality is a translaytion of ego-being into a becoming-otherwise, without return as the same.

14 Hari is a vaiṣṇava epithet used by Sikhs for the Being of beings. It is a masculine name and yet the gendered pronoun should not be read to imply a male godhead; the GGS has both personal and impersonal namings of the absolute.

15 Cosmically, haumai is the process of individuation of all forms; psychologically it is the formation of the individual human ego. See Bhogal (2018; 2020) for an elaboration of

this distinction. That is why it is not to be completely eradicated but reformed: "Ego is a chronic disease, but it contains its own cure as well" (GGS 466, M2).

16 "The cynic does not know the taste of Hari's Nectar; the thorn of egoism being embedded within" (GGS 13, M4); "Attached to love of the other, they [manmukhs] do not savor the taste of the Word" (GGS 28, M3).

17 Such passages are clearly reminiscent of chapter 37 of the Dao De Jing, where the Dao does nothing but leaves nothing undone, and were we to attune ourselves to it, things would transform by themselves. Cf. Nietzsche's "Child" which we only return to after the stages of the "Camel" and "Lion".

18 Four sources of creation: *utabhuj*: "soil," plants etc.; *setaj*: "heat/sweat," insects etc.; *aṇḍaj*: "egg," birds etc.; and *jetaj*: "womb" animals etc.

19 "Everyone begs for a long life, no one wishes to die" (GGS 63, M1).

20 Cf: "We are like gold and the bracelet, or water and the waves" (GGS 93, Ravidās).

21 "The Perfect Primal Being (of being) is Ever-fresh and Ever-young" (GGS 240, M5).

22 Though the Word and Name are different, they are often used synonymously in the GGS. See Bhogal 2001. The Name is like the unnamable Dao and the Word its infinite instantiations/communications.

23 See Bhogal 2023, 212–16 for an elaboration.

24 For a discussion of this concept see Bhogal 2023, 217–20.

25 Cf. The Buddhist theory of Two Truths: conventional (reflection) and ultimate (diffraction), or Sikh saraguṇa and niraguṇa.

26 This is taken from Dogen's fascicle called Learning Through the Body and Mind. John Daido Loori, Roshi cites it discussing Dogen's 300 Koan Shobogenzo, Case 129: "Yoashan's Non-Thinking." See Loori n.d.

27 Thiyagarajan et al. 2012.

28 See Thiyagarajan et al. 2012 (34–36mins).

29 Here, where animal studies meets paranormal studies, such phenomenon as "remote viewing," or "extra sensory perception" (ESP), was not dismissed by the CIA. Rather they funded and ran their own programs since the 1970s—and partly declassified its documents in 1995—where they explored "anomalous cognition" (AC), "precognition," "clairvoyance," and the like, as well as the "non-local" nature of consciousness (Mumford et al. 1995; Schnabel 1997; Tart et al. 1979; Targ and Katra 1999).

30 Importantly, liminality points in other directions of how the world is inherently always already (recorded as) the Word, not just of the present but for all time, past and future. For example, one cannot separate animal communicator's affective and esoteric contacts from the enormously influential and extensively documented cases of the paranormal reading of akashic records by the likes of Edgar Cayse, Helena Blavatsky, Annie Besant, Alice Bailey, Rudolph Steiner and many yogis and Buddhists before them. Here world as Word can be directly experienced by contacting the dimension of *ākāsh* "space, sky, ether," the finest element in Hindu philosophy beyond air, fire, water, and earth that escapes the law of impermanence. Can one separate Barrett's and Breytenbach's clairvoyant "mini-films," or "energetic signatures" from Rudolph Steiner's "picture thinking" that is simultaneously a perceiving that precedes thought-based language? This infinite recording of every being is understood as Chitra-Gupat in GGS from the Indic God Chitragupta (lit. "hidden picture"). Cf. Temple Grandin's *Thinking in Pictures* (2006).

31 "*Nisarga-datta*": lit. "gift" of the "natural state," "innate disposition"; therefore a synonym of *sahaj*.

32 A shift to the post-epistemic soteria is strangely echoed within the decolonization of western forms of knowledge—pivoting on the loss of self. I elaborate this point in Bhogal 2021.
33 I am referring to the "Pre-Trans fallacy" debate in transpersonal psychology (herein I suggest PRE-personal=animal, TRANS-personal=mystic). See Daniels 2004.
34 Carnophallogocentrism combines logocentrism and phallocentrism (Derrida 1991).
35 "[T]he metaphor of darkness and light (of self-revelation and self-concealment) [is the] founding metaphor of Western philosophy as metaphysics . . . [I]n this respect the entire history of our philosophy is a photology, the name given to a history of, or treatise on, light" (Derrida 1978, 27).

Bibliography

Primary Sources

GGS *Guru Granth Sahib.* 2016/1604. Online. www.sikhitothemax.org/
DG *Sri Dasam Granth Sahib: Text and Translation.* 2 vols. 1999. Jodh Singh, Dharam Singh. Patiala: India: Heritage Publications.

Secondary Sources

Aurobindo, Sri. 2005/1914–19. *The Complete Works of Sri Aurobindo*, Volumes 21 and 22. Pondicherry: Sri Aurobindo Ashram.
Barad, Karen. 2001. "Scientific Literacy→Agential Literacy=(Learning+Doing) Science Responsibly." In *Feminist Science Studies: A New Generation*, edited by Maralee Mayberry, Banu Subramaniam, and Lisa Weasel, 226–46. New York: Routledge.
Barrett, M. J., Viktoria Hinz, Vanessa Wijngaarden and Marie Lovrod. 2021. "Speaking with Other Animals through IIC: Towards Cognitive and Interspecies Justice." In *A Research Agenda for Animal Geographies*, edited by Alice Hovorka, Sandra McCubbin, and Lauren Van Patter, 149–65. Northampton: Edward Elgar Publishing.
Bennett, Jane. 2010. *Vibrant Matter: A Political Ecology of Things* Durham: Duke University Press.
Bhogal, Balbinder Singh. 2001. "Nonduality and Skilful Means in Guru Nanak: Hermeneutics of the Word." Ph.D. diss., School of Oriental and African Studies (SOAS), University of London, UK.
———. 2007. "Ghostly Disorientations: Translating the Adi Granth as the Guru Granth." *Sikh Formations: Religion, Culture, Theory* 3, no. 1: 13–31.
——— 2012. "The Animal Sublime: Rethinking the Sikh Mystical Body." *Journal of the American Academy of Religion* 80, no. 4: 856–908.
———. 2018. "Cosmogony, Cosmology, Consciousness: The Meaning and Purpose of Life." English version of Bhogal 2020. *Academia.edu*. Accessed February 1, 2024. www.academia.edu/35852158/Cosmogony_Cosmology_Consciousness_The_Meaning_and_Purpose_of_Life.
———. 2020. "The Sikh" [in Greek]. In *The Book of Simple Questions* (*Το Βιβλίο Των Απλών Ερωτημάτων)*, edited by Petrou Konstantinos. Athens: Forms (Εντυποις).
———. 2021. "Undressing Political Theology for an Animal-Saint Redress." In *Political Theology on Edge: Ruptures of Justice and Belief in the Anthropocene*, edited by Clayton Crockett and Catherine Keller, 183–213. New York: Fordham University Press.

———. 2023. "Death and Dying in the Guru Granth Sahib." In *The Sikh World Encyclopedia, Part III: Articulating Sikh Ethics*, edited by Pashaura Singh and Arvind Mandair, 212–22. New York: Routledge.

Bucke, Richard. 1905. *Cosmic Consciousness: A Study in the Evolution of the Human Mind.* Philadelphia: Innes and Sons.

Cabot, Zayin. 2018. *Ecologies of Participation: Agents, Shamans, Mystics and Diviners.* Lanham: Lexington Books.

Chen, Mel Y. 2012. *Animacies: Biopolitics, Racial Mattering, and Queer Affect.* Durham: Duke University Press.

Daniels, Burton. 2004. "The Trans/Trans Fallacy and the Dichotomy Debate." *The International Journal of Transpersonal Studies* 23: 75–90.

Daston, Lorraine. 2005. "Intelligences: Angelic, Animal, Human." In *Thinking with Animals: New Perspectives on Anthropomorphism*, edited by Lorraine Daston and Gregg Mitman, 37–58. New York: Columbia University Press.

Derrida, Jacques. 1991. " 'Eating Well', or the Calculation of the Subject: An Interview with Jacques Derrida." In *Who Comes after the Subject?*, edited by Peter Connor, Eduardo Dadava, and Jean-Luc Nancy, 96–119. New York: Routledge.

———. 1978. *Writing and Difference.* London: Routledge and Kegan Paul.

Forman, Robert K. C., ed. 1990. *The Problem of Pure Consciousness: Mysticism and Philosophy.* New York: Oxford University Press.

Grandin, Temple. 2006. *Thinking in Pictures and Other Reports from my Life with Autism.* Second Edition. New York: Vintage Books.

Hunter, Jack. 2015. "'Between Realness and Unrealness': Anthropology, Parapsychology and the Ontology of Non-Ordinary Realities." *The Journal of the British Association for the Study of Religions* 17, no. 2: 4–20.

Ingold, Tim. 2000. *The Perception of the Environment: Essays on Livelihood, Dwelling and Skill.* London: Routledge.

James, William. 1987/1910. "A Suggestion About Mysticism." In *William James: Writings 1902–1910*, 1272–80. Edited by Bruce Kuklick. New York: Library of America.

———. 1898. *Human Immortality: Two Supposed Objections to the Doctrine.* Westminster: Archibald Constable and Co.

———. 2002/1902. *Varieties of Religious Experience: A Study in Human Nature.* Foreword by Micky James, Editors Eugene Taylor and Jeremy Carrette. Centenary Edition. London and New York: Routledge

Katz, Steven T., ed. 2013. *Comparative Mysticism: An Anthology of Original Sources.* Oxford: Oxford University Press.

Loori, John Daido (Roshi). n.d. "Thinking Non-thinking." *Lion's Roar: Buddhist Wisdom for Our Time.* www.lionsroar.com/thinking-non-thinking/#:~:text=In%20his%20fascicle%20called%20Learning,through%20which%20you%20view%20phenomena

Maharaj, Nisargadatta. 2005/1973. *I Am That: Talks with Sri Nisargadatta Maharaj.* Translated by Maurice Frydman. Durham: The Acorn Press.

Mumford, Michael D., Andrew M. Rose, David A. Goslin. 1995. *An Evaluation of Remote Viewing: Research and Applications.* Austin: American Institutes for Research.

Neyrat, Frédéric. 2017. *Atopias: Manifesto for a Radical Existentialism.* Translated by Lindsay Turner and Walt Hunter. New York: Fordham University Press.

Postma, Arjan. 2013. "The Incredible Story of How Leopard Diabolo Became Spirit – Anna Breytenbach, "Animal Communicator." 13.16 mins. Accessed February 1, 2024, www.youtube.com/watch?v=gvwHHMEDdT0&ab_channel=ArjanPostma.

Russell, Bertrand. 1910–11. "Knowledge by Acquaintance and Knowledge by Description." *Proceedings of the Aristotelian Society* 11: 108–28.
Sadhguru. 2018. "Memory, Consciousness and Coma." Sadhguru at Harvard Medical School. *YouTube*, July 17. www.youtube.com/watch?v=w7irEcQHChw.
Salaman, Clement. 2007. *Introduction to Asclepius: The Perfect Discourse of Hermes Trismegistus*. London: Duckworth.
Santo, Diana Espírito and Jack Hunter, ed. 2021. *Mattering the Invisible: Technologies, Bodies, and the Realm of the Spectral*. New York: Berghahn Books.
Santos, Boaventura de Sousa. 2014. *Epistemologies of the South: Justice Against Epistemicide*. Boulder: Paradigm Publishers.
Schaefer, Donovan. 2015. *Religious Affects: Animality, Evolution, and Power*. Durham: Duke University Press.
Schnabel, Jim. 1997. *Remote Viewers: The Secret History of America's Psychic Spies*. New York: Dell Books.
Shaviro, Steven. 2014. *The Universe of Things: On Speculative Realism*. Minneapolis: University of Minnesota Press.
Singh, Ganda, ed. 2000. *Bhai Nand Lal Granthavali (Punjabi)*. Patiala: Punjabi University Press.
Singh, Puran. 2000/1927–30. *Spirit of the Sikh: Meditations on Religion and the Spiritual Experience*. Patiala: Punjabi University Press.
Targ, Russell and Jane Katra. 1999. *Miracles of Mind: Exploring Non-Local Consciousness*. Novato, California: New Word Library.
Tart, Charles T., Harold E. Puthoff and Russell Targ, ed. 1979. *Mind at Large: IEEE Symposium on the Nature of Extra Sensory Perception*. New York Praeger Special Studies.
Thiyagarajan, Swati, Craig Foster, Damon Foster, and Vyv Simson. 2012. *The Animal Communicator* [Anna Breytenbach]. 52 mins. www.cultureunplugged.com/documentary/watch-online/play/11936/The-Animal-Communicator
Vásquez, Manuel. 2011. *More Than Belief: A Materialist Theory of Religion*. New York: Oxford University Press.
Viveiros de Castro, Eduardo. 2014. *Cannibal Metaphysics*. Translated by Peter Skafish. Minneapolis: University of Minnesota Press.
Weber, Andreas. 2013. *Enlivenment: Towards a Fundamental Shift in the Concepts of Nature, Culture, and Politics. An Essay.* Publication Series Ecology, Volume 31. Cologne: Heinrich Böll Foundation.
White, David Gordon. 2009. *Sinister Yogis*. Chicago: The University of Chicago Press.
Žižek, Slavoj. 2022. "The Stupidity of Nature." *Compact*, March 22: 1–7.

14 Faḫr al-Dīn al-Rāzī's Multispecies Epistemology

Bligh Somma

Al-Rāzī's Historical Situation on Animal Knowing[1]

The attribution of knowledge to animals presented a challenge to account for animal rationality in the philosophy of the Islamic world. Particularly in Islamic thought that was influenced by Aristotle (d. 322 CE),[2] the account of animal knowing developed in the philosophical tradition did not easily sit side by side with the description of animals in the Qur'ān. It is potentially for this reason that we find an unusual amount of discussion about animal knowledge from both Ibn Sīnā (Latinized as Avicenna, d. 428/1037—the premier Aristotelian philosopher of the eastern Islamic world) and from one of his heirs, Faḫr al-Dīn al-Rāzī (d. 606/1210). Faḫr al-Dīn al-Rāzī was one of the most influential figures in the history of Islamic intellectual history, and historian Ayman Shihadeh has concluded that his place in theology was "somewhat comparable to that of Ibn Sīnā in *falsafa*," that is, Greek-inspired philosophy (2005, 179). This situation holds in particular for the topic of animal knowledge, an issue al-Rāzī returns to again and again throughout his career. Although his exact position on animal knowledge changes over the course of his work, he repeatedly presents a case in favor of animal knowledge that goes beyond the allowances of his predecessors, in particular, of Ibn Sīnā.

One reason that Islamic philosophers and theologians alike were so interested in animals in general, and in animal knowledge and rationality in particular, is that the Qur'ān itself contains many depictions of animals.[3] In many of these instances, animals appear to act and to know in ways that go beyond what is explicable by more minimalist accounts of their intellectual capacities, such as what Aristotelian philosophy offers (more on this below). Though this issue may seem relegated to the sphere of the animal, it is in fact a very human concern. Rationality underlies human-animal difference, on which rests both human identity and moral issues of self-perfection and the afterlife. Upholding intellect as a uniquely human trait, Islamic scholars in fields such as philosophy, theology, and law were able to maintain a strict line between human and nonhuman animals, which both guarantees a clear human definition and explains moral capacity while limiting it to human animals.[4] The two features go together because intellect is responsible both for the higher-level cognition human beings alone seem to perform and for the capacity to know right and wrong and, therefore, to act morally, something also apparently

DOI: 10.4324/9781032659985-18

proper to human beings alone. Further, according to Greek-inspired philosophy, an intellect requires an immaterial soul, and an immaterial soul is the sort of thing that can survive the body. The question of intelligence in animals was especially important to those who were working directly within or who were directly influenced by Aristotelianism, wherein the human being is defined as a rational animal.[5] As a result, when animals exhibit what appears to be rational behavior, thinkers needed to explain that behavior by means of capacities other than intellect. They did so with recourse to psychological capacities that, while sophisticated, are ultimately restricted to the animal soul, namely, the soul consisting of all powers or capacities that are restricted to the body and responsible for an animal's survival, initially developed in Aristotle's *De Anima*.

The animal soul connected to the body is contrasted to the completely immaterial rational soul proper to the human, which according to Aristotle and Ibn Sīnā the human animal alone has (on the celestial bodies and their complicated status as animals according to Ibn Sīnā, see Alpina 2021b). The power of estimation (*wahm*) that Ibn Sīnā developed as part of the inner powers of the animal soul is perhaps the most famous development of the powers of the animal soul aimed to account for animal rationality (1970, 43).[6] Another, earlier explanation of animal rationality would be the Islamic philosopher Miskawayh's (d. 421/1030) claim that animals who are capable of receiving training have a trace of the rational soul. This account relies on a more Neoplatonic worldview wherein the human-animal split, though also occurring along the lines of intellect, can be breached through an illumination of the rational soul upon the animal's soul (1966, 46).[7]

It is becoming more well-known in literature that Faḫr al-Dīn al-Rāzī discusses animal intelligence throughout his philosophical works.[8] However, he discusses animal intelligence in his Qur'ānic commentary, titled *al-Tafsīr al-Kabīr*, as well (Tlili 2012, 176–96). Not only does al-Rāzī return to this issue throughout his career, but he routinely either presents a case by which animals might be intelligent (or, at least, cannot be dismissed as unintelligent) and takes an aporetic stance, or offers strong evidence in favor of animal intelligence. Insofar as he argues in favor of greater shared rational abilities between human beings and animals, his work is fertile ground for reflecting on multispecies epistemology in Islamic thought. I will show that throughout al-Rāzī's work, he presents various accounts of shared knowledge between human beings and animals; these different accounts also raise several questions, which I aim to answer here.

The most fundamental of those questions is what the shared knowledge between human beings and animals amounts to for al-Rāzī. Specifically, although it seems he would agree *that* we share knowledge, and it is not clear whether we as human beings can know with certainty that animals share both kinds of knowledge items (universals) *and* the meaning of that knowledge. If we use Ibn Sīnā's definition of universals, he claims that a universal as such is "that whose very conception does not preclude it from being predicated of many" (Ibn Sīnā 2005, 149). The universal circumscribes an essence shared by many particulars, like horseness, bone, or vegetation, that is abstracted and apprehended by a mind, since universals themselves never exist outside the mind according to Ibn Sīnā (158–59).

While a universal captures an essence that works on the metaphysical and epistemological levels, something's meaning, in the sense I am using it here, works only on the epistemological level. Though the difference between these two options will become clearer in the following, what I mean in broad strokes is this: if we grant that animals know universals (and al-Rāzī will argue in favor of doing so), when, for example, my dog Waylon watches a squirrel and when I watch the same squirrel, we both identify this animal as the kind of animal it is—even if Waylon is not capable of scientific familiarity with squirrels in the way I am. He cannot do as much with the essence that he identifies out there in the world, but it seems clear to me that he *can* identify it. Moreover, even if we are both able to identify a squirrel, it is clear that the furry animal Waylon and I each look for on our walk *means* something different to each of us. To Waylon, a squirrel might be "the thing that runs up into the tree with a bushy tail that I like to chase." To me, the squirrel is a somewhat cute, tree-occupying rodent, instances of which I look out for when I walk with my dog to avoid losing my balance and my dog. When I use the term "meaning," it is not to alter the content of the universal, but to highlight an additional layer of information that guides what the individual in question does with that knowledge and what response the knowledge elicits. The meaning refers to the way the information falls into a network of importance and fills out our inner world. So even if Waylon and I share knowledge of the squirrel in the sense that we grasp the universal—or however one wants to define knowledge in this instance—what that knowledge means to us is different, in part, by being situated within different networks of meaning.

I say "in part," because I neither want to commit myself to saying that absolutely all animals have networks of meaning, nor do I want to commit myself to saying that al-Rāzī thought that all animals have networks of meaning. Some animals might not have such sophisticated of intellectual capacities, and al-Rāzī might not have thought all animals have these networks. For example, Abū l-Barakāt al-Baghdādī (d. c. 560/1165) cites moths (*farāš*) as lacking sophisticated epistemological capabilities (that is, several powers of the animal soul) because they repeatedly run into lamps (1955, 2.313). But al-Rāzī clearly thought that *some* animals do have these networks, as I will argue below.

Returning to our example, the object of knowledge "squirrel" is situated within different networks of meanings. This much is clear from the way I just explained the difference between Waylon's understanding of squirrels and my understanding of squirrels. And that is important on the issue of multispecies epistemology and shared knowledge, since if we are asking whether knowledge, and the world built by it, is genuinely shared between human beings and animals, then we might want to know whether the knowledge shared by human beings and animals is knowledge that is *actually shared* in meaning in a holistic way or is shared knowledge that allows only for *parallel* knowledge grounded in essential likeness but without shared networks of meaning. The latter would be a much weaker account of a multispecies epistemology.

A note about al-Rāzī's historical situation and the history of his works is germane here. According to our current understanding of the chronology of al-Rāzī's works, he started writing his *Tafsīr* while he was in the service of the Ghūrid dynasty starting around 591–2/1195–6 until 602/1206 (Shihadeh 2022, 288). Allied with the ʿAbbāsids, this regional ruling group gave al-Rāzī patronage for over a decade, as part of their program of political, theological, and juridical legitimacy, as Shihadeh has argued (257–63). In turn, al-Rāzī "align[ed] his scholarly production with the Ghūrid project of self-fashioning," in that his works from this period are "devoted entirely to religious subjects" (272). During these years, he did not produce any philosophical works. In fact, Shihadeh highlights the intellectual constraints al-Rāzī was under during his "Ghūrid interlude," saying, "while he could practice *kalām* [speculative theology], philosophy was beyond the pale" (266). The theological works that al-Rāzī does produce "exhibit... an overall conservative tenor that is overall concerned with advocating for a normative orthodoxy than with breaking new ground" (275).[9]

Turning back to the matter at hand in this essay, it is possible that the external factors coming from the Ghūrid project, which focused al-Rāzī's attention on theological issues and on the composition of the *Tafsīr* were, in fact, an occasion for certain developments in his account of animal knowing and multispecies epistemology. Although his philosophical account could not identify an object of knowledge that could guarantee the shared meaning between human and animal knowledge, a theological account could. Working within the bounds of a *Tafsīr* and on a revealed religious text had other certain advantages as well. While working on his commentary, al-Rāzī developed a way to explain how we can account for truly shared knowledge among human being and animals in the rich sense and ameliorate the parallel knowledge problem, simply by placing God at the center, as the knowable object that all animals know, human and nonhuman alike. This object of knowledge has unique qualities which avoid the earlier resultant issues. Of his philosophical works, I will consider primarily the earlier *al-Mulaḫḫaṣ fī l-manṭiq wa-l-ḥikma* ("Epitome on Philosophy and Logic"), while I will take into account some of his distinctions regarding the relation of animal and human souls from his later *al-Maṭālib al-ʿāliya* ("Exalted Topics of Inquiry"), leaving lengthy consideration of his analysis of animal knowing there to others.[10]

Al-Rāzī's Philosophical Account

Animal Intelligence: Apprehension of Universals

I begin by highlighting the result of al-Rāzī's position in his philosophical works, and in particular, the position he puts forward in the *Mulaḫḫaṣ*.[11] There, he argues that there is clear evidence that human beings and animals share many epistemological features and, in fact, that animals have access to much of the same epistemological data as human beings. The terms of the debate were inspired by the philosophical accounts of knowledge and cognition of al-Rāzī's time—staunchly denied to nonhuman animals, as discussed above.[12] Al-Rāzī offers an overview of arguments in favor of the position that animal souls are intelligent, *ʿāqila*—a strong position, given that *ʿaql*, intellect, was characteristically restricted to the realm of the *human* animal alone. Here, as in the *Maṭālib* (7.303) where he offers another extensive philosophical examination of animals, he goes on to consider whether animal souls are intelligent and separate (from matter), which would make them similar to human souls, but not, he is careful to say, identical in essence. The *Maṭālib*'s final words are undeniably aporetic, as he closes by saying "God knows best!" (7.311), while in the *Mulaḫḫaṣ* he ends without a last remark, only calling the finishing proof persuasive (240r 7). In both texts, al-Rāzī closes the discussion only after offering evidently stronger cases in favor of acknowledging that animals have intellectual, separate souls. Such souls would, in turn, grant animals the capacity to work with much of that same epistemological data as human beings do, with the important caveat that they are incapable of reflecting upon that information in the way required for scientific inquiry.

Nevertheless, in both works, and in the *Mulaḫḫaṣ* in particular, al-Rāzī casts doubt on whether this data *means* the same thing to animals as to human beings. As a result, while animals might use the same intellectual capacities to navigate the world, the way they process the information into a network of related, self-centered organized meaning is not the same.[13] Although in this essay I am drawing a rather idiomatic distinction to highlight the problem at issue, I think that the difference al-Rāzī sees between having access to certain epistemic data and working with that data in different ways shows he sees the following basic issue I am highlighting. The fact that different animals can identify the same object as what it is does not guarantee their knowledge—and the world they build around it—is actually shared with human beings in a rich way. While, according to al-Rāzī, human beings and animals share similar souls (not the *same* souls—he is emphatic on this point), they can easily end up with *parallel* worlds of knowledge instead of the *same* world of knowledge. Al-Rāzī is keenly aware that we cannot verify whether the knowledge animals have means to them what our analogous knowledge means to us. Similarly, he is aware of the limitations of our inferences regarding animal epistemological capabilities. So, when we ask al-Rāzī whether there is *shared* knowledge in this richer sense between human beings and animals, he will have to give us an open answer in his philosophical works.

To illustrate this point, I will adduce two primary examples here, which I take together to form al-Rāzī's strongest philosophical defense of animal intelligence from his *Mulaḫḫaṣ*. First, al-Rāzī argues that animals are capable of apprehending universals. While this ability had been the linchpin of human intellectual function according to Ibn Sīnā's Aristotelian account, al-Rāzī argues that animals possess this ability as well. He offers three sub-arguments, and I will focus on one regarding utilizing food sourcing (239v 18–240r 7). Specifically, he argues that each species of animal has a certain species (*nawʿ*) of food that it seeks, for example, bone insofar as it is bone or vegetation insofar as it is vegetation (239v 21).[14] Without this capability, each animal type would not be able to distinguish the food it seeks from that it detests. But this clearly is not the case; consequently, al-Rāzī concludes that animals are able to identify universals and, in fact, have access to more shared epistemological data with human beings than philosophers had hitherto acknowledged. Moreover, wrapped up in this argument is an acknowledgment that the different universals—which each animal species accurately identifies—naturally do not mean the same thing to different species. Al-Rāzī is not merely talking about something like use here—he discusses an animal finding grass per se as something sought and bone per se as detestable. It is true that bone is food to jackals, while vegetation is food to sheep. But bone is more than that. To the sheep, bone is a detestable item and something to be avoided, as evidenced by the sheep's behavior. But to the jackal, it is a desired item and something actively sought (240r 1–7). It is thus clear that the bone *means* something very different to the sheep, even while the sheep and the jackal are both able to apprehend the universal "bone" consistently and accurately, because the apprehension of the bone takes place within a broader network of meaning. This explanation might sound quotidian (of course sheep don't like bones!). However, on the one hand, that is exactly al-Rāzī's point (animals clearly can apprehend universals), and on the other, it is germane to the issue of shared knowledge. If various animals differ amongst one another on basic issues on the meaning of food sources, we as human beings are even more limited in our assurance that the knowledge we share with animals means to them what it means to us.

The use of meaning here is different from something like valence, since the bone to the jackal is thing-I-eat, while the bone for the sheep is thing-I-do-not-eat and thing-I-detest. Both are different from an affective response of attraction or flight. Al-Rāzī shows this distinction in a remark from his third argument from the *Mulaḫḫaṣ*, which he presents in favor of the separability of single, unified (animal) soul. The first premise goes as follows: "Whenever animals sense the form of something pleasant, they have an appetite [for it], but whenever they sense the form of something harmful, they are frightened" (240v 20–21; al-Rāzī 2022, 28). This would be a more direct articulation of valence, via animals' affective responses to what they perceive. Both of these sorts of apprehensions, al-Rāzī argues, go back to a single thing, which must be immaterial. A basic issue here that al-Rāzī's analysis is skirting is the limitations presented by the observation of animal behavior and the reality of being unable to communicate with animals about their inner world (and not having any other access to that world). Al-Rāzī addresses this nexus of problems in the second issue I would like to discuss: self-awareness.

Self-awareness

Al-Rāzī's second argument aims to prove that animals have separate, immaterial souls, rather than intelligent ones (240r 7–20). With this emphasis, self-awareness is the natural activity to defer to, since true self-reference is, canonically, something achievable only by immaterial subjects. In fact, al-Rāzī does not begin here, but, aiming to show that there is some reality (*ḫuṣūṣiyya*) that underlies, in his example, a horse's very being, al-Rāzī has recourse to self-awareness to show the existence of this reality. He argues, for instance, that a horse's awareness of his own friends and foes reveals that horse's self-awareness, since the horse must be aware of (1) the relation the various individuals had toward the horse himself in the past, (2) that the horse himself existed in the past, and (3) that the horse existed in a certain relation to those various individuals in that past. Since horses do indeed run to their friends and flee their foes, we can affirm that they are self-aware and have immaterial, and thus separable, souls. Animals thus share these features of their souls with human souls as well. Further, this argument reveals an even richer understanding of meaning separate from epistemic data, since the horse's inner world not only contains an idea of his own self but at least a loose map of social relations as well.

However, in his argument that animals are self-aware, al-Rāzī is careful to avoid claiming that this awareness is *reflexive*. In the *Mulaḫḫaṣ*, he even echoes a point he will make explicitly in his *Maṭālib*, where he argues that we cannot claim that animals self-reflexively point to their selves, since we do not have access to their minds: "If it were established that animals apprehend their own proper selves, even while being oblivious of the reality of the space around them, then the objection would be telling. *But there is no way to know this premise*" (al-Rāzī 2022, 13; emphasis added). These remarks are in reference to, and in fact use the exact language of, Ibn Sīnā's so-called flying man argument, found in the first book of the *Kitāb al-Nafs* of his encyclopedic *al-Šifāʾ*. This argument requires the one described to have *reflexive* self-awareness and be capable of intellectual intentionality, since in the first book of *al-Nafs*, Ibn Sīnā uses the argument to establish the soul's separability from the body through one's awareness of their soul apart from any bodily awareness (Ibn Sīnā 1970, 15–16).[15] As a result, al-Rāzī recognizes that animal self-awareness, whatever it is, is not of the same type as human self-awareness. This passage from the *Mulaḫḫaṣ* is further telling, because we see, in al-Rāzī's own words, a recognition of what he takes us to be able to know about, and thus to be entitled to claim about, other animals. The subject of self-awareness even more explicitly reveals al-Rāzī's observation that the knowledge one has—human or animal—can be subjectively oriented, that is, situated within a subject-centered network of meaning.

We have seen that while al-Rāzī appreciates that humans can infer that animals must have a certain set of epistemological skills (apprehension of universals, self-awareness), we can only know so much about them *and* we can know little about what their knowledge means to them, in part, because we simply cannot ask them what their knowledge means to them. One way to state the problem is that the knowledge is species-oriented—the sheep knows the bone as a sheep (something to flee), the jackal knows the bone as a jackal (something to eat), and the human being knows the bone as a human being (something to carve). Thus, the universals are oriented to each of the species that apprehends them, and as a result, it seems that we have sets of parallel knowledge rather than shared knowledge. When the jackal, sheep, and I all look at the same bone, we recognize it *as* a bone, but I do not know what that bone means to the jackal and the sheep without watching their behavior (and even then, I am pretty limited in what I can infer, and what I know will be only by grounds of inference). We can affirm now that al-Rāzī is aware of these limitations, and they put a damper on assumptions that his worldview offers an evident case for shared knowledge, in spite of his clear multispecies epistemology.

Completing the Philosophical Account: Shared Knowledge of God

I have argued so far that in his philosophical work, al-Rāzī established that although human beings and animals share many epistemological powers, there is no guarantee that the information gathered by those powers means the same thing to human beings as it does to any of the various animals that share that knowledge. Al-Rāzī is consistently careful on this point, and even after strong proofs in favor of animal rationality, he ends the discussion without comment. This is where al-Rāzī's remarks in his Qur'ānic commentary are pertinent.

God offers a unique opportunity on the topic of shared knowledge, since God is an object of knowledge that *cannot* be subjective—that is to say, the meaning of which cannot be subjectively altered. We will also see that al-Rāzī's source guarantees animals' knowledge of God, and thus solves the uncertainty issue facing al-Rāzī's philosophical discussions of animal knowledge. Since God is an unchangeable, non-subjective object of knowledge, He regulates epistemological frameworks around one object of knowledge (Himself) so that they overlap on one point. God would thus serve as a revolutionarily different type of item of knowledge on the topic of multispecies shared knowledge—if al-Rāzī can establish that animals do, in fact, have knowledge of God. Were animals indeed to have knowledge of God, and were human beings to know that animals have knowledge of God, we would know that they know what we know and that their knowledge means to them what it means to us, namely, that God must be worshipped, praised, and so on (more on the details below).

This statement comes with some qualifications, of course. Animal knowledge of God is not *equivalent* to human knowledge of God, according to al-Rāzī, in the same way that the jackal's knowledge of bone was not equivalent to human's knowledge of bone. That qualification is the most important one, since the sophistication of human knowledge entails things like the capacity for a sophisticated understanding (of the sort required for scientific knowledge) owing to the power of intellect, *ʿaql*, and that sophistication is normatively binding, in the religious tradition and in the philosophical one alike.[16] The capacity for normative obligation is something al-Rāzī is concerned about, as I will discuss in just a moment.

A touchstone passage of multispecies knowledge in al-Rāzī's *Tafsīr* connects to the issue of incommensurable knowledge, which I would like to mention first. This passage is from the 16th *sūra* of the Qur'ān (al-Naḥl, or "the bees"), where we read that God inspired bees with a certain kind of knowledge, which includes things that God dictates for them:

> And your Lord inspired to the bee, "Take for yourself among the mountains, houses [i.e., hives], and among the trees and [in] that which they construct. Then eat from all the fruits and follow the ways of your Lord laid down [for you]." There emerges from their bellies a drink, varying in colors, in which there is healing for people. Indeed in that is a sign for a people who give thought.
>
> (16:68–69, Sahih International[17])

Bees, with their construction of hives, are frequently taken to be intelligent within this tradition (and others), and al-Rāzī comments on this point in his remarks on the above passage. He says bees build their hexagonal homes naturally, while human beings can build such shapes only with toil and the use of scientific instruments (al-Rāzī 1981, 20.71). Thus, while both bees and human beings can construct hexagonal shapes, they do so by different means and with alternative purposes and methodologies. And while, to bees, making hexagons (meaning "those things we bees live in and walk through") does not mean the same thing as it does to human beings (meaning "a six-sided plane figure"), bees genuinely do make hexagons all the same. This discussion serves as an explanation of the more basic fact that bees have sophisticated knowledge at all, in this case, knowledge via God's direction.

Thus, this passage lays out the way that al-Rāzī can affirm knowledge of God to animals in a later verse. Here, bees gain their knowledge through inspiration—*ilhām* (which al-Rāzī highlights is "a state resembling" *waḥy*, a term generally used for human inspiration)—which God gives directly ("God settles [these things] in the [bees'] souls").

We read of animal knowledge of God Himself later in the Qur'ān at 24:41, where we read that *all* animals know both God and the appropriate praise of God, again by way of inspiration. However, in this case, the verse in question does not itself mention inspiration. Rather, al-Rāzī introduces it in his explanation of the verse's contents: "Do you not see that Allāh is exalted by whomever is within the heavens and the earth and [by] the birds with wings spread [in flight]? Each [of them] has known his [means of] prayer and exalting [Him], and Allāh is Knowing of what they do" (24:41).

Although the passage singles out birds in particular, al-Rāzī's commentary makes clear that he takes the general relative pronoun (*man*), translated as "whomever," to entail *all* animals. In his commentary, he claims that many scholars will concede that we witness that God inspired (*alhima*) animals to many apparently rational activities that cause wonder in human knowers (al-Rāzī 1981, 24.10). In turn, al-Rāzī pushes further, asking the reader and scholars with whom he is debating that if they accept that animals can do these apparently rational things, why not accept that God inspired them to knowledge, supplication, and praise (in translation of *al-aṯnā*) of God as well (al-Rāzī 1981, 24.11)?[18] Our author clearly takes the group of animals to be broader than birds and vermin, since he follows this rhetorical question with a list of proofs, which includes a litany of different types of animals, from bees to bears (24.11–12). All of these examples spur wonder among intelligent human beings; in this *Tafsīr*, al-Rāzī is, in fact, highlighting the difficulty of the problem of shared knowledge. This recurrence emphasizes the degree to which both animal knowing and shared knowledge concerned him and that, in the theological context of the *Tafsīr*, they could be traced, in part, to a philosophical background (24.12).

In the end, al-Rāzī actually uses the fact that there is an incommensurability between human and animal shared knowledge (in the sense of meaning)—using the bee hexagon example, among others—to build a case for an example of shared animal knowledge of God that *is* mutually translatable across species, one that is available for human comprehension. Once he established inspiration as a legitimate source of divinely-granted animal knowledge in the building of bee's hexagons (16:68–69), al-Rāzī can apply the concept of inspiration later in his *Tafsīr* (24:41) to explain how it is that all animals in the heavens and earth know God Himself, including how to praise and exalt Him. This solution, in turn, solves the two deeper problems of *multispecies* epistemology, by securing for the human beings reading the Qur'ān—God's word—that animals do know God Himself and in what way they as humans know Him.

A Parallel Concern: Do Animals Have Ethical Obligation Toward Others (*taklīf*)?

Though knowledge of God resolves the two uncertainties left open in the philosophical analyses of shared knowledge by resolving the parallel knowledge concern and eliminating the possibility of subjectively- or species-oriented knowledge (concerning this particular object of knowledge), it presents a major issue. At a certain level of intellectual sophistication, one becomes able to tell the difference between right and wrong. And if one can do that, one is *mukallaf*, or subject to *taklīf*, that is, normative ethical obligations toward others. But this is not something the Qur'ān acknowledges as being the case with animals. In fact, in the Qur'ān, it is human beings of all animals who are willing to take on the Trust, that is, obedience to God together with the obligation of right and wrong, in Qur'ān 33:72, which grants them the normative ethical obligations that rationality supports.[19] Thus, there is a firm line between human beings and animals on this issue.

Al-Rāzī seems to run afoul of this division between human beings and animals in his discussion of a famous series of anecdotes about Solomon's interactions with animals in Qur'ān 27:17–26, where it appears animals *are* granted the burden of ethical obligation. I will focus on the first two stories. Prefaced by Solomon being taught the language of the birds as one of his inheritances from David, in the first story, Solomon gathers forces that include human beings, *jinn*,[20] and birds—the first two groups canonically thought to be rational creatures, but the third not typically thought to be so. The second story highlights the foresight of a particular ant in warning her fellow ants to make way for Solomon's army as it marches through the valley of her residence, lest the ants be crushed—a warning that Solomon also understands. Though it is beyond the bounds of this paper to tackle, speech, too, correlates with rationality—something as apparent in Arabic as in Greek (*nāṭiq*, *nuṭq*; λόγος, λέγειν). That is why al-Rāzī needed to be careful in accounting for the exaltation and praise above in Qur'ān 24:41, since exaltation and praise are usually thought to be verbally expressed.

Al-Rāzī must explain in all these instances both how the facts of the stories can be true *and* how, if they are true, the animals are not, as a result, bound by normative ethical obligations, since the truth of the stories would seem to rest on a sophisticated level of intellect in the animals involved. In his response, al-Rāzī has recourse to a specific stratum of intellect, namely, he claims that the stories could not be true were the birds and ants not either (a) in possession of intellect, and thus *taklīf*, or (b) *near* the border of intellect, and thus on the border of *taklīf*.[21] Being near the border of intellect and being near the border of *taklīf* are mutually entailing, since the two go hand in hand. In this sense, says al-Rāzī, the animals are like adolescents—they are on the verge of being rational, moral agents, but they are also not quite there yet. Thus, they can be reliably called upon for deeds, but they cannot be held entirely responsible for what they do. Nevertheless, in the case of the birds who are part of Solomon's forces, al-Rāzī concludes that in Solomon's day (in contrast to al-Rāzī's own), God brought the birds about *with* intellect, as a result of which there is no conflict, since they would, in that case, be correlatively normatively obligated (al-Rāzī 1981, 24.187). He concludes that in the first story, the birds were not merely on the border of intellect (although this was an option open to him), *but were, in fact, created with it* ("God exalted brought about the birds of His day among what has intellect") (my translation, 24.187). In al-Rāzī's day, by contrast, birds were only *inspired* with the sort of knowledge at issue in the verse ("That isn't the case for the state of the birds in our day, even if there is in them subtleties that God exalted inspired") (my translation, 24.187).

With respect to the moral concern, al-Rāzī also emphasizes the non-equivalence of human and animal knowledge, even while the knowledge is genuinely shared and connotes the same thing. Since he accounts for animal knowledge of God through inspiration, not intellect, he need not worry that *this* knowledge be normatively binding, for in this case, the historical exception withstanding, animals do not have intellect (and the concomitant abilities, like telling right from wrong). Al-Rāzī's emphasis in the *Tafsīr* that animal knowledge of God is not normatively binding—while still being genuine knowledge of God—is consistent with his discussions of the issue in philosophical texts. To remind us, in his *Maṭālib*, he argues that if animals are rational, they have souls that are separate from matter, and he concludes that even if animal souls were separate from matter, they are not identical to human souls in essence (al-Rāzī 1987, 7.303). Together with this point, he emphasizes that animal souls would also *not* be the same in terms of knowledge (*ʿulūm*) or character (*aḫlāq*), since animals cannot undertake scientific inquiry, since they do not have intellects (7.311). On this point, he is consistent across his work.

Summarizing al-Rāzī's Layered Multispecies Epistemology

I have aimed to show that al-Rāzī undertook sustained investigation not only into animal rationality, but into the way that animal knowing and human knowing related to one another. We have also seen that al-Rāzī is consistent across his work when it comes to a realistic skepticism about what human beings can know about animal knowledge. In spite of his bold, if tentative, conclusions in the *Mulaḫḫaṣ*, al-Rāzī clearly understood that showing that human beings and animals alike share an epistemological world takes more than showing they share certain psychological capacities. Since, in al-Rāzī's view, we cannot (and al-Rāzī himself could not) discuss with animals what their knowledge means to them, we cannot know whether we relate to our knowledge in the same ways. In fact, given the behavior of many animals, we have good reason to believe that on many fronts, our knowledge does not mean to us what any given animal's knowledge means to them.

Al-Rāzī is able, in a way, to clear this philosophical hurdle of parallel-but-unshared knowing on one remarkable point, when he takes up the composition of his *Tafsīr* and great theological opus under the administration of the Ghūrids. In al-Rāzī's speculative theology, God is an unchanging object of knowledge, and to the extent that one truly knows God, God will always mean the same thing—even if one's affective responses to God are different and even though what praising God will require will differ for every creature.[22] God will always have the same attributes (whether or not one knows those attributes), require worship, be unchanging, and so on. And, unlike other possible examples of shared items of knowledge, we can *know* that animals know God, since we both read it in the Qur'ān (the revealed text and words of God Himself), and al-Rāzī aims to prove it with the methodologies utilized in the *Tafsīr*. God is only one object of knowledge, but He is the ultimate object of knowledge, and serves as a point of overlap for all knowers. Being a *Tafsīr*, al-Rāzī does not connect this discussion to the philosophical one explicitly (though I suggested above that there are nonetheless connections perceptible in the text). Introducing God as an object of knowledge allows al-Rāzī, nevertheless, to move his investigation forward. After arguing in the *Mulaḫḫaṣ* that animals share epistemological capacities with human beings that give them access to universals and self-awareness, al-Rāzī can confirm in the *Tafsīr* that animals share one primary and overarching referent within the network of meaning, ultimately joining them all together at a central point.

Notes

1 This paper was written with the support of the project "Animals in the Philosophy of the Islamic World," which has received funding from the European Research Council (ERC), under the European Union's Horizon 2020 research and innovation program (grant agreement No. 786762). My thanks also go to Peter Adamson for his constructive comments.
2 With this one exception, the death dates given always offer the date on Hijri calendar followed by that on the Gregorian calendar.
3 See, for example, El Maaroufi 2021; Foltz 2006; Tlili 2012.
4 While here and in select other places, I will use a modifier with the noun "animal," in general, when I use "animals" without a modifier, I always mean nonhuman animals.
5 One of the famous passages for this claim is *Politics* I.2, 1253a7–18, where Aristotle claims that "man alone among animals has speech (λόγος)," speech (λόγος) indicating the rational capacity to produce it, the capacity also indicated with the same word (λόγος). However, Aristotle's *Politics* was not translated into Arabic, and, while its influence was nonetheless circulating in the commentators' works that were translated, we may also look to the well-known passage of Aristotle's *History of Animals* I.1, 488b24–25, that the human being alone among all animals has deliberation and counsel (*ra'y wa-mašwarah*, βουλευτικόν) together with *Nicomachean Ethics* I.7, 1098a1–3 (*nāṭiq*, λόγος).

6 Much has been written on Ibn Sīnā and animals. For an incomplete list on Ibn Sīnā on animals, see Musallam 1989; Kruk 2002; Alpina 2022; and Somma 2022. More on Ibn Sīnā on animal knowing will be cited below.
7 On Miskawayh and animals, see Adamson 2022; Adamson and Somma, forthcoming.
8 Adamson forthcoming; Adamson and Somma 2022; and Virgi 2022.
9 Regarding the *Tafsīr*, we might perhaps hold this latter point in balance with the fact that al-Rāzī completed it after the dissolution of the Ghūrid dynasty, as Shihadeh highlights that the latest dated part of the *Tafsīr* is 603/1207, which is after the Ghūrid period ended (2022, 273).
10 See, for example, Adamson forthcoming.
11 On the nature of this text, see Eichner 2009.
12 On the account of animal intelligence given by Ibn Sīnā, see Alwishah 2016; Druart 2021; López-Farjeat 2012, 2016.
13 There may be rare exceptions to this dissimilarity. As I will argue below, God, according to al-Rāzī, is an example, since God cannot be subjectively interpreted—He must always be He who must be worshipped, He who must be praised, and so on. A more deflationary example is of the second case, namely of something that may be life-threatening to *any* animal such that an animal, qua animal, would find it repulsive and flee from it (whatever that means in each case)—for example, predators, habitat invasion, or extreme temperatures.
14 Here in the third subargument, al-Rāzī is implicitly substituting species (*naw*ʿ) for universal (*kūllī*), which he uses in the previous two sub-arguments. Later in the proof he again uses universal in the bounds of the example.
15 For a selection of secondary literature on the flying man argument, see Adamson and Benevich 2018; Alpina 2018; Druart 1988; Hasse 2000; and Kaukua 2020.
16 For an investigation of the connection between understanding sophisticated knowledge and understanding right from wrong, see Somma 2024.
17 All Qur'ānic verses come from the Sahih International version.
18 Al-Rāzī here may be getting around another problem that he had regarding animal knowledge in his discussion of 17:44, on the concept of *taṣbīḥ*—praise. There, he concluded animals are not knowers in the rich sense, because then they would be ethically obligated. Here at 24:41, he uses a different term, *aṯnā*. See al-Rāzī's *Tafsīr*, 20.219–20.
19 "Indeed, We offered the Trust to the heavens and the earth and the mountains, and they declined to bear it and feared it; but man [undertook to] bear it. Indeed, he was unjust and ignorant" (Qur'ān 33:72).
20 *Jinn* are intelligent beings mentioned in the Qur'ān who are made of smokeless fire or air (MacDonald et al. 2012). In fact, *sūra* 72 is named after them.
21 He discusses the border of *taklīf* in the *tafsīr* on the birds and that of intellect in the *tafsīr* on ants.
22 For example, a sinner's affective response to God is different from a saint's and different from a cow's.

Bibliography

Primary Sources

***

Abū l-Barakāt al-Baghdādī. 1955. *Kitāb al-Muʿtabar*. Edited by ʿA. al-ʿAlawī al-Ḥadramī. 3 vols. Hyderabad.

Aristotle. 1957. *Politica*. Edited by W.D. Ross. Oxford: Clarendon Press.
———. 2013. *Aristotle's Politics*. Translated by Carnes Lord. Second edition. London: Oxford University Press.
Ibn Sīnā [Avicenna]. 1970. Avicenna's *De anima (Arabic Text), Being the Psychological Part of Kitāb al- Shifāʾ*. Edited by Fazlur Rahman. 2nd edition. London: Oxford University Press.
———. 2005. *The Metaphysics of the Healing*. Translated by Michael E. Marmura. Provo: Brigham Young University Press.
———. 2021. *Kitāb al-Nafs*. Selected translations by Tommaso Alpina. In *Subject, Activity, Definition* 190–238. Berlin: De Gruyter.
Miskawayh. 1966. *Miskawayh, Tahḏīb al-Aḫlaq*. Edited by C. Zurayk. Beirut: American University of Beirut Centennial Publications.
al-Rāzī, Faḫr al-Dīn. n.d. *al-Mulaḫḫaṣ fī al-ḥikma wa-al-manṭiq*. Ms or. oct. 623. Berlin: Berlin Staatsbibliothek.
———. 1981. *al-Tafsīr al-kabīr wa-Mafātīḥ al-ghayb*. 32 vols. Beirut: Dār al-Fikr.
———. 1987. *al-Maṭālib al-ʿāliya min al-ʿilm al-ilāhī*. Edited by Aḥmad Ḥiğāzī al-Saqqā. 9 vols. Beirut: Dār al-kitāb al-ʿarabī.
———. 2022. *Mulaḫḫaṣ al-ḥikma*. Partial translation by Peter Adamson and Bethany Somma. In "Faḫr al-Dīn al-Rāzī on Animal Cognition and Immortality." *Archiv für Geschichte der Philosophie* 106(1): 23–52.

Secondary Sources

Adamson, Peter. 2022. "Miskawayh on Animals." *Recherches de Théologie et philosophie médiévales* 89: 1–24.
———. Forthcoming. "Faḫr al-Dīn al-Rāzī on Animal Intelligence." In *Mobilität des Denkens/Mobility of Thought*, edited by Robert Seidel et al. Leiden: Brill.
Adamson, Peter and Bethany Somma. 2022. "Faḫr al-Dīn al-Rāzī on Animal Cognition and Immortality." *Archiv für Geschichte der Philosophie* 106(1): 23–52.
Adamson, Peter and Bligh Somma. Forthcoming. "Virtue Ethics as Animal Ethics in Philosophy in the Islamic World." *Journal of Islamic Ethics*.
Adamson, Peter and Fedor Benevich. 2018. "The Thought Experimental Method: Avicenna's Flying Man Argument." *Journal of the American Philosophical Association* 4, no. 2 (Summer): 147–64.
Alpina, Tommaso. 2018. "The Soul of, The Soul in Itself, and the Flying Man Experiment." *Arabic Sciences and Philosophy* 28: 187–224.
———. 2021a. *Subject, Activity, Definition*. Berlin: De Gruyter.
———. 2021b. "Is the Heaven an Animal? Avicenna's Celestial Psychology between Cosmology and Biology." In *Cosmology and Biology in Ancient Philosophy: From Thales to Avicenna*, edited by Ricardo Salles, 261–78. Cambridge: Cambridge University Press.
———. 2022. "Translating Method: Inference from Behavior to Anatomy in Avicenna Zoology." In *Premodern Experience of the Natural World in Translation*, edited by Katja Krause, Maria Auxent, and Dror Weil, 136–52. New York: Routledge.
Alwishah, Ahmed. 2016. "Avicenna on Animal Self-Awareness." *Arabic Sciences and Philosophy* 26: 73–96.
Druart, Thérèse-Anne. 1988. "The Soul and Body Problem: Avicenna and Descartes." In *Arabic Philosophy and the West: Continuity and Interaction*, edited by Thérèse-Anne

Druart, 27–49. Washington DC: Center for Contemporary Arab Studies, Georgetown University.

———. 2021. "Animal Cognition According to Some Philosophers in the Islamic World: Aristotle or Galen?" In *Europa Cura Te Ipsam! Essays in honor of Rémi Brague*, edited by Elisa Grimi, 186–203. Rome: Stamen.

Eichner, Heidrun. 2009. "The Post-Avicennian Philosophical Tradition and Islamic Orthodoxy: Philosophical and Theological summae in Context." Unpublished Habilitationsschrift. Halle, Germany: Halle.

El Maaroufi, Asmaa. 2021. *Ethik des Mitseins*. Grundlinien einer islamisch-theologischen Tierethik. Freiburg im Breisgau, München: Karl Alber.

Foltz, Richard C. 2006. *Animals in Islamic Tradition and Muslim Culture*. Oxford: Oneworld.

Hasse, Dag. 2000. *Avicenna's* De Anima *in the Latin West*. London: Warburg Institute.

Kaukua, Jari. 2020 "The Flying and the Masked Man, One More Time: Comment on Peter Adamson and Fedor Benevich, 'The Thought Experimental Method: Avicenna's Flying Man Argument.'" *Journal of the American Philosophical Association* 6, no. 3 (Fall): 285–96.

Kruk, Remke. 2002. "Ibn Sina on Animals: Between the First Teacher and the Physician." In *Avicenna and His Heritage*, edited by Jules Janssens and Daniel De Smet, 325–41. Leuven: Leuven University Press.

López-Farjeat, Luis Xavier. 2012. "Self-awareness (*al-shu'ūr bi-l-dhāt*) in Human and Non-Human Animals." In *Oikeiosis and the Natural Bases of Morality. From Stoicism to Modern Philosophy*, edited by Alejandro G. Vigo, 121–40. Hildesheim: Georg Olms Verlag.

———. 2016 "Avicenna on Non-conceptual Content and Self-Awareness in Non-human Animals." In *Subjectivity and Selfhood in Medieval and Early Modern Philosophy*, edited by Jari Kaukua and Tomas Ekenberg, 69–81. Berlin: Springer.

MacDonald, D.B., H. Massé, P.N. Boratav, K.A. Nizami, and P. Voorhoeve. 2012. "D̲jinn." *Encyclopedia of Islam. Second Edition*, edited by P. Bearman, Th. Bianquis, C.E. Bosworth, E. van Donzel, W.P. Heinrichs. Brill. Online edition. http://dx.doi.org/10.1163/1573-3912_islam_COM_0191

Musallam, Basim F. 1989. "Avicenna X. Biology and Medicine." In *Encyclopaedia Iranica*, edited by Ehsan Yarshater, 3: 94–99. London: Routledge.

Shihadeh, Ayman. 2005. "From al-Ghazālī to al-Rāzī: 6th/12th Century Developments in Muslim Philosophical Theology." *Arabic Sciences and Philosophy* 15: 141–79.

———. 2022. "Fakhr al-Dīn al-Rāzī and Ghūrid Self-Fashioning." *Afghanistan* 5, no. 2: 253–92.

Somma, Bligh. 2024. "The Brethren of Purity on Justice for Animals." *Journal of the History of Philosophy*, 62, no. 1: 25–48.

Somma, Bethany. 2022. "Avicenna on Animal Goods." *Journal of Islamic Ethics* 6: 19–52.

Tlili, Sarra. 2012. *Animals in the Qur'an*. Cambridge: Cambridge University Press.

Virgi, Sarah. 2022. "The Mouse's Tale: Al-Jāḥiẓ, Abū Bakr al-Rāzī, and Fakhr al-Dīn al-Rāzī on Animal Thinking." *British Journal for the History of Philosophy* 30, no. 5: 751–72.

Section IV
Experimental Responses

PRAXIS D

MAKING MUSIC WITH MYSTERIOUS SPECIES IN PONDS

David Rothenberg

David Rothenberg presents his latest interspecies musical investigation, not simply with individual birds or whales or bugs as he has done in the past, but by demonstrating music made together with entire underwater environments, the pandemic-directed investigation of the underwater sound world of what he thought were ordinary ponds right nearby his upstate home. The results, musical and philosophical, are anything but ordinary.

Undersound

From above, their depths seem mute. But for a long time, I had known that ponds were full of subterranean sounds. Completely silent above the water, beneath the surface all kinds of rhythms and songs could be heard, if one had the right equipment. I had a few underwater microphones that I used to listen to whales, in the deep and noisy ocean, but whenever I dipped these hydrophones in ponds, I heard nothing. What was I doing wrong? I had heard recordings full of swirling sounds infused with possibility. Why wasn't I hearing this in the ponds around me.

One day, I went to the Wave Farm art park upstate, which features sonic artworks in nature. The Brooklyn artist Zach Poff had made an incredible sound installation, a worn wooden raft permanently floating in a mucky pond, and this thing was continuously broadcasting all kinds of crazy underwater noises. Later I asked him how he did it.

"Different technology," he wrote back, "Kind of like a waterproof contact mic. Picks up only what happens really close to it, as if it could touch the sounds."

"Can you make me one?" I asked, and he obliged. Then I went back out to listen. Here's what happened when I dropped it into a crystal lake near the tip of Cape Cod.

DOI: 10.4324/9781032659985-19

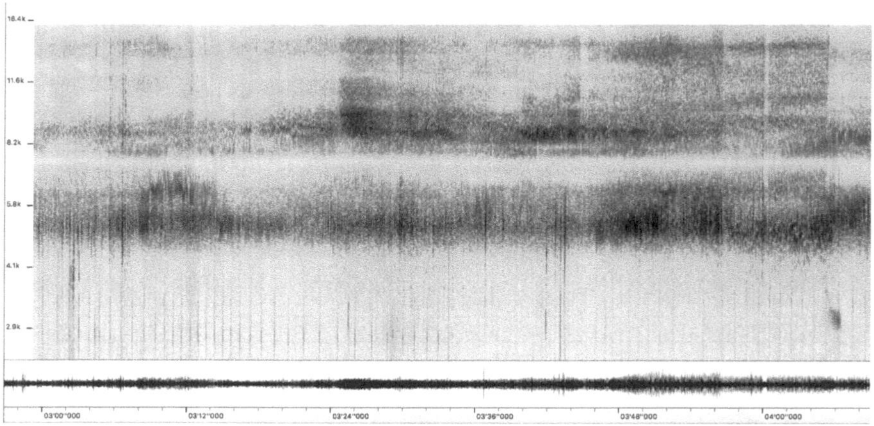

Figure D.1 This is what a busy pond sounds (and looks) like, once you've got the right kind of hydrophone. This segment is from 3:00 to 4:00, one minute total.

Cape Cod Long Pond in Provincetown, Massachusetts

Keep in mind that above the water all is silent. It is only while listening deep down that this music appeared. It was midsummer, a kettle pond in Provincetown, over the first weekend in 2020 that we were actually allowed to go anywhere after the onset of the coronavirus pandemic. The Cape was not crowded, and this pond on the bike path was ignored. We sat down and listened. Took a nap even; midafternoon on a warm day.

I've given you a sonogram of this moment (Figure D.1). This is an image that maps frequency against time, used by scientists, and others, to visually make sense of temporal events that we may have a hard time talking about without having a clear image of them to see. Although these images throughout this reflection appear the same size, they often show different amounts of time, sometimes zoomed in, sometimes out. I'll list the number of seconds or minutes for each sonogram.

The dark chaos above suggests high-frequency complexity, a wash of many insects stridulating, similar to crickets on land in thickets, but these are underwater. The regular vertical lines are the *pop pop pop* bubbles of plants photosynthesizing, exchanging oxygen through the water with the air—that's where regular rhythms come from in this underwater world. The attempt to distinguish plant from animal will be a big part of the decoding of the secret sounds of ponds. Sometimes, it is easy to tell, at other times not so clear.

The QR code below (Figure D.2) is the pure original pond, nine minutes of it so you can really tune in. Later, you'll hear how this piece is used as a musical background, a sonic element to inspire human play.

Composer David Dunn, in his chapter within my *Book of Music and Nature* (2000), describes how the sounds of ponds move in and out around each other;

Figure D.2 Hold your phone camera up to the code to be taken to a link that plays the sound example described above.

in listening, he felt an order, a logic, a beauty emerging, with no one single force behind it all. This is how nature works, he thought, excited to discover something beautiful, musical, and unknown. He writes:

> After a couple of years of listening to these small ponds and marshes, I came to understand a pattern to their underwater sound making. The one consistent factor is how beautiful and complex these miniature sounds are. I have finally reconciled myself to the gut feeling that these sounds are an emergent property of the pond: something that speaks as a collective voice for a mind that is beyond my grasp. I know that this is not a scientific way of thinking, but I can't help myself. Now when I see a pond, I think of the water's surface as a membrane enclosing something deep in thought.
>
> (Dunn 2000, 106)

In early 2020, I was supposed to travel the world playing live with birds. But then a global crisis hit, and no one was allowed to go anywhere. The world got quieter, no more airplanes in the sky. No more commuter traffic, no more buses and trains. The old sounds came back. The birds seemed louder; Carolina wrens next to my window in the morning, waking me up in springtime. Or was that a mockingbird faking the Carolina wren? One never knows. Oh, one knows, all right. Depends how much you know. Depends how long you listen. Our lives soon fell into static routines. Each day the same: Get up, take a walk outside before too many others get the same idea. But everyone already has the same idea—there is nothing else we're allowed to do.

If you do toss a hydrophone, by the way, be sure to hold one end as you swing the other like a lasso out into the wet. Otherwise, you'll soon find yourself wading out after it and scooping up muck, hoping to grab the wire in your messy hands. I've done that more than once, and it's easy to feel ridiculous as the bottom feeding ducks and Canada geese, given up on migration to sit out the winter, seemingly laugh as you paw fruitlessly at the mud, hoping to retrieve the lost and precious device that lets you hear impossible sounds such as tiny vibrating penises with such glory. Yes, technology is a wonderful thing, isn't it, if it lets us hear incredible mysteries right down the line. Why didn't I know the water boatmen were right here, hiding in plain sight? Same reason you didn't. I had read, researched, written, but I hadn't bothered to listen.

Listening is a whole other animal; it reveals things alive before we can claim them. I know that those who are fascinated by these sounds love what they hear. They listen for the tune of a healthy body of water, a thriving pond, not a dying pond. The clicks, pops, and noises lie beneath the great silence of the brackish waters behind our homes in the woods. Suburbia contains multitudes. I am reminded of that ant war in a California suburb where massacres were occurring just a few feet away from innocent people enjoying a barbecue. Nature waits for no one. It does not care if we notice it.

Soundfishing

Listen to these wholly other animals and plants. The rhythmic bubblings that easily frame a beat are like the tiny round bubbles in a fizzy drink, the even-tempo documents of carbonization. A tiny trail of pops blinks upon the water's surface, so I dunk the microphone and hear the *pop pop pop* of a living beat. The water lilies are breathing. Carbon dioxide. The green growing things sigh of relief and desire to save the world. The tap-tap-tapping of a blind man's cane, all indicators in the deeps. We want it to tell us something important, and will keep measuring as long as we hear a sign that the water's alive. Still, it is damn hard to tell which plant or animal is making which sound. In the pond, we hardly ever see what we are hearing.

I want to hand out hydrophones to little kids and watch them toss them in and start to listen. Soundfishing! Let the next generation invent names for what we hear. I imagine hundreds of children and teens tossing these waterproof baubles into the muck at the end of twenty-foot lines of cable. Plug them in to whatever device you've got, and, *voila!*; you will hear a booming, buzzing confusion of remarkable sounds. Even in this century where everything seems possible, morphable, changeable, hearable, or findable at a moment's thought, there are still sounds around us in a time when we are forbidden from going anywhere, immediate sounds that we still don't know.

"Shift your attention to immediate sounds," says Crow to Hoss in Sam Shepard's 1974 play *Tooth of Crime* (1980, 107). "Razor leathers, very razor." The characters in his play make up their own language in hopes that music will save the world. No matter how hard I try, I can't forget those lines 40 years since Sam came to my university to try to whip us undergrads into shape. I think he lasted about one week. Now today, students might figure out Crow's directive on their own if we would let them loose on the secret sounds of ponds.

I collect all these soundfish and try to organize them. I tag them on the machine with the label "Undersound" and when the assembly approaches three hours, perhaps I have enough. Hours stream along, symphonies of the unknown. Just as the performance grounds of the Nashville symphony become overrun with hundreds of screaming purple martins, my normally silent ponds offer up their endless arias of the season, the reach of alien music into our world.

These sounds evoke a place, a place where few have been. When science tells you if you are right or wrong, do not deny your own experience or ideas, but try to blend information together with fiction. Make no mistake. I do not deny the fact of what can be known about the invisible underwater world. Like hearing music in the wind, I simply want to find out what it is we can all *do* with the reality of this immediate but unknown sound.

Bug Music

In much of my professional life, I have performed for the invisible and forgotten originators of music, from birds to bugs, to whales and now pond creatures. Sure of their path in life in the march of evolution, these insects and tiny pond creatures know exactly what kind of sound they must make. Humans are never so sure…of anything. That's why we are the ones who become philosophers.

In my book *Bug Music: How Insects Gave Us Rhythm and Noise* (2013), I argued that insects were the first drummers. For millions of years before humanity, these myriad tiny creatures evolved a world of complex rhythms to further their kind. Human music emerged in forest and veldt, figuring out beats and cries in the midst of a longstanding, thrumming insect world.

Existence is pervaded by all these different species doing the same things, living the same way, with just *tiny* differences. If you look at our species as if you were another kind of creature, we'd likely all look the same, just like a murmuration of starlings looks to us. But, I too want to be original; that's why I sometimes call myself an artist. I seek my inspiration in places no one looks, like beneath the surface of a muck-filled pond. But I know you are all listening for the same thing: something new! But not *too* new, or you wouldn't even want to hear it.

Sometimes I have dreams of these strange impossible, but accessible, sounds. Maybe in dreams a fluorescent glowing green luminous pond creature from the dark lagoon will emerge like the shape of water and stare right back at me. No, I don't need to see anyone like that, however beautiful or beguiling they might be. It's enough for me to listen, to hear rhythms familiar but unfamiliar enough to take notice, to welcome the crackling mysteries into my music. Toss, listen, repeat, loop, remix, tune, resonate: I do whatever it takes to turn the natural magic into music and then I make something new. Never too new, but new enough, different enough, but still something I like enough to free myself from what I like or dislike.

I cast the microphone into the muck with a lasso-like throw, though usually inelegant, an oft-botched ring toss. Plop! It goes down. Old professor hydrophone revealing a techno-haiku. If the recordings are not precise enough for science, they are certainly fine for art. I take them home to the studio, play them back, and tweak them. I take out the noise and put in more noise. I filter unwanted frequencies, massage those that remain or resonate; I add some pitch or tone. The flickering beats of breathing plants invoke some kind of electronic gamelan. There *is* music here, a music few before have ever noticed or chosen to feel.

Little Pond Beasts With Clarinet: Pond Creatures and Plants Recorded at John Allen Pond in Fahnestock State Park, New York

The orchestra, surely, is complete. It certainly doesn't need me, but I take it as a place from which to begin. The human tone engages with the secret world of pond sounds. Interspecies musicians want to ensnare music out of the surrounding sounds. I want to combine my very different clarinet sound with this music of ponds. We've heard animals, then plants, then a human instrument. It's something else.

This soundworld is busy, beating, thrumming, pulsing, all on its own without any transformations. This is the first piece where I add my own instrument to the pond sounds and feel that it works. So much goes on beneath the water. Its bustle implores me to slow down, relax, get lyrical for a change. A minute and a half in, that pitched *güiro* beats under the water. Before writing about it, I play around it. I'm happy being here. This is the pond of ponds. See Figure D.3.

When in doubt, as autumn burns, far from the booming and buzzing of spring, I click a button and let those pond sounds play. Each time they get more familiar, whether in their raw original form or transformed and transposed. I am learning, over time, to distinguish plant from animal. The plants keep time, and the beasts carry a tune. See Figure D.4.

The possibility of these sounds sometimes wakes me in the middle of the night. Maybe they can be, for me, those essential sounds that go on inside the mind *when all other sounds are cut out*, in those midnight hours of almost silence. My Tibetan

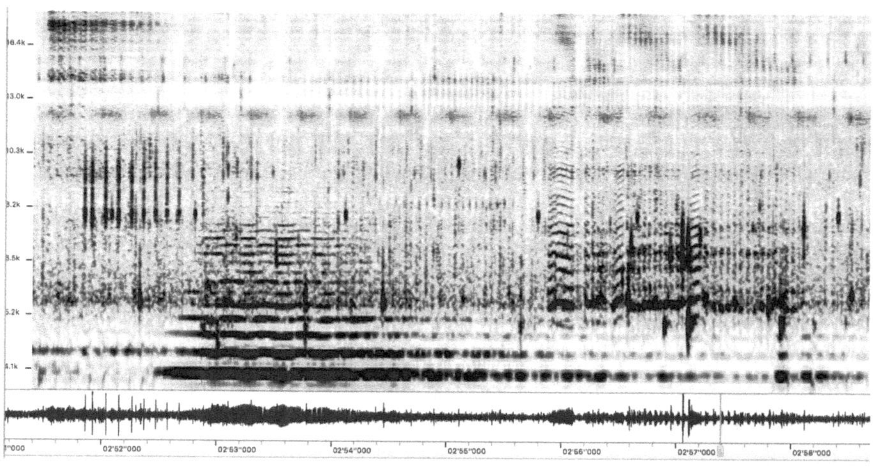

Figure D.3 2:51 to 2:58, seven seconds.

Figure D.4 Hold your phone camera up to the code to be taken to a link that plays a sound sample from John Allen Pond in Fahnestock State Park, New York.

monk teachers in Nepal taught me the circular breathed tones of the Buddhist oboe, the *gyaling*, were the inner musing of a reflecting mind. John Cage describes something similar as the hum of his nervous system and the beat of his heart, inside the anechoic chamber where no sound could survive (1961, 8). For me it's the surprise one hears whenever a hydrophone is tossed into a pond. Will there be something rather than nothing?

I remember visiting the great educational psychologist Howard Gardner at his office at Harvard University, when I was a fairly young student. He was investigating how elementary school kids would invent their own musical notation, without knowing much about real musical notation. The kids would draw lines, up and down, and some would draw individual notes to connect with what they could play. Some were more advanced, others less.

I told him I was interested in something else, the aspects of the sounds they wrote down that musical notation does *not* contain, like the timbre, the color, some of the things Paul Klee drew on in his *Steps to Parnassus* painting (1932), where he imagined a future abstract art as stylized and strategized in the abstract as music. I have been exploring such imaginative notations ever since then, and my fascination with the sonograms you see here surely descends from those days. I want to play unknown pond noises for kids and let them imagine the monsters that make such sounds. Why not? I want to say, "Grownups don't know what's going on down there—*you* have a chance to figure it out." This is one more small thing that the next generation has the chance to explore, along with fighting climate change and ending inequality. We adults sure haven't solved those problems, and I feel ever lost just thinking about them.

The Rhythm of Photosynthesis

As autumn descends, I already look forward to next spring's surge of sound in the ponds. But now things are getting quieter, as the birds and bugs calm down and the leaves fall red, yellow, brown, and green; the season of natural quiet soon comes. It's time for me to hunker down and delve into the stream of sounds. How much can be found in a single minute, a single second? How does an animal take its rhythm from a plant? Everyone has to breathe and to make a world out of light. Photosynthesis in the pond—it has its own tonality. As you listen to them, you shall hear.

The tingle from the insect world makes me shiver a bit. This is the chill that grabs us when certain sounds and images are taken in, a feeling that even the flat internet can present. Not everyone thinks such a feeling is real, and if they do, they often consider it a cheap trick, the aural equivalent of paintings of kittens with really big eyes. I don't think my favorite sounds have anything cheap. I want them to be deep enough to pay close attention to. I have always said that my work playing music with nature might increase our sensitivity and joy with nature, and our sense that humans are always interacting with the natural, even as we try to figure out what counts as art.

At an entomology conference in Missouri some years ago, I heard a Russian expert speak about the sounds of one particular beetle, explaining with heightened emotion how unique the sounds of one particular critter were. In fact, she may have been the only person to ever hear such a sound. I forget the details now, but remember clearly that she was conveying the deep satisfaction she felt when first hearing this sound. It is unlikely that any other human besides her will study this same sound because there are literally millions of other insect species out there waiting to have their habits heard and appreciated. See Figure D.5.

In my recordings, I want to convey to you what it *feels* like to know these subaqueous rhythms exist, that these sounds seem musical, though we rarely get to hear them even though we know they are there. Rhythm is repetition, and as things repeat, a groove is born. This text I write now repeats itself and, in fact, all my books repeat themselves, saying the same thing all over again, more relaxed and with more conviction over time, by no means necessarily better or more fresh: *there is music in nature.* You will hear it in the most unexpected places. Each species I approach ends up transforming my own rhythms—those in music, those in words. The scientific listener in me wants to figure all this out. The musical listener wants to feel the artistic possibilities in what is found. Here's a sonogram of a few of these *Corixidae* species (Figure D.6) that I heard at Lost Pond in Russel Wright's estate, now known as Manitoga. Over several minutes a slow, detectable uneven rhythm clearly emerges.

Lost Pond Mystery: Strangely Regular Sounds From Water Bugs, Fish, and Maybe Turtles

Compared to most insect beats, this tempo is *really* slow, visualized in the sonogram's gaps—though nothing like the seventeen years cicadas have to count before coming in. Each rolling beat is 40 seconds away from the next! Complex, incredibly loud rhythmic pulses inaudible from above the surface, a rhythmic dirge deep down, almost an imperceptibly slow march. It reminds me of the superslow beat of fin whales, each of whom makes a low *whoomph* every two minutes; when sped up, it becomes a beat we can dance to. In this sonogram we clearly see the rhythmic arrangements of these pulse trains, they also sound like the echolocation

Figure D.5 Photograph of an individual from the *Corixidae* species taken at Lost Pond in Phillipstown, New York, not far from my home.

clicks of foraging dolphins. But bugs do not use echolocation…or do they? See Figure D.6.

If you zoom in on one of these loud knocks in the sonogram, it looks even more mysterious. The successively slowed down *thrum*, the bug goes ever *largo*, starting, stretching, and expanding like a single morphing phrase of a mockingbird, exploring the possible beat he is able to achieve. Perhaps it's not an insect, but a

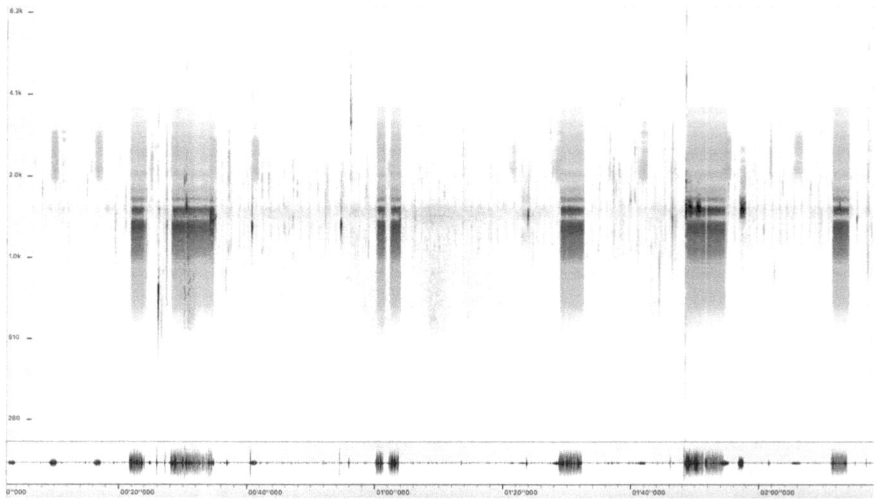

Figure D.6 0:00 to 2:20, nearly two and a half minutes.

Figure D.7 Hold your phone camera up to the code to be taken to a link that plays a sound sample from Lost Pond in Phillipstown, New York.

fish; or a turtle. I smile to realize that even in this nearby and accessible pond, no one really knows the answer. See Figure D.7.

When you sit long enough by this midsummer pond all kinds of creatures appear. The head of a skink peers up from between rocks. I know these reptiles are common, but I have never before seen one at this spot. In the field, the sound is rough and hollow, often silent, empty, repetitive mostly over the longform time. This is a thirty-minute recording in which I also join with the pond music, mostly inhabiting space between sounds, but the pond is very much alive. Mists rise inside me and that's why I add the swirls in the mix, the resounding tones, echo, and resonance; you can't place it because it doesn't come from an actual object in my own mind that wants to find meaning in the emergent, in the patterns

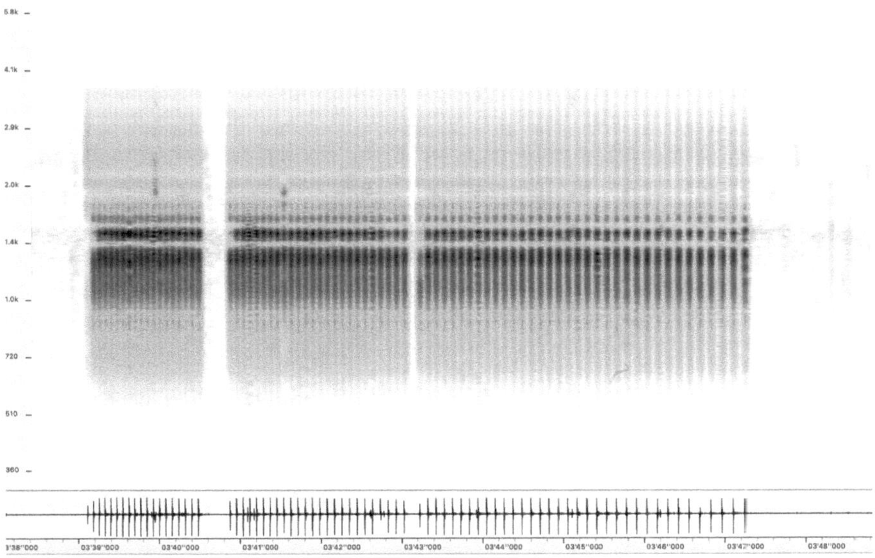

Figure D.8 Lost Pond Zoom-in: 3:38 to 3:48, ten seconds.

of nature that might only be there because we want them to be. Or they might indicate a yet-to-be realized order of the pond ecosystem.

This recording is quite spacious in its original form. Now, I want to fill it in with drip-drop drips of the mystery. I invite you, readers, to look at it, to listen to it, and *you* decide if there is any grand plan there, or just the beat of ambling noise—*three,* not two. The crackles in the pond go on, the shake, the separation, the silence. There *is* a rhythm down there I *knows* it. I *feels* it. I wants to be *in* it. So, I toss on some human effects and add the thrum—an *unidentifiable* hum, not a useful hum. (Though why is it not useful to engage?) Don't ask me what pitch it is, what note, what *rezz*. Resonating the click with a hum, I add recognizable tone. A quick breath, I want to add also. Put a clarinet on it. That's how I plan to make it my own. See Figure D.8.

Multispecies Harmonies in a Time of Forbidden Music

In October 2020, it had been eight months since the performing of live music has been forbidden due to a global pandemic. Friends in Iran over the years sent me photographs of billboards there announcing the illegality of music. I always laughed and then cried inside that somewhere in the world this is what authorities decry. I never thought it could happen here, and of course when it did it was for a different reason, a sense of distress, but also fear. Music was no longer safe. At least not indoors, in a quiet place, surfeiting over the silence. That was gone now, from a world we used to know.

This was the new reality that sent me out to the ponds nearly every day, as often as I could. Nothing so glamorous as an encounter with an octopus, like that film of the moment *My Octopus Teacher* (2020)—so interesting to learn it was made from footage shot for a David Attenborough production a decade previous in 2010. Filmmaker and diver Craig Foster thought about his year with the octopus for nearly a decade before figuring out exactly how the story should be told. The close encounter of one person with one animal, the same kind of thing people describe when staring into the eye of a whale. Me, when I'm making music with nature, I usually don't feel that. There is no me and him, me and her, me and it, but my sounds together with nature's sounds, live and leaping into the lap of the world. It's not that I wouldn't want such a one-on-one contact with someone in nature; it's just that I don't think that's the best way to find music there. It is not an I-Thou situation, but instead the music itself is the unity, the identity, the process that is more real than the various species who produce it together. It cannot be made by humans or ponds alone, but only when we decide that underwater, there is no such thing as silence, and that the sound world we hear there is on the path toward music. Our sense of what can be called music *expands*; that's always the reason I want to make art together with nature.

I thought this pond-derived music would sound impossibly abstruse, far beyond the likes and dislikes of most people, just a complete and utter obscurity, but actually these sounds seem to inspire as the audience starts to feel the rhythm. There are beats out there far beyond the simplicity of human time—the clicks of the plants breathing in, breathing out, make it seem like they are striving toward the need for music, without being aware of what they are doing. Nature is the teacher, underwater lies the source.

By the Awosting Dam, the water is shallow, nothing looks alive. Indeed, the water is supposed to be too acidic to encourage life, but life somehow remains. I hear them even if I don't see them. I listen to the captured hours over and over again. It becomes the natural context for my clarinet dreams. Bass clarinet. *Contra*bass clarinet. Just low, low as you can go. I try to unplay my instrument, to turn myself into a pond creature. I know nothing, absolutely nothing, outside the ringing rhythm of my rippling water.

Awosting Dam Busy With Clarinets: Water Bugs and Water Lilies Resonated With Clarinet Added

Lake Awosting, a famous destination at the end of the carriage roads in New York's Minnewaska State Park, is a fabulous place to swim, unflinchingly wild. I had always heard the water was too acidic to harbor any fish, but it is clearly full of photosynthesis and water bugs, though different-sounding ones than I found elsewhere; there were more high shakers and maracas than thumps and burps. I record the submerged sound, play it back, think about it, massage it gently into music, and wait until inspiration strikes to join along. See Figure D.9.

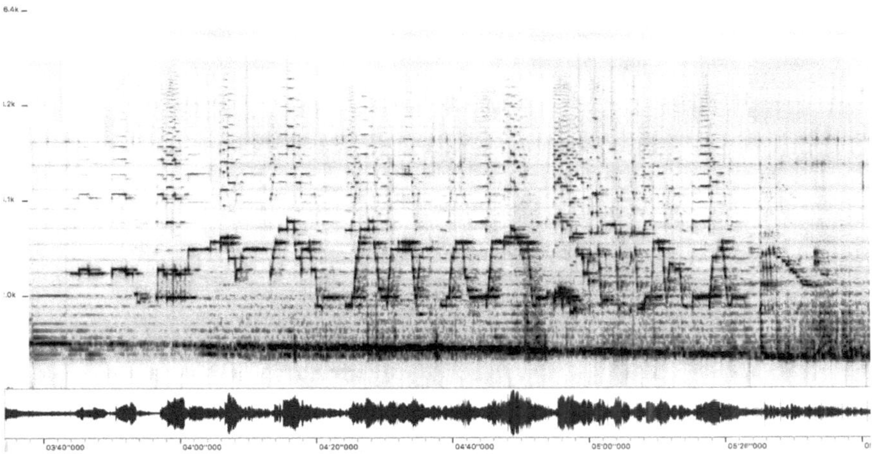

Figure D.9 This sonogram shows the entire final clarinet intervention, a rising and falling melody under the intensified thrum of this plant and animal pond world. The instrumental alternative to what we just heard above, when words were interred into the deep…here I was only ready to play after I had made the pond strike its own bells; 3:40 to 5:40, two whole minutes.

Figure D.10 Hold your phone camera up to the code to be taken to a link that plays a sound sample from Awosting Dam in Minnewaska State Park.

The piece is long. It has to be long because inside of the pond nothing so clear happens. It's as if the microphone is being dragged through the sand, scraping the bottom. But it is not. This is just the sum of what is down there, from the invisible silence above. I have nothing to say, and I am saying it. I am becoming the pond.

The world is bucking up against its ability to cope, and I am in the warming spring woods listening to ponds. These sounds will send me on all kinds of journeys, virtual and real. Listen to the results and ask me how I got here. You'll want a quest too. You deserve one. Ask and you shall answer. Open the door, walk out, head for the faraway nearby, and listen . . . See Figure D.10.

Bibliography

Cage, John, *Silence*. 1961. Middletown, CT: Wesleyan University Press.
Dunn, David. 2000. "Nature, Sound Art, and the Sacred." In *Book of Music and Nature: An Anthology of Sounds, Words, Thoughts*, edited by David Rothenberg, 95–107. Middletown, CT: Wesleyan University Press.
Shepard, Sam. 1980. *Four Two-act Plays*. New York: Applause Theatre Book Publishers.

15 A Multispecies Cosmovision

Knowing Nonhumans Beyond the Cognitive Imaginary in Buddhist Thought and Practice

Daniel M. Stuart

Challenges to Crafting a Buddhist Multispecies Epistemology

Thinking about Buddhism from the angle of multispecies studies, and within the framework of multispecies epistemologies, is challenging for a number of reasons. This is the case, first and foremost, because the referents of premodern "Buddhism" are diverse and remain contested among scholars with different methodological and theoretical commitments. So, while we might look at various case studies from a range of Buddhist contexts to query notions of Buddhist multispecies epistemologies, a normative perspective on how Buddhism or Buddhists might provide resources to think at the edge of multispecies studies remains somewhat elusive. Second, the two most dominant methodological approaches to the study of Buddhism—the doctrinal/philosophical and the socio-historical approaches—tend to project a default humanism or a metaphysical naturalism (often in the guise of a methodological naturalism) onto the foundations of the Buddhist tradition. A default humanism in these contexts tends to minimize the importance of nonhuman agencies within traditional thought systems, while metaphysical naturalisms tend to remove such agencies entirely unless they can be accounted for naturalistically. It is my contention here that the onto-epistemic commitments of these approaches often constrain the ways in which the various species depicted within traditional Buddhist cosmologies get understood and figured, at times obscuring the traditional onto-epistemic relationalities of such cosmologies in their Asian contexts.[1]

In this chapter, I approach the issue of multispecies epistemology within Buddhism by focusing on a specific cosmovision presented in a particular text-complex from around the third century CE in South Asia.[2] This text-complex is comprised of two voluminous and interrelated texts—the *Saddharma-smṛty-upasthāna-sūtra* ("The Teaching on Setting up Awareness of Real Phenomena") and the **Kāya-smṛty-upasthāna-sūtra* ("The Teaching on Setting Up Awareness of the Body")—both of which are attributed by tradition to the historical Buddha. This is a text-complex that I have worked on in a variety of capacities over the past decade. These textual sources are important because they present a picture of the cosmos as envisioned by one community of Buddhist meditation practitioners, known as *yogācāras*, a monastic community that appears to have been specifically

DOI: 10.4324/9781032659985-20

interested in theorizing how a visionary meditative monasticism might be cultivated gradually out of a basic commitment to foundational lay ethics. Unlike many other sources from this time period, these texts stand out in articulating with great specificity the ways in which human practices—be they everyday ethical commitments or the cultivation of rarefied states of meditation—are intimately enmeshed in an elaborate cosmic order populated by a wide range of human and nonhuman entities.

Perhaps most importantly, these are taxonomic texts that delineate a range of relationships within the cosmos by outlining various modes of being and embodiment, modes that are constituted by and constrained by vectors of action (*karma*) within a cycle of continued rebirth. They incorporate many foundational ideas and textual formulations from the earliest Buddhist traditions while at the same time demonstrating an onto-epistemic perspective of relationality in which all beings in the cosmos are inter-implicated in a flux of being, the structure of the cosmos—and its concomitant social worlds—in a process of entropy. This cosmovision appears to represent a somewhat particular form of relationality that is unique to the authors/compilers of our text-complex.

In what follows, I explore a few elements of these texts in an attempt to illuminate some of the ways in which attention to their multispecies taxonomies provides insight into the intertwined thought and practice systems of the Buddhists who produced them. But before getting into the details of the texts, I would first like to touch briefly on a theoretical issue in the field of Buddhist Studies that lurks behind some of my interests in the primary materials I will be presenting.

Methodological Reflections

We might begin with a Buddhist idea and a Buddhist practice that is perhaps as old as the Buddhist tradition itself. In the *Suttanipāta*, we find the following teachings:

> One refrains from sexual intercourse, abandoning sensual pleasures, be they low or high. He is [thereby] free from hostility or attachment towards living beings, be they movable or stationary.
>
> Analogizing [all beings with] oneself [by knowing that] "they are just as I am, and I am just as they are," one should not kill or cause harm.
>
> (*Sn* 709–10; trans. Stuart)[3]

And later in the same text:

> One should not steal nor speak falsely, but should suffuse movable and stationary beings with loving-kindness. When he recognizes turbidity of mind, he should cast it out, as it is 'the faction of the Dark One.'[4]
>
> (*Sn* 973)

In these passages, we find reference to a foundational outlook that runs through the entire Buddhist tradition into the present, a vision of reality in which human beings are kinsmen of plants, animals, and deities in an intertwined cosmic animacy shot through with ethically tinged cosmic forces. However, many of the peculiarities of the relationalities described here become fuzzy under the gaze of the metaphysical naturalist and socio-historicist perspectives of modern scholarship in Buddhist Studies.

I'd like to focus on two ideas at the center of these archaic passages. The first involves the relationship between the Buddhist practitioner and all other beings within the cosmos. The second involves the relationship between actions of the body, speech, and mind of practitioner persons, and the ways in which such actions are implicated in a set of cosmographical dynamics, figured as the interplay of light and dark forces at work throughout the cosmos.

Lambert Schmithausen drew attention to the above-quoted passages in his monograph on *The Problem of the Sentience of Plants in Earliest Buddhism*, noting that, "It would seem that at least oppressing and suffusing with benevolence do not make sense except with regard to living, nay sentient beings, and that hence both the mobile and the stationary beings must be living, sentient beings" (1991, 60). He goes on to note that in another passage from the same archaic textual source, grass and trees are explicitly referred to as "living beings" (*pāṇā*) and are set down at the beginning of a taxonomic list of modes of birth (*jāti*) that additionally includes various insects, four-legged beings, snakes, fish, and birds. As the final item on this list of living beings, the human is noted as distinct from other beings in that no clear differentiating distinctions in mode of birth occur within the human species in the way that they can be found among nonhuman beings (*Sn* 605–16). While we might note that the taxonomy presented here articulates a criterion for commonality between humans and nonhuman species through the notion of life, breath, or sentience (*pāṇa*), alongside a criterion for difference through the notion of mode of birth (*jāti*), I do not presently seek to construct a normative early Buddhist understanding of the implications of this taxonomy. Rather, I seek to call attention to a particular aspect of Schmithausen's methodological approach, which is emblematic of normative doctrinal approaches.

In elucidating his understanding of the early Buddhist sources, Schmithausen provides the reader a glimpse into his positionality as a scholar: "It would seem," he writes, "that at least oppressing and suffusing with benevolence *do not make sense* except with regard to living, nay sentient, beings . . ." (1991, 60; emphasis added). While I think that Schmithausen's conclusions are generally sound in this instance, I find myself concerned by the methodological assumption embedded in the notion that we can adjudicate the onto-epistemic frames of a cosmovision from the distant past by way of how they "make sense" from the perspective of modern Western scholarship.

The problematic implications of such assumptions become even more apparent in the interpretations of more socio-historically oriented scholarship. For example, in a recent series of articles exploring the articulation of "Buddhist" identity as, in part, emergent from a need for the Buddhist community to distinguish itself from other ascetic communities, Claire Maes has drawn on the Buddhist monastic legal/disciplinary literature (*Vinaya*) to suggest that the presence of some non-Buddhist ideas in these texts are the result of a sociological "process of othering" (2016, 2).

That is, precisely because so much was common to the various ascetic communities of early South Asia, the establishment of a normative Buddhist monastic code of law involved the need to delineate specifically between what was formally enjoined by the Buddha in practice and what other ascetic communities enjoined. Maes notes in the overview of her analysis that such interpretations should perhaps be limitedly ascribed to the projects of "the monk-editors of the *Vinaya*," but in the end—as is often the case in such instances—her interpretations exceed such bounds and appear to characterize the process of the construction of "Buddhism" writ large in early South Asia.

Under the gaze of such methodological approaches—which are filtered through analyses of very particular textual projects of the early tradition—Buddhist practice and its concomitant cosmovisions come to look like a series of strategic maneuvers in the service of identity. Take, for example, the following story from the *Vinaya* of the Theravādins:

> At that time a householder, the supporter of the venerable Channa, said to the venerable Channa: "Find the site for a monastery, good sir, and I will have a monastery built for the noble one." Then the venerable Channa, while clearing the site for the monastery, had a tree associated with a shrine (*cetiyarukkha*) cut down. The shrine-tree was worshipped by the [people of the] village, the town, the city, the region, the country. People were upset, troubled, and enraged: "How can these ascetics, Sons of the Sakyan, cut down a tree associated with a shrine (*cetiyarukkha*), which is worshipped by the [people of the] village, the town, the city, the region, the country. These ascetics, Sons of the Sakyan, harm one-facultied forms of life (*ekindriyaṃ… jīvaṃ*)."
>
> (Vin III: 156)

Both Schmithausen and Maes, despite their different scholarly approaches, take this passage and others like it as evidence that the concerns described in it about the harming of plant life are not a reflection of a real onto-epistemic commitment on the part of the Buddhists who transmitted the disciplinary literature (Maes 2011, 86–95). Instead, the passage reflects the politics of early Buddhist identity and the need to assuage a broader lay community that may have had commitments and sympathies beyond the Buddhist fold.[5] While such interpretations "make sense," they also do not do justice to likely scenarios on the ground, where the so-called "Buddhist" community was almost certainly never composed of a monolithic group committed to a single mode of identity. It seems highly likely, I would suggest, that quite a few Buddhists—monastics and lay people alike—would have participated in the norms that are reflected in this story. To his credit, Schmithausen draws on materials beyond the disciplinary literature to complicate such interpretations. But because he—like many others—works under the premise that text-historical sources (and what they tell us about Buddhism historically) can be adjudicated by recourse to a putative normative doctrinal Buddhism, the scholarly "problems" he constructs tend very quickly to morph into reified perspectives, from which possible cross-over practice contexts and onto-epistemic commitments get obliterated under the haze of the normatively doctrinal readings.[6]

Such obliterative modes of interpretation take their most distorting form in gestures such as those of the venerable Bhikkhu Anālayo in his recent collection of essays on *Vinaya Studies* (2017). In an appendix on the concept of *āsava*—a term that refers to karmic forces or "fluxes" that perpetuate the flow of rebirth for individual living beings—within the normative ethics of Buddhism, Anālayo draws on and critiques earlier scholarship that suggests that the Buddhist form of the concept owes its origin to the Jain form (2017, 328, n. 7, citing Jacobi 1914, 472). The Jain notion of *āsava* takes such karmic forces to be material "influences" or "influxes" that flow in upon a practitioner and bind them within processes of rebirth. In the Buddhist tradition, the final state of a liberated saint is—as in the Jain tradition—described as a state in which the *āsava*s have been destroyed. Noting key points of conceptual overlap and important contributions on the topic by Hermann Jacobi and Lambert Schmithausen, Anālayo nonetheless concludes that:

> Rather, sense-restraint [in Buddhism] appears to be concerned with avoiding the 'influence', as a psychologically conceived 'flowing in' that need not involve a material substance, of external detrimental factors on the mind. Such avoiding of detrimental influences on the mind would be very much in line with a general tendency of early Buddhist soteriology not to approach the notion of purification in a material sense, but rather to give overall importance to purity of the mind. There seems therefore to be no real need to assume that the term *āsava* must be a maladapted borrowing of Jain terminology.
>
> (2017, 328)[7]

In one fell swoop, Anālayo psychologizes away the notion of *āsava* as it was likely conceived among a range of ascetic communities in early South Asia, concomitantly associating "the mind" exclusively with the "immaterial" in line with the conceptual contours of the Cartesian philosophical tradition. While there certainly are some key ontological differences between normative doctrinal Jain and Buddhist conceptions of the mind (*mano*) and life (*jīva*), it is difficult to decide when and how such normative distinctions carried weight historically, particularly within the ascetic practice communities that remained the dominant social contexts for Buddhism and its various others into the first half of the first millennium of the Common Era. Perhaps most importantly, even the normative doctrinal concept of "the mind" in mainstream Buddhist thought does not hew to Anālayo's characterization of it. Rather, the term *mano* (Sanskrit: *manas*)—one key aspect of a broader conceptualization of "the mind"— is in many instances conceived of as being made up of an intermediary subtle form of matter, a force that constitutes the realm of *rūpa* (subtle materiality), and that can be discerned in deep states of meditation. The concept of the mind (*mano/manas*) as a form of subtle materiality also shares affinities with the onto-epistemic perspective (*darśana*) of another tradition of thought in early South Asia, that of the *Sāmkhya/Yoga* tradition.[8]

We thus find ourselves in a situation where the goals of modern scholars to construct a normative understanding of a doctrinal "Buddhism" condition their (text-historical and metaphysical naturalist) regimes of interpretation such that they tend to gloss over the complex practice contexts, practice communities, and inter-community co-creation of the teachings as they have been handed down historically. As a consequence of such approaches, it is often the case that complexifying details need to be dismissed, ignored, explained away, or addressed as "problems."

To come back to the notion of *āsava* within early Buddhist ascetic communities, when the extant descriptions of practice contexts we have at our disposal are taken at their word, it becomes very difficult to imagine that the early Buddhist *āsava* is anything other than a karmic force in the universe that tangibly "flows in" upon a meditation practitioner when he is not controlled in his senses. And to render this force as a "psychologically conceived 'flowing in'" levels entirely the ontological complexity of the early (and later) tradition, since the tradition allows for gross and subtle modes of materiality, some of which are explicitly understood to be mind-produced (*manomaya*) and thereby karmic.[9]

We might bring the dynamics of this interpretive issue back down to earth by returning to the passage from the *Suttanipāta* cited above enjoining the foundational early Buddhist practice of *mettā* or "loving-kindness." The teaching (*Sn* 973) instructs the ascetic to "suffuse movable and stationary beings with loving-kindness" and goes on to suggest that "turbidity of mind" should be recognized and "cast out," since it is a manifestation of the "faction of the Dark One." Two complementary aspects of practice are emphasized here: (1) the active cultivation of a positive relationship with all beings, in which the practitioner has the capacity to "suffuse" (*phasse*)—literally, to touch—them with the mental force of loving-kindness (no need to infer an "immaterial" or "disembodied" notion of the mental here); and (2) an associated negative cultivation, the removal of dark aspects of the so-called inner workings of the mind. Practice in relationship to externalities and practice in relationship to internalities are inter-implicated, and a practitioner is thereby at all times inter-implicated in the things and various species they interact with.

We might additionally note here that, in this simple set of prescriptions, a peculiar species of being—not included in the taxonomy mentioned previously—gets invoked. The "faction of the Dark One" refers here to a nonhuman of a very particular sort, an agentive force at work in the cosmos. Such forces can take form as specific deities, intent on the ascetic practitioner's failure—the most famous of which is the Buddha's nemesis, Māra—but they can also manifest within all beings as mental forces of various sorts or other forms of agency. That is, they are figured as both internalities and externalities. Like the elusive karmic force of *āsava*—that can flow in upon a practitioner from outside yet in some sense also becomes an integral part of the makeup of a habitually unrestrained practitioner—such forces perhaps provoke animistic and analogistic vectors of analysis. Yet, I am hesitant to reduce such relationalities to a theory of ontology, since at times such theories can—despite their analytical utility—lead to the leveling of the complexities of the relationalities of a particular system of thought.[10]

Instead, I think it is fruitful to explore the ways in which some of the relationalities implicit in this archaic articulation of Buddhist ascetic *practice* have onto-epistemic implications and reveal a broader set of relationalities that run through the Buddhist tradition, even into the present, in various ways.

The *Saddharma-Smṛty-Upasthāna-Sūtra* and the **Kāya-Smṛty-Upasthāna-Sūtra*: A Buddhist Cosmovision and Its Multispecies Perspectivism

In what follows, I would like to explore the implications of such relationalities and how they find expression in a much more developed set of Buddhist teachings from the Northwest of the Indian subcontinent during the early centuries of the Common Era. By exploring understandings and relationships that a practitioner is encouraged to develop with a variety of human and nonhuman beings—and the conceptions of such beings and their modes of knowing/embodiment within a broader theory of karma—I hope to shed light on a particular Buddhist cosmovision from the distant past and how its multispecies taxonomies might generate productive dissonances in the present in conversation with a variety of ideas about speculative futures emergent at this current moment.

Looking briefly at how these teachings depict the contemplative visionary exploration of modalities of knowing for humans on mythological continents of the cosmos, modalities of being for colonies of minute "worms" or microorganisms (*krimi*) that reside within the human body, distinct life forces for a variety of animal embodiments, the agencies of spirits in the production of natural phenomena, and prescriptions for not harming various possessive spirits, I seek to interrogate the mainstream assumption within Buddhist Studies that "action" (*kamma/karma*) as a capacity of mind within Buddhism ought necessarily be conceptualized in purely (psychological) cognitivist terms. Rather, I suggest that within the cosmovision of the *Saddharma-Smṛty-Upasthāna-Sūtra* and the **Kāya-Smṛty-Upasthāna-Sūtra*—both of which are encyclopedic and taxonomic texts on how the practice of a Buddhist *yoga* practitioner is inter-involved in a grand and particular Buddhist cosmography—the mental karmic forces of each and all beings are conceptualized as intimately bound up with all activities of the cosmos and are figured as largely the result of automatic processes.

If we take this picture of the cosmos seriously, and carry the relationalities presented in these texts to their final conclusions, these teachings appear to present a highly elaborate perspectivism linked to a theory of differential modes of knowing-being among various humans and nonhumans. The guiding focus of the theoretics within which this cosmogram emerges is, however, focused in the main on the implications of human action. That is, the texts' speculative visionary perspectivism can't be separated from a depiction of the path of human Buddhist practice that begins with a commitment to normative lay ethics and eventually gets refined into ascetic ethical modalities that can serve as a foundation for cultivating a visionary relationality that allows for the enaction of the variety of modes of knowing-being that configure the texts' perspectivism.

Species and Sensorium: Body and Sense as World-Constructing Containers

An integral facet of these encyclopedic sources is their focus on a meditative exploration of the six senses and the way in which different modes of sense experience across species construct different kinds of experiential worlds. In this connection, we might first analyze a description of the yogic practice of awareness of the body that is set out at the very beginning of the **Kāya-smṛty-upasthāna-sūtra*, "The Teaching on Setting Up Awareness of the Body," which is worth quoting at length:

> Monks who inspect the body correctly will not be carried off by unskillful attention. They become free from the defilements. The body comprises the eyes, ears, nose, tongue, body, and mind that are of an internal nature, along with external visual objects, tastes, smells, tangibles, sounds, and phenomena. When one dwells observing the body in the body internally, one approaches *nirvāṇa*. In this regard, when a monk sees a visual object with the eye, because he does not cogitate, he correctly inspects the object [as follows] with a state of mind that does not involve lustful desire: 'There is in this body hair, body hair, nails, bones, skin, blood, sinew, consumed food, eyebrows, phlegm, bile, air, intestines, the colon, the anus, lungs, flesh, fat, semen, lymph, pus, the brain membrane, and the brain… .
>
> How does a meditation practitioner correctly inspect the physical eye that is a transformation of the four great elements? The physical eye that is of an internal nature has a slight degree of solidity and firmness. That is what is known as the earth element of the physical eye that is of an internal nature. The physical eye that is of an internal nature also contains a distinct element of water, which is wet and fluid. Thereby the eye retains moisture. That is what is known as the water element of the physical eye that is of an internal nature. The physical eye that is of an internal nature also contains a distinct element of fire, which is warmth and heat. Thereby the eye retains warmth. That is what is known as the fire element of the physical eye that is of an internal nature. The physical eye that is of an internal nature also contains a distinct element of wind and air, thus yielding movement. That is what is known as the wind element of the physical eye that is of an internal nature. In the same way that one understands the eye as it is, so also a monk discerns the ear, nose, tongue, and body… .
>
> (Kāsu 5.3–4)[11]

The text then goes on to depict a practice in which a meditation practitioner speculates on the sensorium of humans of a different species (or "class" or "embodiment"; *ris* [Sanskrit, **nikāya*?]) that live on (mythical) continents other than Jambudvīpa, the continent that is the homeland of Buddhist tradition. In the traditional cosmography of the systematic Buddhist traditions, there are four continents: (1) Pūrvavideha in the East, (2) Jambudvipa in the South, (3) Godānika in the West, and (4) Uttarakuru in the North. I present here a description of the meditation on the ear sense-faculty of persons living on the northern continent of Uttarakuru:

> Further, the meditation practitioner examines the bodies of humans in Uttarakuru by means of his own body in Jambudvīpa: 'The sounds that my ears can hear may be unpleasant or clear, but I do not register sounds when there is an obstruction. I hear sounds when they are loud. But is it the same for humans in Kuru in the north?' When he thus examines the ear as it is, he will notice that due to their unique situation, the people on Kuru in the north hear sounds whether they are nearby, distant, weak, pleasant, or unpleasant. They hear in a way that resembles the way the sun shines on all things, whether they are nearby, distant, common, inferior, pure, or impure. Humans in Kuru in the north hear sounds in a manner similar to that.
>
> (Kāsu 5.8)[12]

What is going on here? We have in this description a unique mode of practice cum theorization, one that demonstrates how a peculiar multispecies speculation is part of a broader cosmological project of the early systematic Buddhist traditions. In this version of such a project, traditional cosmographic ideas become avenues for speculations on differential modes of embodiment—and thereby on different onto-epistemic perspectives. These speculations then get enacted as visionary practices in which a complete world-image or cosmogram—and the various modalities of knowing and being within it—can be traversed by a meditation practitioner.

The Enlivened Human Body

This cosmographic perspectivism gets brought down to earth—and extended to the heavens—in various ways in our sources. For example, the **Kāya-smṛty-upasthāna-sūtra* additionally demonstrates a unique repertoire of medical knowledge, perhaps most clearly delineated in the text's system of 80 winds (*vāyu*) and 80 microorganisms or "worms" (*krimi*) that interactively support the functioning of the human body. What stands out about this system are the specific descriptions and conceptualizations of the microorganisms, and the ways in which they interact with the bodily winds in agentive modalities to create balance or imbalances in human bodies.[13]

The meditative discernment of this system of microorganisms in the **Kāya-smṛty-upasthāna-sūtra* follows upon the discernment of the external cosmography partially detailed in the above discussion of the theorization of humans from different continents, a meditation that concludes with a taxonomy of modes of rebirth in the animal realm. This depiction begins as follows:

> In this way, brahmins, householders, and monks of the town of Nālati, a meditation practitioner dwells observing the body within the body, having correctly discerned the external body by means of the internal body. How, then, brahmins, householders, and monks of Nālati, does one dwell observing the body within one's own internal body? Discerning as they really exist the various impure

> elements in one's own body internally, one dwells observing the body within the body by repeatedly examining and investigating [it] from head to toe.
>
> (D Śa 117)[14]

The text then goes on to detail the microorganisms and winds of the body in ways that suggest that a practitioner should discern modalities of knowing and being of the bodily microorganisms in order to fully comprehend the life of the body. The description of the contemplation of the "blood-dwelling" microorganisms is exemplary:

> The meditation practitioner continues to dwell observing the body within his own internal body: 'How do my microorganisms influence the condition of the main channels that run through my body, starting from the tail bone?' He sees with knowledge derived from scripture or with the divine eye that when ten types of microorganisms produced in the gut enter the spleen and liver, they give rise to disease. Those are the hair eater (**romāda*), the driller (**sosura*), the mother of beings (**jantumātara*), the *audumbara*, the one that flows (**prasrāva*), the hair protector (**romadvīpa*), the microorganism infuriated by blood, the blood-consumer (**romabhakṣa*), the trembler (**simisimi*), and the *aṃdhūka*. All these microorganisms have no legs and they writhe freely. They are no larger than the most subtle particle, have no eyes, and they cause itching and disease.
>
> Further, the meditation practitioner continues to dwell observing the body within his own internal body: 'What are the activities of each of those microorganisms?' He sees with knowledge derived from scripture or with the divine eye that when the hair eaters become aggressive, they consume the eyebrows and eyes, or one's facial hairs and body hairs. They also cause warts. When the driller microorganism becomes aggressive, it enters the bloodstream and causes disturbances throughout the body. The mother of beings microorganism gives the bloodstream a foul smell. Being a blood microorganism, the *audumbara* is provoked and agitated by blood, thus causing pustules, pimples, and ulcers in the mouth and nose. [F.124.b] When the one that flows becomes aggressive within the blood stream, it causes disease of the liver, blood, and bile. When the hair protector enters the blood, it becomes aggressive and causes an aggravation of those conditions, thus giving rise to itching, jaundice, and pustules. When the microorganism infuriated by blood enters the bloodstream, it causes warts and diseases associated with desire. In women, it also causes an itch of the blood and skin ruptures. The blood-consumer causes dizziness, an unsteady neck, and inflammation of the anus. When the trembler enters the veins, it causes fatigue, paralysis, shaking, and discomfort of the mouth. Such are the effects of the blood-dwelling microorganisms and their retinues.
>
> (Kāsu 5.69–70)[15]

We might note here how these depictions present a dynamic picture of the physical body as active and animated by a variety of agencies. The winds and microorganisms function interactively as the body in a process of life, and to contemplatively discern and understand that process involves understanding those agencies, some of which are figured as extremely subtle. In the case of the microorganisms in particular, the explicit meditative discernment of these subtle interactions also involves the discernment of how such minute beings themselves know and act in dynamic relationalities within the body. It should also perhaps be pointed out that within this system of winds, it is the wind of karma, the *karmavāyu*, that has paramount importance, since it is ultimately the arbiter of the things that truly matter for the spiritual life.

Taxonomies of Intentionality, Sentience, Life-force

Another meditative modality found in our text-complex—this time in the *Saddharma-smṛty-upasthāna-sūtra*, or "The Teaching on Setting up Awareness of Real Phenomena," emerges from early Buddhist canonical systematizations wherein various beings get classed according to their modes of sustenance or "food" (*āhāra*). In the *Saddharma-smṛty-upasthāna-sūtra*, we find a peculiar taxonomy along such lines:

Further, that monk dwells observing phenomena among internal phenomena:

> '[Being] bound by which bonds do beings run on, fare on, in the flow [of existence]?'
>
> He sees with knowledge produced through scripture, or with the divine eye: 'This world is bound by two bonds: [1.] the bond of food (*āhāra*) and [2.] the bond of contact.
>
> 'In that respect, [the world is bound] by a fourfold bond of food: gross material food, the food of mental volition, the food of meditation, and the food of contact.
>
> 'Which [beings] subsist on gross material food? The four abodes of men, the six [types of] deities of the realm of sensuality, [beings in] the eight great hells, as well as some hungry ghosts and animals [subsist on gross material food].
>
> 'Fish and those of that type subsist on mental volition (*manosañcetanā*).
>
> 'Deities of the realm of subtle materiality subsist on meditation (*dhyāna*).
>
> 'Birds and snakes subsist on contact, as do others of that type, who are beguiled by mating through contact.
>
> 'All foolish worldlings of the sensual sphere—men, hungry ghosts, animals, or denizens of hell—comprise the sensual sphere because [they] indulge in sensual pleasures.
>
> 'Those of the immaterial realm subsist on the objects of the [immaterial] meditative attainments.
>
> 'In this way, the world is bound by two [bonds]…'.
>
> (Saddhsu II 9.5–9.6.9 in Stuart 2015, 520–23; modified)

As far as I am aware, this taxonomy of modes of sustenance is unique to this particular text-complex, but a range of associated taxonomies can be found in the earliest layers of the oldest Buddhist literature.[16]

There's a lot going on here, and little space to fully unpack all of the relationalities at work. We might, for present purposes, simply single out several aspects of this taxonomy. First, the species described here include a variety of humans and nonhumans. The nonhumans include nonhuman animals, and a range of supernormal beings found in various realms of the cosmos, such as hungry ghosts and immaterial deities, among others. Second, and perhaps most importantly, "food" or sustenance here has a wide semiotic range. We can understand it in concrete terms as the gross material food humans subsist on, but in the case of deities and certain animals who subsist on "meditation" or "mental intention," or "contact," we have to understand food as more basic sorts of life-force at work in the cosmos, life forces that are inherent to those modes of existence and that also function as modes of sentience.

I would like to focus in particular on the notion that fish and other animals like them subsist on "mental intention" (*manaḥsañcetanā*). This technical term is the active modality of another important technical term within the Buddhist tradition, *cetanā*, which often gets translated as "intention," "will," or "volition." But through much of this text, and also in the oldest Buddhist scriptures extant, *cetanā* mostly appears as an inherent force that is active in every moment of mental life, constitutive of the most basic forms of consciousness or awareness along with sentience (*vedanā*) and percipience (*sajñā*). The translation of *cetanā* as "intention" and *mahaḥsañcetanā* as "mental intention," therefore, may in fact lead away from a proper understanding of the way that these texts—and perhaps other Buddhist philosophical traditions as well—thought about mental action. Here *mahaḥsañcetanā* appears as a life-force deeply intertwined with a more basic sentience, a largely automatic impulse that supports existence for certain kinds of beings. *Cetanā* elsewhere appears to work in this way—as a basic and automatic force associated with sense contact—for the mental life of humans, even though such beings also must be nourished by gross material food.

Against Naturalism: Discerning Karma Beneath the Natural

A number of other relevant agencies and modes of knowing-being get invoked at various points in our text-complex, and I would like to turn to a couple of them in closing as we consider the relationship between the perspectivism depicted in these sources and their stipulative teachings on how to act in the world as a human. One aspect of the *Saddharma-smṛty-upasthāna-sūtra* that comes to the fore as it develops its picture of the cosmos involves the incorporation of a mythical narrative of a perennial cosmic battle—common to many strands of premodern tradition in South Asia—that is constantly being waged, and in which human actions are implicated.[17] In the *Saddharma-smṛty-upasthāna-sūtra*, this narrative is also accompanied by a somewhat strident anti-naturalism. We find in the text, for instance, an elaborate description of the importance of wholesome human action in conditioning the agencies of deities, earth spirits, sky spirits, and celestial snake persons (*nāga*)—classed as a form of nonhuman animal—battling supernormal counterforces at work in the universe. These forces are also figured variously as the material substance of what some now (and then) consider(ed) natural phenomena, such as earthquakes, eclipses, and meteor showers:

> The monk who has knowledge of action (*karma*) and the ripening of its fruit will further survey the activity of *nāga*s and evil *asura*s: 'How does a meteor shower occur?' He sees with knowledge derived from scripture how those spirits with great power and great force inform the wandering deities, as previously [described]...Those wandering deities of the Retinue of the Four Great Kings will bristle with excitement: 'Don't fear, don't fear! The faction of the *asura*s is diminished and our deity faction is strong. In Jambudvīpa, there are righteous people who follow the Dharma, people who respect their mothers and fathers, are ascetic and brahminic, have taken up restraint, and are caretakers of the leading clans. [Because of this,] we shall vanquish the aggressive *nāga*s and evil *asura*s!'
>
> Hearing such speech from the wandering deities, the spirits with great power and great force bristle with intense anger. With anger directed at the *asura*s, [they proclaim] all that as was previously [proclaimed]...so as to inform the righteous *nāga* kings who follow the Dharma, beginning with Takṣaka and Vāsuki. If these deities—bodies shining bright—advance [during the day], they can't be seen because the rays of the sun outshine them. If, with their bodies blazing, they advance into the sea at night to inform the righteous nāgas who follow the Dharma, they are visible to all.
>
> Prognosticators think of that vision (*dṛṣṭotpāda*) [in various ways] as follows: 'A meteor has fallen. Harvests will be good! Harvests will be bad! There will be [continuity of] rule! There will not be [continuity of] rule! There will be war!' 'There will be peace!' 'Cows and brahmins will flourish!' 'Cows and brahmins will decline!' In this way, they imagine things to be so without having direct visionary knowledge of action (*karma*)."
>
> (Ms 112a1–5)

In addition to this being another instantiation of the text's perspectivism, which here can be seen in the monastic's visionary discernment of the modes of knowing-being of various immanent spirits, material deities, and mythical non-human animals whose activities form a central structuring element of the positive and negative forces of the cosmos, we also see here an argument for the idea that human activity powerfully influences the order of things in all realms of the cosmos. What is more, in this perspective on reality, a true visionary meditation practitioner will come to have direct knowledge that phenomena considered to be natural by non-Buddhist prognosticators ultimately have supernormal referents, referents that are conditioned by the forces of human action.

These cosmic relationalities point to an overarching perspective of our text-complex: the idea that the health and welfare of the entire cosmos hinges on whether humans will undertake the righteous ascetic lifestyle stipulated by the Buddha. The renunciatory outlook of the texts is perhaps most emblematically embodied in its stipulations on the benefits of giving the "gift of life" (*jīvitadāna*). We see one instantiation of such a stipulation in a depiction of how one might develop karmic qualities that lead to rebirth in a particular heaven where righteous deities dwell:

> And how does one remain abstinent from the taking of life? He performs good actions here and is fearful of the next world, with complete conviction in the law of action and fruit. Innocent by nature and harmless, he avoids bad friends and seeks happiness. [Such a person], upon being possessed by a possessive spirit or taken over by an ancestor while going into rivers or crossing a river, relinquishes the self. He does not want to kill that possessive spirit by administering poison. [He understands] that if poison is administered to that spirit of his, that possessive spirit will die. Though cognizant of such a remedy, he does not kill that possessive spirit. Though cognizant of such a remedy for an ancestor, he does not kill [it]. Its death [can be brought about] by submergence in water, by looking into a mirror, or by gazing at the sun. Though cognizant of a means to combat [the ancestor], he does not kill [it]. He instead relinquishes his own life. He performs good actions. In this way, one who carries out good actions, upon the breaking up of the body at death, is reborn in a good heavenly destination, in a level of the deities of the Heaven of the Thirty-three, among deities known as Having Bodies Produced by Freedom of Want.
>
> (Ms 209a3–5)

This ascetic willingness to relinquish life for a larger cosmic order, an order involving relationalities that implicate all human activity in the welfare of every facet of the cosmos—from microorganisms to trees to nonhuman animals to earth-bound spirits to exalted deities—emerges from a very particular cosmovision. It is one that calls on all humans and nonhumans to look very closely at every action they find themselves involved in and take responsibility for their part in the state of affairs in the broader cosmos.

Pluriversal Reworlding? Seeking Buddhist Multispecies Relations Beyond Doctrine, Caricature, or Critique

To return to the initial framing of this essay, I would also suggest that this cosmovision departs radically from some of the normative modern interpretations of the Buddhist teaching on karma as "intention," which can perhaps be gleaned from a narrow focus on the sources of the legal/disciplinary and early doctrinal Buddhist literature. For example, we find the following account in the *Vinaya* of the Theravādins:

> At one time a certain monk was possessed by a spirit (*amanussa*). Another monk struck him [to try to remove the spirit]. The monk died. The monk who had struck him got worried [as to whether he was guilty of an offense]… . [The Buddha declared that]: 'It is not an offense, monk, since there was no intent (*adhippāya*) to kill him.'
>
> (Vin III: 187 [85])

These legal scenarios highlight one modality of the early Buddhist tradition, and they provide important material for thinking about certain aspects of the social life of the tradition. They have also served historically as resources for caricature and critique by non-Buddhist critics. Doctrinal sources likewise do similar work in a different sphere of discourse. But it is my sense that such sources—and the normative perspectives that scholars often derive from them—lead us to overlook some of the more fundamental onto-epistemic commitments and relationalities that undergird many of the practice contexts of Buddhist communities in history. Engaging the speculative visionary perspectivism of the cosmovision I have analyzed in this chapter provides an opening for a loosening of the grip of normative doctrinal/philosophical and socio-historical methodologies in Buddhist Studies. It also possibly provides a pathway for a consideration of how distant and dissonant modes of knowing life and life-knowing might provide resources for the cultivation of novel modes of existence in the present.[18]

This cosmovision from the distant past may be alienating to some, and it is perhaps for this reason that textual sources like the ones we have explored here have been largely ignored or scorned by modern scholarship. Yet, it is my sense that the recovery of material of this sort—with a view to a pluriversal reworlding through alternative multispecies epistemologies—may, in fact, be catalyzing in our present circumstances. In invoking the notion of a "pluriversal reworlding" here, I do not mean to suggest that we can simply recover an ancient cosmovision and live it. Nor do I necessarily suggest that we might want to do so even if we could. Many facets of the cultural regimes of ancient ascetic South Asia will likely have little purchase in our current cosmopolitical moment. Rather, I gesture here toward what Martin Savransky refers to as the "pluralistic problematic" (2021a, 145, 150, and 155–56; 2021b, 8, 11). Savransky develops this idea as one strategy of living with the fundamental tensions inherent in the pluralism of material and cultural worlds, and how they stand in relation to experiences and posits of a singularity—a singularity that might at times appear immanent and at others transcendent. Whichever world we find ourselves in as we move into possible futures, I hope that our collectives can learn to take responsibility for their contributions to the structure of the cosmos. Perhaps an ascetic orientation rooted in a radically embodied loving-kindness and an epistemically open perspectivism might be one place to begin.

Notes

1 See, for example, the fairly recent works of Ambros 2014 and Ohnuma 2017.
2 For previous work on these materials, see Stuart 2015, 2017, 2018, 2019, and 2020.
3 All translations are mine.
4 I follow Bhikkhu Bodhi (2017, 319 [verse 967]) in rendering the phrase *"kaṇhassa pakkho" ti* as "faction of the Dark One."
5 In a recent article on Jain nonviolence, Brianne Donaldson has incisively pointed out that normative perspectives often emerge from polemical contexts that, "by nature, mischaracterize and simplify rival positions. Any close reading of early Buddhist and Jaina texts, including the Jaina concepts of *ārambha* and *parigraha* described above,

reveal that both communities consider action and intention as entangled phenomena" (Donaldson forthcoming).

6 For a useful overview of some of these dynamics across traditions in South Asia, in a similarly normativizing—though still useful—register, see Ferrari and Dänhardt 2016, xi–xxxi.

7 It is noteworthy that Anālayo cites (and appears to be following) Richard Gombrich's take on this issue—which echoes Schmithausen language of sense-making: "the term [*āsava*] makes no sense, as in Buddhism there is nothing which 'flows in' on one" (Gombrich 2009, 56). It seems clear to me that this is an instance of problematic interpretive work, and yet such interpretive approaches remain normative in the field of Buddhist Studies.

8 On the classical metaphysics of the Sāṅkhya tradition, see Larson 1998.

9 On the concept of the "mind-made body" in early Buddhist and and Upaṇiṣadic thought, see Notaris 2019.

10 Here I have in mind ontological theories of the type that Philippe Descola has developed in *Beyond Nature and Culture* (2013) and elsewhere.

11 Dharmacakra Translation Committee 2021: 5.3–5.4 (cf. T17, no. 721: 379a18–25). This translation has been heavily modified based on my own reading of the Tibetan translation [D Śa 110a4–b7], the Chinese translation, parallel terminology in Sanskrit, and my preferences of technical terminology.

12 Dharmacakra Translation Committee 2021: 5.8 (slightly modified) (cf. T17, no. 721: 379b27–c4). On the lack of egoism among the inhabitants of Uttarakuru, see Stuart 2015: 414–15.

13 On the eighty winds, see Ishikawa 2002a and 2002b. On the eighty "worms," see Kritzer 2023. These pieces present schematic and partial analyses of the system of winds and microorganisms in the **Kāya-smṛty-upasthāna-sūtra*.

14 Cf. T17, no. 721: 381c11–14. This translation is informed by Dharmacakra Translation Committee 2021: 5.37. The depiction of practice here is an elaborate expansion of the traditional Buddhist meditative practice that takes as its object "impurity" (*aśubhā*). On the early form of this practice, see AN V 109 (trans. Bodhi 2012: 1412).

15 This translation is based on that of the Dharmacakra Translation Committee, but it has been heavily modified with reference to Chinese and Sanskrit parallel materials (cf. D Śa 124a2–b4; T17, no. 721: 384a6–28; Ms 71b6–7). On the worms in this passage, see Kritzer 2023, 320–21.

16 See, for example, SN (12.3.5) II 37–41 (trans. Bodhi 2000, 559–62) and SN II (12.7.3) 97–100 (trans. Bodhi 2000, 597–99).

17 For a brief version of such a mythical battle from a "Hindu" source, see Dimmitt and van Buitenen (trans.) 1978, 299–303. On the broad theme of Buddhist cosmography and its connection to Buddhist practice and other traditional South Asian cosmologies, eschatologies, and timescapes, see Kloetzli 1989. For a speculative reflection on the possible connections between this narrative of a perennial cosmic battle, mythological South Asian cosmography, and Ptolemy's *Geography*, see Kloetzli 2010.

18 See, for example, Bollée 1999, in which Bollée details "the oldest extant dispute between Jains and heretics" as a critique of Buddhist theories of "intention." In that account from *Sūyagaḍa* 2.6, the Jain mendicant Adda, upon encountering some Buddhist monks, caricatures a normative Buddhist approach to ethical responsibility in which "intention" is the benchmark for karmic culpability: "If someone puts a man or a child on a spit and roasts it on a fire taking it for a lump of oil-cake, it would be fit for Buddhists to end their vow of fasting with" (Bollée 1999, 413).

Bibliography

Primary Sources

AN *Aṅguttara Nikāya* (see Hardy [ed.] 1958/1900).

D Derge Edition of the Tibetan Tripiṭaka, published by the Tibetan Buddhist Resource Center 2002 (based on a scanning of the photomechanical reprint of the *par phud* printing published in Delhi by Karmapae chodhey gyalwae sungrab partun khang, 1976–79).

Kāsu **Kāya-smṛty-upasthāna-sūtra* (see Dharmacakra Translation Committee 2021 [D Śa 109b7–227b7 (T17, no. 721: 379a10–417c19)]).

Ms A photocopy of the single known Sanskrit manuscript of the *Saddharma-smṛty-upasthāna-sūtra* kept in the collection of the China Tibetology Research Center (Box 12, No.1), the original of which is held at Norbulingka in Lhasa.

Saddhsu *Saddharma-smṛty-upasthāna-sūtra.*

SN *Saṃyutta Nikāya* (see Feer [ed.] 1989/1888).

T Taishō edition of the Chinese Buddhist canon (CBETA 電子佛典集成光碟 2011).

Vin *Vinaya Piṭakaṃ* (see Oldenberg [ed.] 1993/1881).

Secondary Sources

Ambros, Barbara. 2012. *Bones of Contention: Animals and Religion in Contemporary Japan.* Honolulu: University of Hawai'i Press.

Anālayo, Bhikkhu. 2017. *Vinaya Studies*. Taipei: Dharma Drum Publishing Corporation.

Bodhi, Bhikkkhu, trans. 2000. *The Connected Discourses of the Buddha: A New Translation of the Saṃyutta Nikāya*. Somerville: Wisdom Publications.

———, trans. 2012. *The Numerical Discourses of the Buddha: A Translation of the Aṅguttara Nikāya*. Somerville: Wisdom Publications.

———, trans. 2017. *The Suttanipāta, An Ancient Collection of the Buddha's Discourses, Together with Its Commentaries*. Somerville: Wisdom Publications.

Bollee, W. B. 1999. "Adda Or The Oldest Extant Dispute Between Jains and Heretics (Sūyagada 2, 6), Part Two." *Journal of Indian Philosophy* 27 (1999): 411–437.

De Notaris, Bryan. 2019. "The Concept of *Manomaya* in Early Buddhism and the Upaniṣads: A Study with Particular Reference to the Pāli *Sīlakkkhandhavagga*." *Journal of the International Association of Buddhist Studies* 42: 47–81.

Dharmacakra Translation Committee (under the patronage and supervision of 84000: Translating the Words of the Buddha), trans. 2021. *The Application of Mindfulness of the Sacred Dharma (Saddharmasmṛtyupasthāna)*. https://read.84000.co/translation/toh287.html

Dimmitt, Cornelia and J. A. B. van Buitenen, trans. 1978. *Classical Hindu Mythology: A Reader in the Sanskrit Purāṇas*. Philadelphia: Temple University Press.

Donaldson, Brianne. Forthcoming. "The Hunter, the Bow, and the Arrow: The Development of Intentional Harm as *Kriyā* in the Early Jain Canon." In *(Non-)Violence in Jaina Philosophy, Literature and Art*, edited by Peter Flügel. 1–4. New Delhi: Dev Publishers and Distributors.

Feer, M. Léon, ed. 1989/1888. *The Saṃyutta-nikāya, Part II, Nidānavagga*. Oxford: Antony Rowe Ltd. for the Pali Text Society.

Ferrari, Fabrizio M. and W. P. Dänhardt, ed. 2016. *Roots of Wisdom, Branches of Devotion: Plant Life in South Asian Traditions*. Sheffield and Bristol: Equinox.

Gombrich, Richard. 2009. *What the Buddha Thought*. London and Oakville: Equinox Publishing.

Hardy, Edmund, ed. 1958/1900. *The Aṅguttara-nikāya, Part V, Dasaka-nipāta and Ekādasaka-nipāta*. London: Luzac & Company, Ltd. for the Pali Text Society.

Ishikawa, Mie (石川美恵). 2002a. "『正法念処経』における「風」について, 一チベット訳を中心として一." 東洋学研究 39: 125–42.

———. 2002b. "『正法念処経』の rlung について." 印度學佛教學研究 50 (2): 210–12.

Kloetzli, W. Randolph. 1989. *Buddhist Cosmology: Science and Theology in the Images of Motion and Light.* Delhi: Motilal Banarsidass.

———. 2010. "Ptolemy and Purāṇa: Gods Born as Men." *Journal of Indian Philosophy* 38, no. 6 (December): 583–623.

Kritzer, Robert. 2020. "Meditation on the Body in Chapter 7 of *Saddharmasmṛtyupasthāna-sūtra*." *Religions* 11, no. 6: 283.

———. 2023. "Worms in *Saddharmasmṛtyupasthānasūtra*." In *Burlesque of the Philosophers: Indian and Buddhist Studies in Memory of Helmut Krasser, Part I*, edited by Vincent Eltschinger, Jowita Kramer, Parimal Patil, and Chizuko Yoshimizu, 309–48. Bochum/Freiburg: Projekt Verlag.

Larson, Gerald James. 1998/1969. *Classical Sāṃkhya: An Interpretation of its History and Meaning.* Delhi: Motilal Banarsidass.

Maes, Claire. 2011. "One-Sensed Facultied Life (*ekindriya jīva*) in the Pāli Vinaya: A Camouflaged Debate Between Early Buddhists and Jains." *Bulletin d'études indiennes* 28–29: 85–104.

———. 2016. "Flirtation with the Other: An Examination of the Processes of Othering in the Pali Vinaya." *Bulletin of School of Oriental and African Studies* 79, no. 3: 535–57.

Ohnuma, Reiko. 2017. *Unfortunate Destiny: Animals in the Indian Buddhist Imagination.* New York: Oxford University Press.

Oldenberg, Hermann, ed. 1993/1881. *The Vinaya Piṭakaṃ, One of the Principal Buddhist Holy Scriptures in the Pāli Language, Vol. III, The* Suttavibhaṅga, *First Part* (Pārājika, Saṅghādisesa, Aniyata, Nissaggiya). Oxford: Antony Rowe Ltd. for the Pali Text Society.

Savransky, Martin. 2021a. "Counter-Apocalyptic Beginnings: Cosmoecology for the End of the World." *Tapuya: Latin American Science, Technology, and Society* 4, 1–14. Article number 1914423.

———. 2021b. "The Pluralistic Problematic: William James and the Pragmatics of the Pluriverse." *Theory, Culture & Society* 38, no. 2: 141–59.

Schmithausen, Lambert. 1991. *The Problem of the Sentience of Plants in Earliest Buddhism.* Tokyo: The International Institute for Buddhist Studies.

Stuart, Daniel M. 2015. *A Less Traveled Path:* Saddharmasmṛtyupasthānasūtra *Chapter 2, With a Study on Its Structure and Significance for the Development of Buddhist Meditation*. Vienna and Beijing: Austrian Academy of Sciences Press—China Tibetology Publishing House.

———. 2017. "Unmanifest Perceptions: Mind-matter interdependence and its consequences in Buddhist thought and practice." In *Śrāvakabhūmi and Buddhist Manuscripts*, edited by Jundo Nagashima and Seongcheol Kim, 109–71. Tokyo: Nombre Publications.

———. 2018. “Yogācāra Substrata? Precedent Frames for Yogācāra Thought among Third-Century Yoga Practitioners in Greater Gandhāra.” *Journal of Indian Philosophy* 46, no. 2: 193–240.

———. 2019. “Becoming Animal: Karma and the Animal Realm Envisioned through an Early Yogācāra Lens.” *Religions* 10, no. 6: 363.

———. 2020. “Map Becomes Territory: Knowledge and Modes of Existence in Middle Period Meditation Practice.” In *Mārga: Paths to Liberation in South Asian Buddhist Traditions*, edited by Vincent Eltschinger and Cristina Pecchia, 277–302. Vienna: Austrian Academy of Sciences Press.

16 How to Love Animals?

Attentive Platonic Love as an Epistemic and Moral Method

Elisa Aaltola

Although love is often viewed as a deeply personal emotion with little to offer for serious societal and moral thought, it is worth noting that many significant figures in Western philosophy have acknowledged the intricate connections between love, politics, and ethics. For instance, Simone Weil and Iris Murdoch have argued that love allows us to understand "the good," and that its potentially powerful role should be recognized in political and moral thinking (Murdoch 1971, 2003; Weil 2002). In this paper, I will examine their take on love in comparison to a philosopher who greatly influenced both, namely Plato. This examination will take place in the context of animal ethics. In short, I will focus on: (1) how Plato's definition of love applies to how human beings relate to nonhuman animals, and (2) how Weil's and Murdoch's suggestions can complement it in important ways that give shape to a politically and morally relevant form of loving animals.

Platonic Love Toward Nonhuman Animals

Plato's *Symposium* offers a classic and nuanced take on love. Albeit written in a humorous style, it offers a serious depiction of love that matches Plato's take on metaphysics. In this dialogue, Socrates and his companions compete over who offers the most viable definition of love. Much wine is consumed, and many triumphant sentences are offered. After others offer definitions that focus on passionate love and yearning for unity with one's "missing half" (whom the gods have cruelly separated from us at the beginning of time), Socrates stands up in order to deliver his take on the nature of love. He paraphrases the priestess Diotima, whom he posits to be his source on this topic, and argues that instead of carnal pleasures or existential longing, love consists of a focus on the transcendental ideas expressed in others. The most advanced variety of love is not about lust or yearning for one's "half," and indeed, it does not even primarily concern itself with corporeal beings. Rather, it is directed toward universal qualities such as the good, beauty, and truth. This again means that love extends to all those who express these qualities; if love is aimed at the good, truth, and beauty, it does not favor one individual over another, but rather loves all individuals capable of manifesting these transcendental

DOI: 10.4324/9781032659985-21

qualities (Plato 1892, 210a-b). Therefore, love concentrates on the transcendental qualities of the beloved, not her tangible embodiment.

Since universal qualities belong in the reality of forms and are thereby eternal, love allows one to connect with and be touched by immortality. Most importantly, via love one can access knowledge of the perfection of beauty, together with the truth and the good which it entails. As such, love allows for metaphysical insights, as we transcend the immediate corporeal level of being, and begin to understand the realest components of existence. Plato, via Socrates, describes this process with the following elevated words:

> Drawing towards and contemplating the vast sea of beauty, [the lover] will create many fair and noble thoughts and notions in boundless love of wisdom; until on that shore he grows and waxes strong, and at last the vision is revealed to him of a single science, which is the science of beauty everywhere.
>
> (Plato 1892, 210c-d)

The depiction of love offered in the Symposium can be called *the quality view of love* (Kolodny 2003, 135). The quality view of love (here called also "Platonic love") focuses on the universal and/or transcendental qualities of the beloved. Instead of loving one particular physical being, we love the qualities she manifests; as such, love concerns physical beings only indirectly, as the vessels of the beloved qualities. Next to Plato, some other classical Western philosophers, including Immanuel Kant, have made references to such love. According to Kant, "Love of human beings" (which he positioned as the highest form of love) resembles moral respect toward persons. Such love is directed toward rationality and moral autonomy, that is, the qualities that render beings "ends in themselves" (Arroyo 2016, 589–92; Rinne 2019). Again, love connects the lover to transcendental realms—in this case it forms an entry point to appreciating the noumenal reality (Rinne 2019, 51).

The quality view depicts love as a morally charged emotion. David Velleman has underscored that, following Kant's definition, love and the recognition of the inherent value of another are closely connected. In fact, since love and respect entwine, love forms an integral part of morality (1999, 365). The same moral charge is found also from Plato's definition, which is critical of "owning" the beloved and highlights how love cultivates virtue. In both accounts, the lover respects the beloved as the beholder of revered universal qualities, and as a result, the latter is never merely an instrument for hedonistic desire or personal satisfaction.

It is quite possible to undergo Platonic love also toward nonhuman animals. When observing a crow, a bee, or a fox, one can experience a primordial sense of connection with a transcendent dimension of existence, often unnoted in the hectic immediacy of everyday life. Suddenly, a seemingly mundane pigeon can manifest herself as perfected beauty, and the usually ignored beetle may seem to express most evident truths, and all of this can appear as something "good," in the sense that the existence of such creatures is just as it should be. In the process, the human perceiver may undergo sudden epiphanies, for instance, in the form of recognizing the mindedness of also nonhuman beings, the interconnectedness of conscious life, and the insignificance of the human species amidst the incalculability and intricacy of the animal world. This, again, can lead to a heightened moral sensibility and concern over the treatment of nonhuman animals (For a discussion on the historical instances of such love, see Aaltola 2022). Kant's version is slightly trickier in this respect. Since Kant argued that animals are not rational or autonomous creatures, and thereby cannot be "ends in themselves" (Kant 1991, 443), they remain beyond the scope of duty and moral respect, and thereby also beyond the limits of "love of human beings." Yet, as Velleman points out, there are also other universal qualities than rationality and autonomy, which tie together with inherent value (1999). Such qualities can be found also from environmental and animal ethics, and include not only sentience, but also life, vulnerability, and diversity. Importantly, these qualities make it possible for the Kantian observer to extend her love toward also pigs, pikes, gazelles, bats, and other nonhuman creatures.

Platonic love helps to ward off, what I have elsewhere called "The Love Paradox" (Aaltola 2021). If we centralize self-gratification or pleasant effects of love, we risk destroying the object of our love. This is the core of The Love Paradox: love can cause destruction. As it comes to loving nonhuman nature and animals, this paradox is regrettably prevailing. When a nature-lover seeks majestic sceneries in order to find elated experiences, they may book flights or drive long distances to the national park of their choice, where they may leave litter and disturb wildlife, all of which causes unnecessary harm to the very nature they profess to love.

Similarly, when a self-professed lover of animals wants to feel affection for nonhuman beings, they may favor the sort of "cute" and neotenic (juvenile; feel free to adjust this explanatory term, but include something) animals prone to spark rushes of oxytocin and other pleasure-inducing neurotransmitters, and thereby go, for example, for big-eyed pugs, who suffer from severe health problems precisely because they have been bred to look appealing. Again, one's desire to feel good in the company of animals causes those animals severe harm (Aaltola 2021, 95; Estren 2012). As pointed out by Soren Kierkegaard, love can be extremely selfish, and sometimes leads to more harm than good (2009, 25). I argue that when it comes to love of nature and animals, this is too often the case. However, Platonic love, in its insistence that love is not about personal gratification but cultivating esteem toward the qualities manifested in the beloved, avoids The Love Paradox. Yet, such love also faces various challenges. After addressing three of them, I will turn back to Weil and Murdoch to see if their own take on love could offer a needed solution.

Challenges to Platonic Love: Idealization, Universalism/Generality, and Expendability

The first of the challenges concerns the manner in which Platonic love centers on positive qualities and, by implication, pushes less perfect qualities beyond its scope. Thereby, one's love of, for instance, unappealing, chaotic, or destructive beings and entities (as opposed to beautiful, truthful, and good ones) becomes a conceptual impossibility. This emphasis on the positive can be questioned, as, for instance, Kierkegaard argued that the true test of love is whether it can flourish in challenging contexts and also attend to those who are troublesome.

Reflecting Christian *karitas*, meaning charity or love, Kierkegaard went on to suggest that, if anything, love is most deserved by the most imperfect rather than the most perfect among us (2009, 77). Kierkegaard's reminder is valuable; there are obvious moral reasons to direct love toward imperfection, for love is often needed most where there is tension, struggle, and brokenness. As Alan Soble (1989) points out, it is tempting and easy to love that which is close to perfection; however, does doing so truly capture the nature of "love"?

This question is noteworthy also in regard to the love of the nonhuman world. When such love is directed only toward pristine national parks, breathtaking landscapes, or majestic species, it appears hollow, as does love of animals that extends only to, say, cute kittens and big-eyed pugs. The risk is that, despite its lofty aims, Platonic love is reduced to feel-good factors, whereby what began as Platonic love collapses into egoistic pursuit of gratification and The Love Paradox becomes relevant once more. Therefore, focusing only on perfection and positive qualities can come with morally counterproductive consequences.

Simultaneously, much of nature is left beyond the scope of love. Nonhuman nature is often messy, aesthetically difficult, violent, and dangerous. Immaculate landscapes are relatively rare, and in place of them stand marshes, thickets, and arid lands filled with the ever-present danger of illness, injury, and death. Moreover, nature comes with immense violence and suffering (Horta 2010). Hobbes may have painted an overtly grim picture of nature as brutish and violent, but it is equally misleading to portray nature as an idyllic place filled with nothing but truth, beauty, and the good. In short, whilst the transcendentalist lover may feel elated at the sight of colossal mountains, intricate flowers, and majestic elks, nature is often ordinary and even unappealing to the senses, and colored by constant cycles of destruction and misery. The same applies to nonhuman animals, who can also be challenging, and for whom conflicts of interest are an everyday occurrence. Animals harm each other routinely and they can also be dangerous to human beings; some of them threaten the survival of other populations and even species, etc. This being so, the danger is that, when following Platonic love, much of nonhuman nature and animals are excluded from moral appreciation.

Ultimately, the underlying issue concerns the idealization of nonhuman nature and animals. When the nonhuman realm is depicted only through exceptionally positive qualities, idealization is the evident end-result. Such idealization leaves no space for imperfection or negative qualities, and it may ultimately serve egoistic motivations. Simply put, idealization can feel elevating and rewarding. Intriguingly, idealization of nonhuman nature and animals can thus walk hand in hand with the destruction of nature and animals, and thereby entangle with The Love Paradox. Further, it may alienate one from practical environmental concerns and arguably also from animal ethics (Coeckelbergh 2015). It is difficult to respect and support nonhuman nature and animals when one does not hold a realistic understanding of them.

Following Kierkegaard, I propose that the test of our love in this setting is whether it can extend to aesthetically and affectively displeasing groves or swamps, to crop-destroying storms, and, for instance, to those animals frequently categorized as "vermin." Can we love other animals and nature also when their qualities appear undesirable from the perspective of human interests, and when they cause economic losses, discomfort and worry, or harm to other species? Can we love rats, who possibly (although not likely) carry diseases, or wolves, who may kill the occasional lamb, and who is often (even if unnecessarily) feared? Can we love so-called "invasive species" that threaten local wildlife?

Extending love to qualities typically considered negative or challenging does not necessarily require us to forsake Platonic love. First, even when nonhuman nature and animals are not idyllic at first sight, their troublesome features can ultimately enable the existence of more perfect qualities, whereby the mosquito-covered thicket emerges as a source of intricate diversity, and whereby violent deaths caused by predators facilitate the birth of new life. Therefore, the qualities of perfection may not always be immediately apparent, but rather linger on the meta-level and become recognizable only after the type of intellectual effort highlighted throughout Plato's philosophy and present also in Murdoch's suggestion that one must train one's "vision" in order to learn the art of loving (Murdoch 2003). In this reading, love is an intellectual process dependent partly on the cultivation of one's ability to observe and understand the qualities of other beings and things, whereby what seemed displeasing may reveal itself in a new light—rats cease being "vermin" and instead manifest as creatures of intelligence and rodentic exquisiteness (we will come back to such cultivation in the last part of the paper). Second, the qualities, which we love, need not be categorized into "positive" or "negative"; what suffices is that they are *worthy* of love (Warner 1979). Similarly, as human imperfection may be worthy of love despite the fact that its fragile edges can cause harm, those parts of the nonhuman world that potentially threaten human interests or the interests of other species may be worthy objects of love. In short, despite their potential destructiveness, a member of an invasive species can also deserve love.

The second challenge concerns universalism or generality, and the category of lovable "nature." "Nature" and related notions such as "wilderness" are famously problematic due to their vagueness (see, for example, Woods 2017). Also the distinction between "nonhuman nature" and "human nature" is problematic for various reasons, such as dualism and the fact that the latter is made of the former; humans are, of course, part of nature) (Soper 1995). Further, if we take nonhuman nature to refer to all of existence beyond *Homo sapiens*, we are left with an enormous, generic category, and the scope of love appears impossibly vast. Are we to love all of existence (black holes and deadly coronaviruses included), and if so, does love not suffer from conceptual inflation to the extent that it becomes morally superfluous? That is, if one loves everything, does one love nothing? The same applies to nonhuman animals; are we to love all animal life, including bacteria? This generality intertwines with universalism. Ultimately, the quality view would necessitate universal love of all the beings and things with the relevant qualities, which would mean that love of nature and animals comes with an exceedingly broad scope.

Now, in his defense of the quality view of love, and addressing the possible request to love all human beings, Velleman has argued that the problem of universalism can be avoided. Although lovable qualities are widely present in the world, love is not universal, but instead focuses on a limited group of individuals. This is because some individuals manifest the lovable qualities more obviously than others; in the Kantian case, some are more evidently moral and rational, whilst others seem hopelessly lacking in these qualities. Therefore, due to differences in how these qualities are expressed, not everyone invested with them is equally lovable. Velleman also adds that love is a finite emotive resource; we only have a limited amount of it, and thereby cannot administer it to all those, who in principle could be its legitimate objects (1999, 372). Yet, these claims are not altogether convincing. If we know that also those, who do not appear particularly moral and rational, still have these qualities on a basic, categorical level, why exclude them? Additionally, more justification is required for the presumption that love is a limited emotion capable of focusing on only a few objects. One option is to postulate that, as an epistemic, intellectual emotion, love comes close to an attitude that can be broadened to include even the entire planet. One cannot affectively feel love toward, for example, all individual animals as our emotional lives are not suited to such affective universalism. However, one can adopt the *attitude* of love toward all animals, whereby one intellectually orientates toward them as lovable.

Therefore, the issue of universalism and generality remains. However, as it comes to the love of the nonhuman world, this issue need not be troubling, but quite the opposite. Love that entwines with a sound animal or environmental ethic would not restrict itself to only given individuals or entities, such as one's companion animals and home forests, but instead would be more general and universal in scope. If we wish for love to fulfill a serious and encompassing moral role, it needs to involve not just this particular elk or river, but all of them. What may sound absurd, such as love of hurricanes or bacteria, may in fact make sense on a broader scale, when considering that all things and beings in nonhuman nature hold their own distinct role in the web of being. For one to truly pay moral heed to nature and animals as holistic, intricate, plural, and interconnected entities, one thereby needs to direct love toward them precisely on a universal, generic level.

However, there are further difficulties ahead. The danger of such generality is that it condenses love into a rather abstract emotion, the moral implications of which are difficult to pinpoint. What would love for vast, universal qualities entail in everyday moral practice? If one is to love most or all of the nonhuman world, how could that love advise moral decisions in any precise fashion? If I love both bacteria and dogs, what am I to do when the two exist in a state of conflicting interests? Universal love can allow one to depict much of nonhuman nature as morally considerable, which is constructive from the perspective of environmental and animal ethics.

At the same time, universal love makes moral decision making arduous in situations, where we have to choose which of the multiple morally considerable entities and beings are to be prioritized. Fortunately, these difficulties are not insurmountable. One option is to distinguish love from ethics and posit that even if love comes with moral implications, the latter should not dictate how we define the former. We, of course, still need moral theory, which can specify how to solve conflicts of interest. The suggestion here is that love only points toward the value of others in a very general sense and is not meant to be used as a method of more precise decision making. In other words, love only offers one moral psychological foundation for ethics, and more detailed decisions rest on other factors, such as rational analysis. This leaves us with the conclusion that the generality of love need not be a problem.

The third challenge concerns expendability. Since the quality view of love can, in principle, be directed toward any being or thing that manifests the relevant qualities, the objects of love are wholly replaceable since anyone expressing the coveted qualities will do (Kolodny 2003, 140–142; Smuts 2014, 520). Indeed, this is what Plato proposes: we should not get stuck on one beautiful person, but rather love the beauty in all. Although to some this may sound refreshingly liberating, it also means that one could easily discard one's dog or home forest, if another dog or a forest with similar qualities was on offer. Such discarding would even be a logical necessity in situations where the newly found dog or forest came with superior qualities.

Here, the quality view goes against a common intuition that love will not treat the beloved as disposable. Now, Velleman has sought to address such critique by arguing that the dignity of our beloved erases the problem of expendability. Love requires respect toward the beloved as an end in herself, and thereby approaches her as a being with dignity. Treating her as replaceable would violate such respect and dignity, and thus doing so is alien to love (Velleman 1999, 366). Therefore, even if Platonic love would otherwise logically lead to expendability, adding Kantian dignity into the picture changes the situation and saves Platonic love from intuitively unappealing consequences. Yet, this defense it not entirely convincing, for it is less than obvious why recognition of dignity would be integral to the quality view. In fact, if love's focus is on qualities, such dignity is difficult to justify since the beloved can be an end in herself, like billions like her, but she cannot stand out as irreplaceable, for the qualities she has are shared by also others.

Another way to address the issue of expendability is to deny its moral relevance. So what, one could ask, if the objects of love are disposable? One option is to argue that it is quite logical for love to change and diminish when the qualities it is aimed at alter, and that this does not necessarily come with great moral costs. Recognition of the inherent value of another is not dependent on love, and when love lessens, other moral emotions and reason can still depict the ex-beloved as inherently valuable. As Amelié Rorty has pointed out, expendability does not necessarily imply lack of moral worth, and the fear of such expendability may limit one's understanding of love as a changing, varying emotion rather than a life-long constant (1987, 403–404). These suggestions are highly relevant to animal and environmental ethics. Exclusivity is alien to the sort of love of nonhuman beings and things that holds relevance from the perspective of environmental and animal ethics; we must love many, adapt to nature's constant alteration, and continually bring new creatures and things into the scope of our love.

Similarly, expendability also poses significant problems for environmental and animal ethics. Some constructivists have argued that authenticity and originality have no relevance in environmental philosophy, as "nature" is ultimately nothing but a cultural construction that varies from context to context (Crist 2004). Such musings are not only theoretical, for biotechnologies have brought forward novel options for both replication (whereby species that went extinct long ago, can potentially be brought back) and modification (whereby new hybrids and even species can be brought to existence) (Gamborg 2014). These options underscore the environmental risks of expendability. Why worry about the destruction of nonhuman beings and things, if they can be replaced (perhaps even with superior versions) through future technologies? Such expendability is evident also on a planetary scale, as some dreams of making Mars habitable so that Earth can be forsaken when its nature has been depleted and destroyed (Cuthbertson 2019; McKay, Toon, and Kasting 1991). Would Earth have any special significance if another habitable and lusher planet was found, and if not, can we still meaningfully proclaim to love Earth and its nonhuman nature? Moreover, if we could clone dogs or use biotechnology in order to bring new and enhanced species into the world, would the original dog or the old species lose their significance? These questions underscore the counterintuitive feature of expendability and, with it, the inadequacy of Platonic love.

Robert Nozick has argued that love concerns primarily individuals, not qualities (1991, 423) and Bernard Williams has contended that querying whether one ought to save one's drowning partner or a perfect stranger is asking one question too many, for special ties act as an obvious moral navigator (1981, 17). Similarly, Irwing Singer (2009) and Harry Frankfurt (2004) have argued that love makes us value the beloved above others by bestowing special worth on its object. These claims highlight the intuition that love needs to focus on factors other than qualities. On these grounds, and toying with the idea of what a Kantian dating ad would sound like ("rational, autonomous individual seeking alike"), Niko Kolodny has argued that love focused on universal qualities remains too far removed from the everyday realities in which our beloved are considered unique (2003, 174).

In particular, expendability points toward the need to complement Platonic love with other notions of love more capable of acknowledging specificity. Although Platonic love comes with merits, it alone does not provide an adequate account of love. Rather, it risks ignoring the particularity of our loved ones, the potential consequence of which would see the objects of our love become but insignificant and easily replaceable containers of worthy qualities. I argue that the above is particularly pertinent in regards to the love of nonhuman animals. If we exude universal love toward the qualities of the nonhuman world (or indeed the whole of existence, human or nonhuman) and discount other factors, do we not risk ignoring something which is fundamental to love, namely, the particular, unique being in front of us?

Love as Attention: Eliminating Expendability

One way to address the issue of expendability comes from Simone Weil's and Iris Murdoch's philosophy. It helps to refine Plato's take on love also on a broader level, as it clarifies not only what the irreplaceability of the beloved is founded upon but, more generally, what it means to become better able to notice the qualities of another, that is to rehearse vision, in the art of loving.

According to Weil and Murdoch, love consists of an epistemic state called attention. In order to understand what attention is, we must first understand the role of "the self" in preventing attention from unfolding. Both Weil and Murdoch underscore that love requires the process of "unselfing." They argue that "the self" has been excessively prioritized in Western thought whereby we tend to think that "the self" and its desires are the primary unit of existence. This has led to a worryingly egoistic way of approaching the external world and its beings, as we tend to notice only those aspects in them which serve our own personal interests. Consequently, we become so focused on our "self" and personal interests that we forget about morality. Yet, "the self" is nothing but a construct of our own doing, which means that we let what is ultimately invented and imaginary govern our lives and prevent us from engaging with morality. As suggested by Murdoch, "the self" or "ego" (she uses these terms synonymously) is "an illusion" (1971, 91) that "veils the world" and prevents us from grasping "the good" (2003, 175). In other words, the self is a mere construction which guides us to perceive things from the perspective of how they serve our own self-interests, and in so doing, it prevents us from understanding morality.

Therefore, self-gratification, a phenomenon also stressed by Plato and Kant, again emerges as a challenge. Yet, instead of simply offering a conception of love that avoids such egotism, Weil and Murdoch underscore that we need to first question the primacy of "the self." The process of *un*selfing requires that we become aware of how our "self" is an elusive construct, and how we are never precisely what we wish to think we are. We are not as intelligent, advanced, or loving we wish to be, nor are we as failed as we may fear. Instead, we are always something different than our preconceptions allow us to recognize, and the task is to approach this discrepancy with compassion: "I am also other than what I imagine myself to be. To know this is forgiveness" (Weil 2002, 9). It is such realism and forgiveness that begin the process of unselfing, whereby our own presumptions of who we are no longer dictate what we understand others to be.

Murdoch underscores that the process of unselfing also requires humility which, in her mind, is rare among human beings because "the anxious avaricious tentacles of the self" are often at work in how we humans perceive reality (1971, 101). In short, humility requires that we learn that we and our interests are not above others and their wellbeing—it reminds us that we are not the centers of existence. Weil stresses the relevance of humility as well. According to her, we are to acknowledge our mediocrity, question the relevance of our desires, and even become "nothing." Indeed, Weil goes so far as to suggest that unselfing requires that we become "voids" lacking an ego such that "the self" eager to triumphantly endorse its own interests is replaced by an empty space (2002, 5–42). It is such unselfing, accompanied by forgiveness and humility, which paves the way for what Weil and Murdoch call attention.

According to Murdoch, the goal of attention is "to come to see the world as it is" (1971, 89). For her, attention is about witnessing the "realness" and truth of another. Through this, we come to recognize the realities of others, uncolored by the otherwise constant self-directedness of the recognizer, which warps perception and biases understanding. Such attention, and its ability to recognize "truths," is nothing less than love: "Love is truth, truth is love" (Murdoch 2003, 293). Weil further describes attention as the ability to set aside one's ego or "the self," and to witness the reality of another. "'I' has to be passive," she writes, "Attention alone—that attention which is so full that the 'I' disappears—is required of me" (Weil 2002, 118). Likewise, she claims that such attention is love: love consists of becoming open to the reality of another being.

The moral implications of love as attention are at least three-fold. First, as just argued, attention urges us to notice the reality of another. As Murdoch famously stated, love is "the extremely difficult realization that something other than oneself is real" (1959, 215). This realization may seem simple, yet it is uncomfortably challenging; if reality is qualitative rather than quantitative, others are *as real* as oneself and they are equally existent as I myself am. In my interpretation, when this realization concerns minded, sentient creatures, it comes with a moral dimension. We not only note that others exist, but we recognize that their existence is *equally relevant to our own*. Since other-minded beings also have felt perspectives onto the world, as in a Nagelian sense, it is *like something* for them to exist; they are *individuals* with their own inner subjective lives. As we tend to be preoccupied with our own perspective and interests—the realities of our own life—recognizing this can be arduous. Yet, such a recognition allows us to notice that the flourishing of another is equally morally pertinent to our own flourishing. Thereby, it motivates us to approach others in the light of "the good." In short, acknowledging that another minded being exists as an equally real individual, is an inherently moral affair – and it may be precisely this, which makes it also so difficult (our egoism finds it hard to fully witness the moral significance of others).

Second, by eroding the walls built around "the self," attention facilitates a recognition of the interconnections and unison between oneself and the other. We are no longer encapsulated into our own individualistic existence, but instead adopt a more relational and open mode of being. Indeed, Murdoch describes how the process of unselfing and the ensuing attention reveals "a natural way of experiencing the interconnectedness of things, their beauty and strangeness, their liveliness in and to our consciousness as 'ours,' and yet also as independent witnesses to reality" (2003, 342). This regard for connections and holistic networks again paves the way for the possibility of relational ethics, wherein it is not only the beloved individual, but also our relations with her and others, which emerge as morally pertinent.

Thirdly, and most importantly from the perspective of this paper, the particularity of another is the focus of our attention. As Murdoch puts it. "[The particular] teaches us love, we understand it, we *see* it" (2003, 497). Attention allows one to notice the details and specificity of another, and thereby it is through attention that the uniqueness of another being comes into a fuller view. Weil explains this via a metaphor of suddenly seeing more clearly in the dark, stating, "We see the same colors; we hear the same sound, but not in the same way" (2009, 100). Via attention, we come to appreciate the particulars of another—their distinctive way of existing, their uniqueness. Therefore, although Weil and Murdoch draw from Plato's metaphysics, which underscores universalism, they combine it with their own brand of particularism. Whilst universal qualities such as "truth" and "good" reserve an important role, what we love in another is those qualities' unique and particular expression. Universalism thereby entwines with exceptionality—one can appreciate both universal ideas and their inimitable manifestations. This, again, provides an answer for the above problem of expendability. The beloved is not expendable, because they are wholly exceptional and as such irreplaceable.

An Under-Explored Ethical and Epistemic Alternative

What could all of this mean when applied to the context of love of nonhuman animals? To begin with unselfing, I suggest that developing attention requires the letting go of anthropocentric presumptions concerning human perfection. Although cultural narratives keep insisting that humans are flawless "images of God" or perfected "pinnacles of evolution," neither is true. Instead, we are one species among a multitude of other species, all perfected and advanced in their own ways. When taking into account the capacities and intricacies found in nonhuman nature, a set-apart and dominating human supremacy is no longer warranted. Releasing these separatist and exclusivist conceptions further brings into question the anthropocentric ethos of always prioritizing human interests at the cost of non-human flourishing. In place of such a tendency for human hubris, we ought to forgive ourselves for not being quite as God-like or perfect as cultural stories suggest, and endorse what I call "species humility," which recognizes both the advantages and limitations found in humanity as realistically as possible. In short, instead of offering grandiose descriptions of human species, it is more advisable to notice how humans—like all species—come with strengths and weaknesses and have a lot to learn from our nonhuman kin. Most importantly, species humility, when applied to animal ethics, requires revisiting the ontological and moral roles of *Homo sapiens* through what I describe as the "un-human-self," inspired by Weil and Murdoch.

When letting go of the strict confines of the "the human self" that can be layered with anthropocentric presumptions of human supremacy, we may begin to appreciate the existence of nonhuman animals—not as instruments for human use, but rather as agents of their own lives. As a result, one may start to appreciate "the truth" of animals, which I understand to refer to not only the noticing of their existence, but also the act of apprehending that animals have their own realities, untouched by the societally-common insistence on approaching them as dull instruments for human use. What is the reality of a pig independent from how human societies make use of her? When spending time with this question without constant references to the human self as separate, exclusivist, and dominating, one may come to see glimpses of "the pig as she is, in herself." These glimpses and the very act of paying attention to the pig's reality are nothing less than love.

Such love of animals comes with the three moral dimensions mentioned above. First, when loving a nonhuman being, we acknowledge that her existence is equally real as our own, and that her felt perspective—her inner realm—is as relevant as is ours. This prevents us from sidelining nonhuman experiences, and instead invites us to pay heed to the viewpoints of pigs, cows, wolves, rats, vipers, eels, cockroaches, and other animal relations. Second, attention toward animals reminds us of the interconnections between humans and their kin—the webs between different ways of being, which also give wider moral meaning to human existence, and urge people to protect our surrounding worlds. Third, love as attention comes with the recognition of the particularity of the animal. When being attentive toward a pig, we open up to her minute details, the distinctiveness of her being. Therefore, attention is not about generic categories, such as the abstract concept "pigs," but about individuals and their particularities. How does this particular pig exist, what may her perspective entail, and how does she make sense of her reality?

Importantly, this means that love toward animals does not involve replaceability, as if one beloved pig could easily be substituted with another. Instead, all individual animals are inimitable and, as such, not expendable. As humans begin to note the details of animals, our vision becomes clearer, and what used to be a murky shape—perhaps a stereotypical category with little resemblance to this real pig—becomes lively and enriched, and we can suddenly envisage nonhuman ways of being in all of their specificity. Here, love as attention evokes empathy and moral concern toward the uniqueness manifested in each animal; the more we notice the details of the pig, the more we can envision her experiences and viewpoint, and the more morally relevant she becomes to us. As Murdoch emphasizes in her analysis of Plato, vision comes with morality, as it allows us to witness the world in light of "the good" (2003, 2425). I argue the same to apply in the context of envisioning what it is like to be a nonhuman animal such that awareness of her uniqueness entwines with the recognition of her worth. In sum, when approached via love as attention, nonhuman animals will not remain faceless representatives of generic categories and as such easily expendable, but rather emerge as unique individuals, each deserving of our moral consideration (Aaltola 2018, 2019).

Therefore, love as attention helps to solve the most serious challenge faced by Platonic love: the problem of expendability. The beloved is not expendable, because attention allows us to recognize her intricate nature, which again makes her unique and irreplaceable. In so doing, love as attention also helps to make sense of the problem of universality: the beloved is not a generic manifestation of esteemed qualities, but rather an intrinsically particular being, who manifests those qualities in her own, singular fashion. As such, Weil and Murdoch offer a more complete definition of love than Plato and Kant.

In spite of this significant theorization of love within and beyond classical Western philosophers, the concept of attention has gained regrettably little attention in animal philosophy. There are some exceptions, such as Josephine Donovan (2007), my own research into the topic (Aaltola 2018, 2019), and a recent book by

Silvia Panizza (2022). Additionally, Weil's relevance to animal studies has been analyzed more broadly by Anat Pick (2011). However, the role of attention in how to make sense of and engage ethically with nonhuman animals remains under-explored. My argument is that it is precisely attention as love, and its potent moral implications, which offer one important future direction for animal philosophy and ethics. In particular, love as attention offers an epistemic alternative to the rationalist stance often favored in animal ethics—it is not only via reasoned analysis but also through attention and more broadly love, and the type of Murdochian "moral vision" which these entwine with, that one can grasp the moral significance of nonhuman beings.

Before concluding, it is worth noting that Murdoch herself refers to nonhuman animals as teachers of attention. According to her, paying focus to other animals can help the process of unselfing, as it draws us away from our egocentric desires and aches, and reminds us of the world outside of "the self." She details a moment of observing a bird flying in the sky: "Then suddenly I observe a hovering kestrel. In a moment everything is altered. The brooding self with its hurt vanity has disappeared. There is nothing now but kestrel…[we ought to] give attention to nature in order to clear our minds of selfish care" (Murdoch 1971, 82). I argue that humans need such moments, not only in order to cultivate attention and thereby love in general, but also in order to refine our ability to appreciate, love, and revere the multitudes of nonhuman individuals, with whom we share this finite and precious planet.

Platonic love (as defined by Plato and later Kant) can easily be applied to also nonhuman animals. Yet, it faces the interlinked problems of idealization, universalism/generality, and expendability, out of which the last is the most difficult to solve. Thereby, albeit potentially morally constructive, Platonic love struggles particularly with the challenge of expendability and the ensuing inability to account for the specificity of our loved ones. Love as attention, as outlined by Weil and Murdoch, helps to solve this problem by reminding us that it is not only the qualities of the beloved that draw us to her, but also and primarily, her unique way of existing which attention as an epistemic and moral method can helps us to appreciate as *real*.

Bibliography

Aaltola, Elisa. 2018. *Varieties of Empathy: Moral Psychology and Animal Ethics*. London: Rowman and Littlefield International.

———. 2019. "Animal Love: Iris Murdoch and Simone Weil." In *The Routledge Handbook on Love*, edited by Adrienne Martin, 193–204. London: Routledge.

———. 2021. "Egoistic Love of the Nonhuman World? Biology and the Love Paradox." *Ethics, Policy & Environment* 26, no. 1: 86–105. https://doi.org/10.1080/21550085.2021.1885245

———. 2022. "Platonic Love of Nonhuman Nature and Animals." *The Harvard Review of Philosophy* 29: 33–44.

Arroyo, Christopher. 2016. "Kant on the Emotion of Love." *European Journal of Philosophy* 24, no. 3: 580–606. https://doi.org/10.1111/ejop.12128.

Coeckelbergh, Mark. 2015. *Environmental Skill Motivation, Knowledge, and the Possibility of a Non-Romantic Environmental Ethics*. London: Routledge.

Crist, Eileen. 2004. "Against the Social Construction of Nature and Wilderness." *Environmental Ethics* 26, no. 1: 5–24.

Cuthbertson, Anthony 2019. "Elon Musk Says Giant Mirrors Will Warm Up Mars and Make it Habitable for Humans." *Independent*, August 21, 2019. www.independent.co.uk/tech/elon-musk-mars-nuke-humans-live-mirrors-spacex-a9072631.html

Donovan, Josephine. 2007. "Attention to Suffering: Sympathy as a Basis for Ethical Treatment of Animals." In *The Feminist Care Tradition in Animal Ethics*, edited by Josephine Donovan and Carol Adams, 174–97. New York: Columbia University Press.

Estren, Mark J. 2012. "The Neoteny Barrier: Seeking Respect for the Non-Cute." *Journal of Animal Ethics* 2, no. 1: 6–11. https://doi.org/10.5406/janimalethics.2.1.0006

Frankfurt, Harry. 2004. *The Reasons of Love*. Princeton: Princeton University Press.

Gamborg, Christian. 2014. "What's So Special about Reconstructing a Mammoth? Ethics of Breeding and Biotechnology in Re-Creating Extinct Species." In *The Ethics of Animal Re-Creation and Modification*, edited by Helena Siipi and Markku Oksanen, 60–76. London: Palgrave MacMillan.

Horta, Oscar. 2010. "Debunking the Idyllic View of Natural Processes. Population Dynamics and Suffering in the Wild." *Télos: Revista Iberoamericana de Estudios Utilitaristas* 17 (January): 73–88.

Kant, Immanuel. 1991. *The Metaphysics of Morals*. Cambridge: Cambridge University Press.

Kierkegaard, Soren. 2009. *Works of Love*. London: Harper.

Kolodny, Niko. 2003. "Love as Valuing a Relationship." *Philosophical Review* 112, no. 2: 135–89.

McKay, Christopher P., Owen B. Toon, and James F. Kasting. 1991 "Making Mars Habitable." *Nature* 352, no. 6335: 489–96. https://doi.org/10.1038/352489a0

Murdoch, Iris. 1959. "The Sublime and the Good." *Chigago Review* 13, no 3 (Autumn): 42–55.

———. 1971. *The Sovereignty of Good*. London: Routledge.

———. 2003. *Metaphysics as a Guide to Morals*. London: Vintage.

Nozick, Robert. 1991. "Love's Bond." In *The Philosophy of Erotic Love*, edited by Robert Solomon and Kathleen Higgins, 231–40. Oxford: Oxford University Press.

Panizza, Silvia. 2022. *The Ethics of Attention: Engaging the Real With Iris Murdoch and Simone Weil*. London: Routledge.

Pick, Anat. 2011. *Creaturely Poetics: Animality and Vulnerability in Literature and Film*. New York: Columbia University Press.

Plato. 1892. *The Dialogues of Plato*. Edited by B. Jowett. Oxford: Oxford University Press.

Rinne, Pärttyli. 2019. *Kant on Love*. Berlin: De Gruyter.

Rorty, Amelie Oksenberg. 1987. "The Historicity of Psychological Attitudes: Love Is Not Love Which Alters Not When It Alteration Finds." *Midwest Studies In Philosophy* 10, no. 1: 399–412. https://doi.org/10.1111/j.1475-4975.1987.tb00548.x.

Singer, Irving. 2009. *Philosophy of Love: A Partial Summing Up*. Cambridge: The MIT Press.

Smuts, Aaron. 2014. "Normative Reasons for Love." *Philosophy Compass* 9: 518–526.

Soble, Alan. 1989. "Analysing Love." *Philosophy and Social Science* 19: 493–500.

Soper, Kate. 1995. *What Is Nature?: Culture, Politics and the Non-Human*. London: Wiley-Blackwell.

Velleman, J. David. 1999. "Love as a Moral Emotion." *Ethics* 109, no. 2: 338–74. https://doi.org/10.1086/233898

Warner, Martin. 1979. "Love, Self, and Plato's Symposium." *Philosophical Quarterly* 29, no. 117: 329–39.
Weil, Simone. 2002. *Gravity and Grace*. London: Routledge.
———. 2009. *Simone Weil: An Anthology*. London: Penguin.
Williams, Bernard. 1981. *Moral Luck: Philosophical Papers from 1973–1980*. Cambridge: Cambridge University Press.
Woods, Mark. 2017. *Rethinking Wilderness*. Broadview Press.

17 Transspecies Selves

Rethinking Species-Being and the Boundaries of the Human

Gabriele Schwab

In recent years, we have witnessed a significantly increasing philosophical interest in transspecies relationships. A flourishing body of transspecies imaginaries is responding to the threatening anthropogenic extinction of species, the so-called sixth extinction. The growing awareness of the dwindling diversity of life forms on Earth and the concomitant ecological imbalance has generated profound challenges to anthropocentric worldviews and given rise to urgent calls for new relationships with other species and the natural environment more generally. As Parker Palmer writes in *Healing the Heart of Democracy*, "We are a profoundly interconnected species, as the global economic and ecological crises reveal in vivid and frightening detail. We must embrace the simple fact that we are dependent on and accountable to one another" (2011, 15).

If we follow Marx in assuming that species-being is always determined by specific social and historical conditions, we might wonder whether we have reached a stage at which the massive intervention into the natural order of things forces us to recognize the entanglement of the fate of humans with that of other species and therefore compels us to extend our sense of species-being to include all living species. This means facing the challenge of thinking of humans as transspecies beings—not in the sense of a posthuman transspecies reproduction but in the sense of the human entanglement with other species. As corporeal, sensuous beings, humans share vulnerability and mortality with other species and are dependent on sustainable environments for their well-being. The extent to which they have threatened the life of a sustainable planet by exploiting its resources generates a particular responsibility and obligation to care. We are, in other words, at a historical conjuncture where our ethics of care needs to include transspecies care.

Theories of Transspecies Selves

The following thoughts on "Transspecies Selves" aim to outline a transspecies ecology inspired by Gregory Bateson's *Steps to the Ecology of Mind* (1972). I articulate a conceptual framework based on multispecies imaginary ethnographies and transspecies relationalities. In response to current challenges to planetary life and the survival of many species, recent debates in critical theory have started

DOI: 10.4324/9781032659985-22

to develop new theories of the human that profoundly reconsider the boundaries between humans and their environment, including other species. Already in the early seventies, Bateson argued forcefully that the most important factor for the survival of the planet is the development of a new ecology of mind. He called for modes and categories of thinking that are mindful of the entanglement of humans with their various and evolving environments, insisting that "the organism which destroys its environment destroys itself. The unit of survival is a flexible organism-in-its-environment" (Bateson 1972, 457).

Let me begin by introducing and critically discussing a few theories that inspired my project and the readings of specific literary, artistic, and cultural objects. Samantha Frost's concept of "biocultural creatures" (2016), Neel Ahuja's concept of "bioinsecurities" (2016), and Merlin Sheldrake's concept of "entangled life" (2020), for example, offer conceptual challenges to familiar definitions of the boundaries of the human. I will draw on these and related concepts to develop my own focus on emergent subjectivities in precarious ecologies, and especially my own concept of "transspecies selves."

Samantha Frost is an interdisciplinary scholar whose theory straddles the boundaries between political science, gender studies, and contemporary biology. Her ambitious goal in *Biocultural Creatures* (2016) is to develop a new theory of the human. Recent critiques of the category of the human and human exceptionalism, she claims, have left us bereft of a category of the human subject that can be mobilized to address today's multiple crises and challenges (Frost 2016, 1). This implicit critique of poststructuralist and postmodern, if not posthuman, theories that proclaim the death of the subject, constitutes a crucial philosophical intervention that is more widely shared in the environmental humanities, including in Sheldrake's theory that I will discuss later. Yet, far from returning to conventional theories that define the human subject as bounded and distinct from its environment, these theories reconceptualize the human with fluid and permeable boundaries toward the environment and other species.

Frost, for example, theorizes humans as "biocultural creatures" to encapsulate the co-constitution of body and environment, biology, and habitat (2016, 4). *Homo sapiens,* she insists, is not a pure species but emerged from variegated entanglements between the *homo* species and their environments. Rather than conceptualizing the human in distinction to other species, Frost focuses on the "conditions of existence that humans share with all living creatures" (11). In this context, she also analyzes today's environmental crises as well as the forms of structural violence that mark human societies. To meet today's challenges, she argues, we need to address what humans do to the world and to each other (12).

Thinking in terms of transspecies entanglements also implies thinking bodies and environments together. "Like social theorists," Frost asserts, "scientists increasingly confirm that there are complex interactions and interchanges between biological and social processes that muddle any distinction we might want to make between body and environment" (2016, 13). At a molecular level, Frost's perspective is attuned to what Nikolas Rose describes as a molecular turn in the life sciences that emphasizes the permeability, rather than boundedness, of biological entities (Rose 2007, 9–52).

The permeability of boundaries also pertains to species more generally. In *Death-in-Life* (1991), Robert J. Lifton coined the term "species self" to describe the fact that many atomic bomb survivors, for the first time, thought about themselves in terms of their species and its survival rather than in terms of familial or national belonging. In *Radioactive Ghosts*, I argued that considering climate change, environmental destruction, and the continued threat of radioactive contamination and nuclear war, all of which threaten planetary life, we must expand our ecology of mind to include other species and therefore think in terms of "transspecies selves" (Schwab 2020, 248–49). What we could call a "transspecies turn" in thinking about the human, in fact, amounts to a paradigm shift that affects many disciplines as well as ways of thinking and being in the world. Frost writes: "the body is conceived as porous or permeable in a way that belies the sense either that the 'self' is distinct from the body or that the body is distinct from the social and material environment" (2016, 18). This does not, however, mean that we give up categorical distinctions. Rather, it means that we categorize differently within fluid epistemologies. Frost proposes thinking of creatures as "constituted by their environs or habitats yet also identifiably distinct from them" (27). In a similar vein, Karen Barad introduced the concept of "intra-action," or action within a field, to capture the processes by which the biological and the social reciprocally work on one another (2007, ix). "Culture, symbolic forms of communication, and imaginative anticipation shape the ways that bodies compose and recompose themselves over time," writes Frost, and concludes that we, therefore need to think not only in terms of the "molecularization of life" as laid out by Rose, but also of the "environmentalization of the molecule" (2016, 19–20).

This relational epistemology is not only grounded in a genetic but also in an epigenetic perspective, mindful of the fact that "a creature's response to a habitat in one generation can shape the development and growth of offspring in subsequent generations via epigenetic markers" (Frost 2016, 29) We may think of the chemical molecules that enter and transform the substance of our bodies and selves, of viruses that replicate within our cells, or even of transgenerational epigenetic changes and transmission.

Importantly, Frost conceptualizes matter as an effect of the flows of energies between organisms and environments. She locates agency on both sides of the equation, tracing "how the processes of sensory and imaginative perception generate energetic chemical reactions within and between cells that transform the functioning of the cells" (Frost 2016, 148). The main point is that there is a biological inscription of social perceptions and that the social, symbolic, and material is transformed into the biology of the body. While the resonance of Frost's theory with psychosomatic medicine is striking, she pursues this knowledge at different scales and uses it to retheorize the human and its boundaries.

Reflecting on the temporal scale, Frost leans on Maurizio Meloni and Giuseppe Testa's argument that biocultural creatures are anticipatory forms "seeded with traces of [their] future" (Frost 2016, 149). While there is a frenzy of biochemical activities on the micro-scale of an organism, these usually do not manifest at the macro-scale of its large-scale function and behavior (150). Frost also addresses the productivity of "constraints" that limit the possibilities of response to the habitat. This claim invites further consideration of how constraints have widely been used to enhance the productivity of creative literary and artistic productions.

According to Frost, embodied existence is constituted by, but not reducible to, habitat (2016, 151). Yet the profound dependence on habitat cannot be overstated. "Identity, affects, norms, germs, nutrition, pollution, and other forces and phenomena have their acculturating effects on our development, growth, and persistence in living" (151–52). This perspective requires a reconceptualization of "culture" not only as symbolic and representational but also as "chemical, spatial, thermal, viral, bacteriological, nutritive and so on" (152). A salient example is provided by the culture of our the Covid-19 pandemic, where virtually all aspects of culture were affected by the proliferation of a virus. In conclusion, Frost insists on a double conceptual move in the theorization of biocultural creatures: we need to denaturalize the body and "deculturalize" culture (152).

Ontological Insecurity

Let me now draw out some implications of Frost's concept of biocultural creatures for theories of subjectivity and the self. Human subjectivity develops in response to habitats, including trees, oceans, deserts, air, the sun, the climate, the sky, the stars, and the planets. Moreover, human subjectivity is also impacted at both an individual and a collective level by an entire environment of nutrients, chemicals, pharmaceuticals, pollutants, etc. This impact can mark an entire generation as seen in the so-called Prozac generation and the Ritalin generation. Subjectivity also develops in response to rural or urban environments, architectures, and living spaces, as well as technologies and modes of communication and information, including literature, the Arts, entertainment, and the culture industry. It develops in response to religion and spirituality but also in response to politics, violence and war, structural and slow violence. Finally, and perhaps most importantly, subjectivity develops in the modes of connecting to other humans as well as other species. Whether we are aware of it or not, we are profoundly "transspecies selves." The challenge is to build this knowledge into our theories and conceptual systems.

The latter requires inter- and transdisciplinary collaboration. In biology, epigenetics is the theory that addresses the impact of environmental forces on the biological organism, that is, what Frost calls "the material and corporeal effect of the symbolic, representational, and political dimensions of biocultural habitats" (2016, 154). To fully theorize biocultural creatures, epigenetics, however, needs to be supplemented by other theories as well as the visions embedded in literature and the Arts. Frost points to the resurgence of interest in ontology in contemporary theories of posthumanism. One of the reasons for this resurgence that Frost does not mention is the fact that the current challenges to planetary survival, including climate change, the threat of nuclear war and catastrophe, as well as the onset of the pandemic age, have generated a profound and pervasive ontological insecurity.

Efforts to retheorize the human and its boundaries are partially responding to this ontological insecurity. Frost ends with her hope that her theory of biocultural creatures may contribute to "the political project of transforming the conditions of our collective existence" (158). In this context, the question "How should we live?" or the age-old philosophical question "What is a good life?" becomes paramount once again, albeit under completely new parameters.

Bioinsecurities and Immunity

Frost's environmental politics is grounded in a relational theory that focuses on reconceptualizing the boundaries of the human. In contrast, Neel Ahuja's book *Bioinsecurities* (2016) focuses on the large-scale perspective of governmental control of the relationship between humans—if not nations—and other species. *Bioinsecurities* is a prime example of work done today in the field of Medical Humanities. Ahuja approaches the issue of transspecies encounters through the lens of U.S. imperial expansion and the related project called "the government of species" through which health and military officials tried to control the interactions between humans, animals, viruses, and bacteria. To return to the example of COVID-19, we currently face efforts by global governments to govern the species of the coronavirus. We also recall that initially, under the Trump administration, the U.S. had gained visibility through its spectacular failure on all fronts.

Ahuja traces the efforts by the U.S. to join with other countries, including Panama, Cuba, the Congo, Iraq, and India, to control the spread of viruses and bacteria such as polio, smallpox, and HIV. Emphasizing the racialization of fears of contagion, he analyzes state uses of pharmaceuticals, medical experimentation, military intervention, and incarceration to regulate immune responses. In the process, biocultural creatures became sites of struggle in the politics of empire. Ahuja also traces the forms of activism and resistance to medical and military authority that developed globally across spheres of U.S. influence.

Ahuja's book has gained new relevance in assessing the efforts at the "government of species" during the current pandemic, particularly concerning efforts to trace the transspecies encounters involved in the process. Ahuja highlights the often-racialized uneven distribution of risk and practices of care during epidemics. Moreover, he opens new perspectives on the racialized panic in the U.S. and elsewhere, mainly from right-wing politics. He coins the term "dread life" to describe the dialectics of racialized fear of contagion as well as hope placed in and enthusiastic embrace of new medical, often life-preserving, technologies: "Dread life," Ahuja writes, "is the term I use to describe this racialized channeling of the fear of infectious disease into optimism regarding the remaking of life through technical intervention" (2016, 6).

Ahuja also draws on literature and the Arts, including novels, short stories, memoirs, films, and photographs, bringing them into conversation with government documents. It is through artistic works that he grasps the implications of the specific conjunction of fear, anxiety, and hope for the making of biocultural subjectivities. In this context, Ahuja's concepts of "immune ecologies" and "disease ecology" are central, especially in connection with Derrida's reflections on autoimmunity and immunitarian logic. According to Ahuja, ecological models of disease explore problems of settlement, urbanization, and networking across colonized space. "Enmeshed in processes of territorial, economic, military and biological expansion, disease ecology gave a specific form to the matter of the immune system that stressed its constant emergence through interspecies entanglement" (Ahuja 2016, 12). Disease, as we are painfully aware, transforms both bodies and environments.

Niels Jerne proposes a network theory of immunity that explores the ecological relationship between antigens and antibodies (Ahuja 2016, 12). The problem is that the immune system does not only recognize external threats like viruses, bacteria, and toxins; it can also react to internal elements as imagined threats and attack an organism's own body. This is the problem of autoimmune deficiency. The immune system's constant re-creation of the borders of the body generates an iterative process of self-recognition at work in the immune system that, according to Ahuja, inevitably involves minor processes of self-destruction (13). The immune system continually reprocesses autoimmune recognition and reproduces the self by managing the complexity of its interspecies relations. This is a very different aspect that supports the notion that we are always already "transspecies selves" at the molecular level of the body's internal transspecies world.

There is an interesting history of transcoding between military discourses and discourses of immunity that can tell us a lot about the political unconscious. In the war on terror, for example, the concept of "networked immunity" became essential to the biological front of the U.S. In this context, "anticipatory containment" is key to the logics of biosecurity and biowarfare (Ahuja 2016, 15). Catastrophic risks generate intense affects and affective knowledge, including biosemiotics accounts of interspecies intimacies. They also call for the management of public fear. The government of species, Ahuja argues, "renders life itself an ongoing site of social and political contestation as disease interventions layer ever more complex technologies of containment and circulation" (26).

Disease interventions, Ahuja posits, are political processes that "securitize privileged forms of circulation against the privileged forms of interspecies contact" (2016, 195). In today's area of mass extinction and ecological crisis, he adds, critical theorists are imagining "the intensification of species war coincident with current forms of empire" (196). According to Ahuja, this shift to the securitization and remaking of biological life coincides with privatization, the dismantling of social welfare and imperialist efforts to contain the effects of decolonization. Health and medicine, he argues, are embedded in logics of intervention that are deeply politicized. This includes state visions and media assemblages that take planetary space and interspecies relations as sites of intervention. Inner and outer spaces are technologized to engineer empire's forms of human security against intensifying precarities (201).

Political "Autoimmunity" and Resistance

While I agree with many of Ahuja's core arguments, I think that the logic of governance is more complex and ambivalent. The coronavirus pandemic has taught us some painful lessons. Governance is not monolithic and not unilaterally evil, nor even always completely restricted to the interests of empire. There is a hidden danger in pushing this logic toward the ultra-right-wing paranoia of the "deep state." The U.S. has become a paradigm of failed governance during the pandemic, not because of too much governance, but because of a lack of good enough governance. Everything depends on how interspecies relations are conducted and whose interests targeted interventions serve.

The pandemic itself is already an outcome of destructive ecological interventions in interspecies contact, most prominently through the deforestation of the rain forest and the concomitant endangerment and displacement of species. Ahuja also invokes the "vast zones of exclusion" as an outgrowth of slow death and the capitalist generation of a risk society (2016, 202). But these processes are not caused by an excess of governance but by a lack of governance that would restrict the murderous free flow of capital and the hegemony and political power of global mega-corporations. It is the latter that have contributed to setting in motion the chain of events that began with rampant deforestation and displacement of species and ended with today's climate crisis and the onset of a pandemic age. In short, it is not governance per se that needs to be the target of critique but empire in collusion with unrestrained global corporate capitalism.

At stake is, as Ahuja accurately points out, the relation between U.S. empire and the development of global capitalism (2016, 204). Where I disagree with him is in his unilateral conceptualization of biodefense, environmental government, and human security as neoconservative. While these institutional forces undoubtedly follow a neoconservative logic under certain historical and political conditions, they are not neoconservative in and of themselves. Ahuja emphatically states, "When empire loses its sole character as the sovereign and takes on a circulatory form, it produces complex geographies of contact that embed the regenerative powers of bodies in intricate and intimate forms of warfare" (205). While I couldn't agree more, I would add that what Ahuja calls "the regenerative power of bodies" also generates intricate and intimate forms of resistance. Alternative non-hegemonic forms of equitable governance belong to these forms of resistance, as do new ecologies of mind that reconceptualize the boundaries of the human along concepts such as the "logic of autoimmunity," "biocultural creatures," and "transspecies selves."

In this context, the logic of autoimmunity warrants special attention. In the nineties, discourses of autoimmunity proliferated in the political philosophies of Roberto Esposito, Jean Baudrillard, Byung-Chul Han, and Derrida. Derrida's biophilosophical concept of the "logic of autoimmunity," for example, theorizes "autoimmunity" as a full ethico-political concept, focusing on epidemiological and immunological tropes of contact and contamination (2005, 155–57). Defining autoimmunity as a quasi-suicidal behavior of a living being to destroy its own immune system, that is, its own system of self-protection, Derrida performs, as many have pointed out, a pointed misreading of the medical definition. According to the latter, the organism fails to recognize its own constituent parts as "self" and triggers an immune response against its own cells and tissues. The immune system thus does not destroy itself but the body's cells and tissues. In biological terms, natural autoimmunity plays a protective function by ensuring the tolerance of so-called "self-antigens" and by regulating the entire immune system. Autoimmune diseases, by contrast, are linked to high level auto-reactive immune cells and the loss of immunological tolerance toward the self.

Using the term "political autoimmunity" to construct his position against nationalism, Derrida emphasizes the relationship between *politikon*, *physis*, *bios*, *Zoe*, life, and death (2005, 9). Relating Derrida's political autoimmunity to Ahuja's reflections on empire opens an interesting angle of critique in support of my claim for more differentiation and complexity in theorizing empire and the vicissitudes of governance. I propose to link "autoimmune disease" to what Chalmers Johnson has analyzed as "blowback," that is, the unintended consequences of empire (2000, 3–33). Autoimmunity would be an unintended consequence of the immune system's governance of the body and its mechanisms of defense against attacks by antigens or hostile foreign agents. If this defensive system goes overboard, so to speak, it begins to attack its own internal agents or cells by mistaking them as enemies. In a similar way, political autoimmunity can be seen as a form of blowback in which the forces of empire generate unintended consequences by destroying its own internal forces of resistance.

In respect to a robust concept of transspecies selves, Ahuja's theoretical reflections mainly contribute to conceptualizing institutional and political attempts to regulate the relationship of humans and viruses or bacteria. Within the human body, it is at the cellular and molecular level that these relationships unfold. Institutionally, however, they mobilize political forces that aim at controlling the threat of epidemics and pandemics and thereby also use mechanisms of controlling not only the fate of harmful viruses and bacteria but also the mind and psyche of those involved.

New Ecologies of Mind

Merlin Sheldrake's *Entangled Life: How Fungi Make our Worlds, Change our Minds, and Shape our Futures* (2020) presents a shift in perspective that is pivotal for theorizing transspecies relationalities and resistance. In the prologue, Sheldrake states:

> [A] fungal network laced out into the soil and around the roots of nearby trees. Without this fungal web my tree would not exist. Without similar fungal webs no plant would exist anywhere. All life on land, including my own, depended on these networks.
>
> (2020, x)

How many of us are aware that our lives, our very survival, and even our minds and subjectivities are entangled with fungal networks? "Fungi are everywhere but they are easy to miss," states Sheldrake (4). Closer attention to fungi and humans came with the molecular turn in the life sciences because, as Sheldrake maintains, "at a molecular level, fungi and humans are similar enough to benefit from many of the same biochemical inventions" (9). Moreover, both fungi and humans suffer from the ecological destruction of forests and soils. Unsustainable agriculture, for example, reduces the ability of plants to form relationships with the beneficial fungi on which they depend. Sheldrake also reports that "the widespread use of antifungal chemicals has led to an unprecedented rise in new fungal superbugs that threaten both human and plant health" (8). On the other hand, the awareness that fungi and mushrooms can be beneficial to health and healing is growing, with environmental scientists reporting that "radical fungal technologies can help us respond to some of the many problems that arise from ongoing environmental devastation" (9).

There is, in short, growing awareness of our entanglement with the more-than-human world, starting with the knowledge of our microbiome and its impact on mind and mood. The microbial sciences subtly challenge the boundaries of the human, making it harder to define where one organism or individual stops and another starts. Sheldrake defines this epistemological and ontological challenge:

> We are ecosystems composed of—and decomposed by—an ecology of microbes… . Our microbial relationships are as intimate as any can be… .'We' are ecosystems that span boundaries and transgress categories. Our selves emerge from a complex tangle of relationships only now becoming known.
> (2020, 17–18)

These new knowledges are beginning to generate a profound ontological and epistemological rupture that Sheldrake interestingly compares to conventional symptoms of mental illness such as loss of a sense of self-identity, delusions of self-identity, and experiences of alien control. In psychological terms, these experiences indicate a state of ontological insecurity. In ecological terms or under the conditions of a new ecology of mind, however, I rather think that we are challenged to inhabit a position of ontological uncertainty. Ecosystems have shaped—and been shaped by—human cultures for thousands of years. But only the destructive environmental impact of human cultures that led to the profound ecological crises we are facing today forced us into a growing awareness that we are, in Frost's terms, "biocultural creatures" whose lives are entangled with that of other species and environments. Seen in this way, the challenge to inhabit a new type of ontological uncertainty is a healthy and future-oriented attitude.

But how can the growing awareness of these entanglements shape the transspecies relationship with mysterious organisms, such as branching fungal *mycelia* roots that appear so alien to us? It is already a profound challenge for humans to relate to many of the so-called non-charismatic animal species such as certain insects because they appear so alien. How much harder is it to imagine a transspecies relationship to fungi or mycelia? "Despite their nearness, fungi are so mystifying, their possibilities so *other.* Should this scare us off?" asks Sheldrake, "Is it possible for humans, with our animal brains and bodies and language, to learn to understand such different organisms? How might we find ourselves changed in the process?" (2020, 22). Understanding radically different organisms pushes the limit of empathy and begs the question of anthropomorphism.

Sheldrake challenges us to explore what we might pass over and not notice if we follow the strict prohibition on "humanizing the world" to understand the lives of organisms on their own terms (2020, 41). As we may recall, this stance resonates with Frost's claim that recent critiques of the category of the human and human exceptionalism have left us bereft of a category of the human subject that can be mobilized to address today's multiple crises and challenges. The challenge is to reconceptualize the category of the human in ways that allow one to account for the rich arsenal of possible human relationalities to environments and other species. The latter includes anthropomorphism as a basis for empathy.

Sheldrake suggests that we might learn from Indigenous languages a "grammar of animacy" that attributes aliveness to the more-than-human world:

> Equipped with this 'grammar of animacy,' it is possible to talk about the life of other organisms without either reducing them to an 'it' or borrowing concepts traditionally reserved for humans… . Might we be able to expand some of our concepts, such that speaking might not always require a mouth, hearing might not always require ears, and interpreting might not always require a nervous system?
>
> (2020, 42)

A "grammar of animacy" is, of course, particularly attuned to transspecies relationships that unfold at the molecular scale at which the notion of transspecies selves includes the awareness that humans are not only transspecies through their relationships to other species but also because they already harbor other species in their bodies. This includes the bacteria without which humans would not be able to live, but it also includes the viruses and bacteria that threaten human life and generate the governmental inventions that Ahuja foregrounded. While molecular perception and thinking heighten transspecies sensitivity and ethics, it might, therefore, also increase awareness for dangers that occur at the molecular scale, including the dangers of the "alien" toxins humans introduce into the environment, among them, most prominently, the radioactive substances that affect the bodies of humans and animals at the molecular level.

Transspecies Animacy

Mel Chen's notion of animacy allows one to foreground the transformative affection of bodies by alien matter. Chen describes such embodiment in terms of a "marriage of bodies and chemicals" (2012, 1). Mutations are an extreme manifestation of such a "marriage," inscribed at the molecular level of cells. As Chen explains, these mutations are concrete embodiments of what Jasbir Puar calls the "bio-necro political" aspect of contemporary life (6). This perspective of animacy compels one to envision entirely new forms of entanglement. Chen describes the challenge in terms of a new ecology of mind, body, and affect:

> Environmental toxicity and environmental degradation are figured as slow and dreadful threats to flesh, mind, home, and state. Myths of immunity are challenged, and sometimes dismantled, by transnationally figured communicable diseases, some of them apparently borne by nonhuman animals.
>
> (7)

The concept of animacy enables a vision that encompasses both large-and small-scale effects. "The concept of 'animacy' can be regarded as some kind of assumed cognitive scale extending from human through animal to inanimate," writes cognitive linguist Mutsumi Yamamoto (Chen 2012, 8). Informed by "philosophical considerations of life, care, and molecularity" (8) the concept of animacy is implicated in political questions of the distribution of power and the recognition of different forms of life, challenging one to "rethink the significance of molecular, cellular, animal, vegetable, or nonhuman life" (10).

Inspiring us to expand the boundaries of our concepts of language and communication, Sheldrake's notion of a "grammar of animacy," I suggest, implicitly works with a notion of anthropomorphism *without* anthropocentrism. This is, I think, a crucial distinction in conceptualizing transspecies encounters, namely, a distinction that allows one to rethink the long histories of animism as well as the emergent animisms in contemporary art, literature, and critical theories. The latter includes a wide range of diverse theorists, including most prominently Gilles Deleuze and Félix Guattari, Eduardo Viveiros de Castro, Bruno Latour, Donna Haraway, Mel Chen, and Sheldrake himself. As Sheldrake demonstrates with the example of fungi, without a certain anthropomorphism, we would not be able to grasp the functioning of these mysterious organisms on which our lives depend, let alone have empathy with other organisms in the first place.

Finding an adequate non-anthropocentric language of description might, Sheldrake suggests, require a compromise of approximation: "Our descriptions warp and deform the phenomena we describe, but sometimes this is the only way to talk about features of the world: to say what they are like but are not. Might this also be the case when we talk about other organisms" (2020, 44)?

In the case of complex networks of decentralized organisms such as mycelium that belong to a microbial ecology, we might even have to expand, if not bend, the notion of organism. "Mycelium is ecological connective tissue, the living seam by which much of the world it stitched into relation," writes Sheldrake, "In practice, it is impossible to measure the extent to which mycelium perfuses the Earth's structures, systems, and inhabitants—its weave is too tight. Mycelium is a way of life that challenges our animal imaginations" (2020, 46). Emphasizing the conceptual slipperiness of mycelium, Sheldrake states that from the point of view of the network, mycelium is a single interconnected entity, whereas, from the point of view of a hyphal tip (a feature of filamentous fungi), it is a multitude.

Such network-based organisms are not only naturally entangled with human life; humans have also begun to use them to solve human problems. Witnessing the supreme skills of maze-dwelling slime molds to solve spatial and geometrical problems, researchers, for example, incorporated slime-mold behavior into the design of urban transportation networks or apply the strategies that fungi or slime molds use to navigate labyrinths to program robots or solve mathematical problems (Sheldrake 2020, 48). Exploring fungal communication, electrical signaling emerged as the crucial way for fungi to convey information. Like brains, mycelium is endowed with complex networks of electrically excitable cells. This led computational biologist Andrew Adamatzky to develop the idea of a fungal computer, that is, of mycelial networks that compute information encoded in electrical activity (Sheldrake 2020, 69–70). In a similar vein, in his exchange with neurobiologists, computational biologist Simon Olsson speaks of his dream to hook up a fungus to a computer to communicate with it and perhaps even change its behavior (Sheldrake 2020, 63–65). Fungal cyborgs would thus join the ranks of and become entangled with human cyborgs.

Sheldrake argues that a mycelial network is a reminder that all life forms are in fact processes not things. He quotes geneticist William Bateson saying, "We commonly think of animals and plants as matter, but they are really systems through which matter is continually passing" (Sheldrake 2020, 53). Comparing mycelium to a "polyphony in bodily form," he points out that to comprehend mycelium, we need to let go of focused attention and approach it through unconscious scanning, asserting, "Several streams of consciousness have to commingle in the mind. My attention has to become less focused and more distributed" (55). Interestingly, this mode of attention corresponds exactly to what art theorist Anton Ehrenzweig describes as the source of creativity and the basic mode of artistic production and reception (1967, 5–70). It also corresponds to current neuroscientific theories of the distributed mind. I would go so far to argue that, in accordance with the molecular turn in the life sciences, we need this mode of unfocused attention or unconscious scanning to attune ourselves to the scale of molecular animacies and networked organisms or mycelial and microbial forms of life.

Transspecies Imaginaries

Let me end with a literary example and play with a reading of Samuel Beckett's *The Unnamable* as a literary response to Sheldrake's vision of speaking without a mouth or hearing without ears. In many ways, the Unnamable envisions himself as a transspecies character, most prominently in one of his most prominent embodiments as Worm. In this context, the Unnamable also envisions an organless hearing and speaking: "Without an ear I'll have heard, and I'll have said it without a mouth" (1958, 134). As the Unnamable's monologue unfolds, he increasingly uses, in Sheldrake's terms, a "grammar of animacy" that de-anthropomorphizes the human. His phantasms of animate organless bodies culminate in the conception of Worm, an experimental transspecies creature never completely born yet buried before his time. Wai Chee Dimock imagines literature as a "contagious site" (2020, 10). Beckett's work figures humans as a finite species, forcing us to inhabit a position of radical onto-epistemological uncertainty, not unlike the one invoked by Sheldrake, who compared this uncertainty to the symptoms of mental illness. Like Sheldrake, Beckett de-pathologizes alternate mental states. Both challenge us to embrace a new ecology of mind grounded in porous boundaries of bodies, distributed minds, and transspecies entanglements.

The transspecies imaginaries and concepts of transspecies selves I have presented here also allow one to rethink the long histories of animism and animacies. Finding an adequate non-anthropocentric language to describe the emergent animisms in contemporary art, literature, and critical theory is not easy. The Unnamable, in his attempts to approximate what he experiences as unknowable, consistently exposes the inadequacy of available forms of expression. More conciliatory, Sheldrake suggests—as stated above—a compromise of approximation: "Our descriptions warp and deform the phenomena we describe, but sometimes this is the only way to talk about features of the world: to say what they are like but are not. Might this also be the case when we talk about other organisms" (2020, 44)?

Inevitably our theories, interpretations, and discourses will be marked by the epistemological uncertainties of understanding and relating to species that are radically alien to us. These gaps in our knowledge about other species confront us with the unknowable that will always be part of our world. Beckett shows that nothing fuels our imagination more than what we experience but do not yet or will never know. Dimock writes, "humans as a species have a track record of bouncing back, turning the endgames of finite individuals into linked chapters in our ongoing life" (2020, 7). So let us create transspecies stories that sustain the survival of the rich diversity of planetary life.

Bibliography

Ahuja, Neel. 2016. *Bioinsecurities: Disease Interventions, Empire, and the Government of Species*. Durham: Duke University Press.

Barad, Karen. 2007. *Meeting the Universe Half-Way: Quantum Physics and the Entanglement of Matter and Meaning*. Durham: Duke University Press.

Bateson, Gregory. 1972. *Steps Toward an Ecology of Mind*. San Francisco: Chandler Publishing Company.

Chen, Mel. 2012. *Animacies: Biopolitics, Racial Matting and Queer Affect*. Durham: Duke University Press.

Derrida, Jacques. 2005. *Rogues: Two Essays on Reason*. Stanford: Stanford University Press.

Dimock, Wai Chee. 2020. *Weak Planet: Literature and Assisted Survival*. Chicago: The University of Chicago Press.

Ehrenzweig, Anton. 1967. *The Hidden Order of Art: A Study in the Psychology of Artistic Imagination*. Oakland: University of California Press.

Frost, Samantha. 2016. *Biocultural Creatures: Toward a New Theory of the Human*. Durham: Duke University Press.

Johnson, Chalmers. 2000. *Blowback: The Costs and Consequences of American Empire.* New York: Henry Holt.
Lifton, Robert J. 1991. *Death In Life: Survivors of Hiroshima*. Chapel Hill: University of North Carolina Press.
Palmer, Parker. 2011. *Healing the Heart of Democracy: The Courage to Create a Politics Worthy of the Human Spirit.* San Francisco: Jossey Bass.
Rose, Nikolas. 2007. *The Politics of Life: Biomedicine, Power, and Subjectivity in the Twenty-First Century.* Princeton: Princeton University Press.
Samuel Beckett, Samuel. 1958. *The Unnamable.* New York: Grove Press.
Schwab, Gabriele. 2020. *Radioactive Ghosts.* Minneapolis: University of Minnesota Press.
Sheldrake, Merlin. 2020. *Entangled Life: How Fungi Make Our Worlds, Change Our Minds and Shape Our Futures.* New York: Random House.

18 Knowing Multiply, Unexpectedly, Imaginatively

Decolonizing Knowledge with Gloria Anzaldúa

Suzanne Bost

Last summer, I attended *Forothermore,* an exhibition of the installation artist and fashion designer Nick Cave at the Museum of Contemporary Art in Chicago. My thinking is still touched by the experience. It was atmospheric, filling the space entirely with shimmer and cacophony. But the highlight was Cave's "Soundsuits": humanoid garments that invoke, reclaim, and multiply overdetermined figures of race and queerness in the United States. These quirky and sometimes monstrous figures welcome otherness, weirdness, and "for-ness" (a word I made up to express the lasting gifts of this installation). The suits are generally shaped like humans, but they incorporate surprising found objects like beaded flowers, fur, twigs, yarn, and racist "Relics" (Cave's term) like Sambo dolls. Cave's archive contains endless possibility, remaking and resignifying each object. Toys and metal birds spring from the upper half of the body in *Soundsuit* 2011, below (Figure 18.1). Black hands and rainbow-colored leggings invoke the human aspects of this figure, while a phonograph horn takes the place of a face. A circular target is positioned over the point where the human genitals would be, assuming that there is a human wearing this suit. Blackness and queerness spring to life in full ambivalence. Is this a posthumanist approach to humanity? These forms obviously integrate multiple ways of being, communicating, and knowing, but their inspiration is more material (and political) than theoretical.[1]

Cave and his collaborators wear some of the suits for performance (which we can watch in screening rooms), even when these suits go well beyond the boundaries of human bodies. We encounter the Soundsuits through multiple senses: the literal and ideological weight of the suit pulling down on a body, the sound of susurrating beads and labored breathing behind a mask; the sparkling light bouncing off reflective pinwheels, sequins, and mirrors. Sound, color, and shape each contribute their own kinds of information. The multiplicity of objects that populate a Cave installation overwhelm the limits of human focus; we have to get "close" to notice details, put on a headset to add the soundtrack (Beckwith 2022a, 53). There is movement in all dimensions, what Cave calls "vastness": an excess of visual and aural noise Cave draws from sources like *Soul Train*, fashion shows, and house dance parties to activate bodies and to ignite our motions towards an unknown future.

DOI: 10.4324/9781032659985-23

Figure 18.1 Nick Cave's Soundsuit (2011).

Cave's multisensory exhibits remind those experiencing them that understanding our world requires inclusive perceptiveness, from the smallest to the largest scales of relation. Humans cannot know apart from their ecosystems, their atmospheres, and the communities that give them perceptual frameworks. Attending to all of these at once might be impossible, but I have abandoned pragmatism as an intellectual limit. This chapter builds from my earlier work on transspecies epistemologies—and from years of studying ontological and epistemological oddities in *The Gloria Evangelina Anzaldúa Papers*—to develop inclusive ways of knowing.

My aim in this chapter is to create a multidirectional approach that will open doors to new kinds of ethical research practice. Scholars generally follow the paths of those who came before them—not just to follow critical trends, but also to join conversations already in process, to have company. Writing from the depths of post-pandemic despair, mass shootings, and persistent racist injury, I am interested in finding as many approaches as possible that will help us break existing molds and reinvent the world in new ways. In an interview in which he discusses his methods as a professor at the Art Institute of Chicago, Cave suggests: "We're teaching to shift our way of existing in the world" (Beckwith 2022b, 263). This approach happens in the classroom, the studio, the museum, and the archive.

My faith in the possibilities of shifting is inspired by the intellectual border-crossings of Latinx decolonial writings and the Black diasporic imaginary of Alexis Pauline Gumbs, self-described "Queer Black Troublemaker…Black Feminist Love Evangelist…aspirational cousin to all sentient beings" (Gumbs 2023). I bet that Anzaldúa loves her from the grave; they would have been allies in bringing about a less violent future.

Archival Relations

Anzaldúa's archive is a maze of treasures. After her death in 2004, more than 200 file boxes of materials were transferred from her home in Santa Cruz, California, to the Nettie Lee Benson Latin American Collection at the University of Texas at Austin, where scholars now encounter hundreds of unpublished stories, poems, essays, and even entire book manuscripts, most in multiple, conflicting drafts. I found stories about jaguars that shift shapes like werewolves, queer human lizards who leave puddles on their chairs, a woman who loses herself in eating, and time travel. I also found a memoir that recounts her multiple deaths and rebirths. How could one not be shifted by this material?

I have already published two articles about the unpredictable and unending processes elicited by the materials that converge in this archival research—"Messy Archives and Materials that Matter" (Bost 2015) and "*The Gloria E. Anzaldúa Papers* and Radical Research Practices: A Collage" (Bost 2022)—but there is always more to say. I am obsessed with the variety of objects we consider to be the content of the archive, the printed texts with their traces of editing and often doodles in the margin, as well as the technologies that constrain and enhance our research, such as the limits and possibilities of digital cameras, document request forms, and the light outside and temperature within the library's reading room. Sometimes my eyes shift outside to the trees, grackles, squirrels. (A friend reminds me that squirrels are also archivists, storing away nuts and crackers in tree shelves we rarely see).[2] Though most of my archival experience has been lively and generative, Monica Huerta's experimental autobiography *Magical Habits* (2021) reminds us that archives can also be places of alienation: "I propose that constantly exhuming archives looking for versions of certainty can also become suffocating; that living as critique can manifest dizziness, distance, loneliness" (iv). The search for certainty in dusty papers and messy handwriting can be alienating as well as inducing various kinds of emotional and physical malaise. One might get stuck in the past. But the dancing squirrels keep me in the present. I talk back and forth with the documents, reading and re-reading, feeling accompanied in my uncertainty.

Anzaldúa's work helps me to think beyond human epistemology, beyond what's been documented and stored on shelves. She accepted mystical encounters and multispecies minglings without judgment. She experienced her own disabling conditions, including Type One Diabetes and a hormone imbalance, in the epistemological fold between modern medical diagnosis (which generally enforces normative corporeality) and sacred Aztec figures who fall outside the shapes of Humanist perception and the requirements of empirical science, like the snake-headed goddesses Coatlicue and her mutilated daughter Coyolxauhqui, whose body was chopped to bits by her brother, the war god Huitzilopochtli, and cast into the sky to become the moon and stars. In this sense, Anzaldúa's ideas are prehumanist, or simply outside the cultural-temporal evolution of Humanism, as much as they are posthuman; for this reason, I step aside from the chronological emphasis on the pre- and the post- to regard Anzaldúa's epistemology as simply other-than-Humanist and her imagery as other-than-human.[3] The dozens of doodles that accompany Anzaldúa's manuscripts reflect this broadening of life and species.

"Two Fishy Cats" (Figure 18.2) (or is it "Fishy Two Cats"?) doesn't look like fish or cats at all. They share a torso, but each has a separate tail. They have antennae and fangs but no mouth? They remind me of a sock with a hole on both ends. Is "fishy" a noun or an adjective here? Another image (Figure 18.3), which might be a cover for a book never published, employs a maze-like drawing style that Anzaldúa often used in her doodles. This one, to me, resembles a brain with white figures (or holes?) contained inside it. Three of the figures seem to have two eyes, one seems to have one eye (maybe in profile?), and another has three. But what makes me assume that these dots and circles *are* eyes? The figures are not necessarily shaped like heads, and they don't seem to have other facial features. Neither of these doodles replicates known species, recalling the elaborate creations within Nick Cave's Soundsuits. Indeed, one of the "figures" in Figure 18.3 looks like a ghost (one of Anzaldúa's favorite symbols), and another looks something like a cupcake. Pinning down these figures to any labels would defy their multiplicity, their whimsy, and their reach beyond "the real." To assign them a species would colonize their imaginative wildness.

Figure 18.2 Doodles by Gloria Anzaldúa.

I spend time on Anzaldúa's doodles because she saved so many of them in her files, so presumably they must have been important to her; maybe she was even planning to publish a book of them. Anyone who has spent time in her archives will have encountered these hand drawings. I am attracted to the idea of "doodles" as such because they are opposed to mastery. Anzaldúa's doodling, like much of her writing, was created in a trance-like or meditative state, often beyond the conscious deliberation. AnaLouise Keating describes Anzaldúa's approach to writing as "dialogic, recursive, democratic, spirit-inflected, and only partially within her conscious control. She relied extensively on intuition, imagination, and what she describes as her 'naguala' [or animal companion]" (2015, xi). Anzaldúa's writing was communal, incorporating the suggestions of her many "writing comadres," as well as shamanistic: Anzaldúa emphasizes that the "ability of story (prose and poetry) to transform the storyteller and the listener into something or someone else is shamanistic…The writer, as shape-changer, is

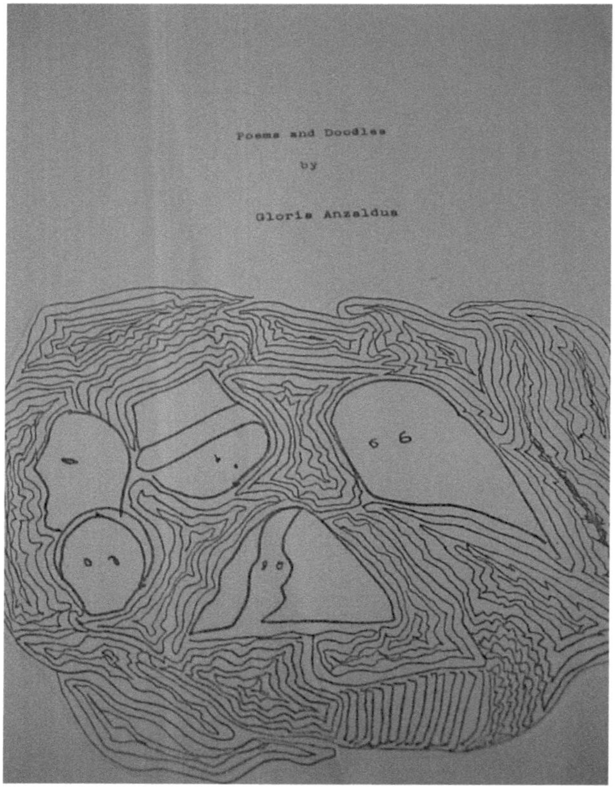

Figure 18.3 Doodles by Gloria Anzaldúa.

a *nahual*, a shaman" (1987, 66). Writing is performed, sacred, enacted for both the writer and for others:

> When I write it feels like I'm carving bone. It feels like I'm creating my own face, my own heart—a Nahuatl concept. My soul makes itself through the creative act. It is constantly remaking and giving birth to itself through my body. It is this learning to live with *la Coatlicue* that transforms living in the Borderlands from a nightmare into a numinous experience. It is always a path/state to something else.
>
> (1987, 73)

Coatlicue, mother goddess, embodies both creation and destruction; she teaches us ambivalence. Anzaldúa draws this embodied form of writing from early Mesoamerican sacred texts and the role of sacrifice in shamanistic performance.

Writing, in this way, generates something else and makes life "numinous"—giving it a supernatural sheen of possibility.

Working outside the frameworks of Humanism enables Anzaldúa to imagine more capacious possibilities for being. Breaking with the rules of reason that limit academic discourse enabled Anzaldúa to have a hopeful perspective: "I have this visionary experience where I'm flying in the sky as an eagle… . [W]e're going to leave the rigidity of this concrete reality and expand it. I'm very hopeful" (1987, 284–85). Whether or not one believes in the possibility of out-of-body experiences or shifting shapes into animal forms, these imaginative propositions take us beyond status-quo thinking. These other-than-human beings are a logical extension of the permeable nations, genders, and cultures associated with the early theories Anzaldúa is already famous for (like mestiza consciousness and the borderlands), though these more imaginative ontologies might alienate some academic readers.

One of the more "practical" outcomes of rethinking human life is that it leads us to re-imagine the kinds of relationships we form in the world. Even if we don't consciously experience states of shapeshifting or boundarylessness, perhaps we should practice our relations with others as if we did. After all, we share an eco-system with a great variety of stuff. Mel Chen's important work, for instance, replaces our perception of human subjects with an ontologically diverse field of "animacy." Chen writes, "In its more sensitive figurations, animacy has the capacity to rewrite conditions of intimacy, engendering different communalisms and revising biopolitical spheres, or, at least, how we might theorize them" (2012, 3). Rethinking not just our sources of knowledge but our approaches to ontology shifts how we move through the world. Anzaldúa developed a theory called "new tribalism" out of the commonality between the different entities in our ecosystem:

> Your identity has roots you share with all people and other beings—spirit, feeling, and body make up a greater identity category. The body is rooted in the earth, la tierra itself. You meet ensoulment in trees, in woods, in streams. The roots del árbol de la vida of all planetary beings are nature, soul, body.
> (2002, 560)

We inhale and exhale the same air as the trees and animals around us. We leave traces of ourselves—not just our ideas, but our skin and hair—on the objects we encounter. Our bodies and movements emerge within and shape the material context that shapes us. These phenomena show how community and environment, as Anzaldúa puts it, carve our very bone.

One is never alone in the archive. One is always enmeshed with the entire atmosphere of the library: toxic chemicals on paper, preconceptions of value, and structures of authority all exert an influence on the library ecosystem. Squirrels outside the reading room windows move in the periphery of my attention. Materials in an archive are conventionally mined for "data" to defend a scholarly argument, but objects in themselves are meaningful, three-dimensional, and weighty. These objects often resist clear translation into information; their impenetrability confounds our research goals.

Though the archivists might shudder to hear it, the transference of the archive is a two-way street. The materials I examine leave a mark on me, and I leave a mark on the materials: airing the pages, forgetting the order of randomly gathered clippings, smudging paper with my barely visible prints. I am marked by what I see, read, and feel with these papers. The archive and I share an intimate experience, passing the cup back and forth, licking our lips with new encounters and new ideas. An archive is a world of intertwined entities that give scholars their ideas and their materials. An archive is an ecosystem, a community. The scruffy squirrels outside the library windows defy gravity as they circle and leap.

One of the many unpublished essays I bumped into in the *Anzaldúa Papers* is called "Com-positioning: Theories of Art" (plural "theories" rather than *a* theory). The gerund evokes processes of which we are in the midst, and the hyphen in "com-positioning" emphasizes the "com," the with-ness and communality of composition. The neologism also resonates with composting and repositioning: "Surprise," unexpected words and loopy images catch readers off guard. Juxtaposing words of different character linked for effect… . Music, rhythm, and soul, words that touch the inner ear and biological roots (Anzaldúa n.d., 3). This approach resonates with what Walter Mignolo calls "epistemic disobedience," delinking thought from established forms of knowledge-making (2009, 162). Our words, both in teaching and in scholarship, are more powerful when they create multiple pathways towards knowledge.

Decolonizing Being and Knowledge

Feminist decolonial critics have expanded the boundaries of what counts as knowledge and what counts as an archive. When Macarena Gómez-Barris describes her own decolonial methods in *The Extractive Zone* (2017), she calls for a broad view that includes submerged environments like the "yellowish green-brown muck" that gathers around piers in the ocean (xiii). Setting aside the Humanist view from above and our aversion to lifeforms that ooze, intermix, and cling, what if researchers submerged ourselves in other worlds? Knowledge is murky. The stuff that gathers under the pier is an archive of intermingled substances: seaweed tangled with shards of shells and dead fish parts. For Gómez-Barris, "Submerged modes flurry in their activity, random, complex, and coordinated systems that are often illegible to those with state and financial power that assume simplicity where complexity actually dwells" (xvi). Rather than following one direction, submerged ways of knowing "flurry." Gómez-Barris pursues a model of development borrowed from Andean anti-imperial, anarcho-feminisms: "the idea of freedom as an operation of political activity that keeps on developing," expressed through creativity and the pursuit of a "future-oriented tense" (127–31). The unfolding of biomatter is opaque, unpredictable; it disturbs our categories. The story of biomatter intertwines across scales and levels in a verb tense oriented towards the future but without any single determining verbs.

María Lugones describes comprehension as layers of understanding that are next to each other, next to comprehension, but not comprehensive. They are characterized by "manyness" rather than clarity (2003, 4–5). Lugones's signature term "playfulness" describes a "vocation not bound by the meanings and norms that constitute one's ground" but as an "openness to uncertainty" (26). Playfulness is like hopping on a train without knowing where the tracks lead—and also knowing that new ideas will come down the road. In her theory of "world-travelling," Lugones suggests that moving between realities elicits epistemic and ontological shifts, new ways of thinking and being.

These shifts resonate with Kelli Zaytoun's analysis of *nagualism*, those sacred Mesomaerican forms of shapeshifting found in Anzaldúa and others. For Zaytoun,

> [A]ccessing and liberating the decolonial imagination offer much untapped potential for refashioning selfhoods, identities, bodies, relationships, social-justice coalitions, and even realities. For writers like Anzaldúa, artistic endeavors are a primary means for such refashionings.
>
> (2022, 121)

Speculation is a frequent mode in Anzaldúa's work, including experiments in shapeshifting, offering new interpretations of Aztec cosmology, and writing extensively in science fiction genres. Speculation shapes this approach to decolonialism, one that imagines new kinds of futures rather than finding evidence, truth, or history. Decolonial speculation does not merely critique the colonial structures that distort and inhibit reality; it brings about ways of doing selfhood and sociality differently. "I become the jaguar. I become the serpent, I become the eagle," says Anzaldúa in an interview with AnaLouise Keating (1998–99, 284).

Crossing, as kin to the shapeshifting and speculation imagined by Anzaldúa, is a theme that links Latinx and Black diasporic thought. Latinx writers like Anzaldúa and Lugones emphasize border-crossing as a material-intellectual activity that decolonizes not just nations or languages but also hegemonic knowledge and hegemonic ontologies. Laura Elisa Pérez, following Marjorie Garber's theories of cross-dressing and her own writings about the permeable boundaries between life and death, emphasizes that "crossing is crisis-inducing when the space of crossing is revealed as such, as a space beyond, as a borderlands (to invoke Gloria Anzaldúa), a third space outside binary logic, a zone operating by multiple codes" (Garber 1992, 172).

Crossing multiplies, muddles, and unfolds in complex formations. In *Pedagogies of Crossing* (2005), M. Jacqui Alexander similarly poses crossing itself—as derived from the embodied and psychic "Crossings" of the Middle Passage—as a form of teaching and a form of learning. She writes, "[P]edagogies that are derived from the Crossing fit neither easily nor neatly into those domains that have been imprisoned within modernity's secularized episteme. Thus, they disturb and reassemble the inherited divides of Sacred and secular, the embodied and disembodied…" (Alexander 2005, 7). The teachings of the Middle Passage are violent, disruptive, and life-altering. Ontologies and epistemologies are undone. Pedagogies, for Alexander, break through, transgress, disrupt, displace, and invert "inherited concepts and practices, those psychic analytic and organizational methodologies we deploy to know what we believe we know so as to make different conversations and solidarities possible" (7). In order to build new knowledges and new solidarities, Alexander, like Gómez-Barris, turns to subordinated sources, deep under the water or deeply buried in memory, in order to destabilize existing practices of knowing. This knowledge is not found in maps, books, archives, museums, or "documents of any kind" (294). This knowledge is found in grief, in the ocean, in Spirit, and in ancestral wisdom.

Living with (Im)Possibility

While Anzaldúa turned to ontologies like shapeshifting in order to get around the many borders that crossed her, Alexis Pauline Gumbs responds to Crossing by telling histories that could never be true according to conventional reason, forging her own epistemology. She charts transnational and transspecies connections rooted in some of the worst violence in world's memory. I'm invested in Gumbs' perspective because her onto-epistemology has shaken my ground. Her sense of Blackness is capacious, global, and in motion, somewhat like José Esteban Muñoz's vague/inclusive phrase "sense of brown" (2020) in the title of the book he was completing at the time of his death: a broad and murky alternative to identity. Gumbs's book *M Archive* (2018) is a "speculative documentary" (xi), telling a true history of what never could have happened, imagining conditions under which history could have been radically different. She speculates: what if those bodies from the Middle Passage that ended up in the ocean developed gills and learned to live underwater? The "dark science" of the Black oceanographers Gumbs invents uncovered evolutionary species adaptations that would never be possible—at least, not in the time scales we currently understand—but that nonetheless present revolutionary possibilities, breathing air and water at the same time.

Gumbs also messes with the periodic table of elements, selecting the elements she needs for particular kinds of knowing. The chapters in *M Archive* are themselves archives, with words and voices from the past as well as the material archives that determine our being: "archive of dirt," "archive of sky," "archive of fire," and an "archive of ocean," along with "lab notes," "baskets," and a "memory drive," each with their own elements highlighted or blacked out (depending upon how you see it) from the periodic table. The archive of dirt, for instance, includes sodium, magnesium, calcium, iron, silicon, and oxygen; the archive of fire includes hydrogen, iron nitrogen, helium, and neon; the archive of ocean, of course, includes hydrogen, oxygen, and possibly sodium or chlorine.

Gumbs even invites her readers to stage a dance version of her "mud mothers," to put ourselves inside her ideas:

> if you could look close enough (or listen carefully enough, the critical geologists would have corrected), you could see the churning planet making herself brown. if you were to choreograph a dance about it (which, i agree, would be an excellent idea), you would need to have everyone cover themselves in mud.
>
> (2018, 23)

A group of artists in Minneapolis and St. Paul did indeed stage their own interpretation of *M Archive* with a dance directed by Signe Harriday (2019), to embody the fire, sky, ocean, and dirt. The performance also included Trinidadian cuisine and installation pieces inspired by the text. Communal, corporeal inhabitation—like the drama of eating and moving according to textual suggestion—leads to literal kind of physical comprehension. As Gumbs puts it, "the individual body is not a workable unit of impact" (2018, 6).

Gumbs describes *M Archive* as "ancestrally co-written," seeking to "create textual possibilities for inquiry beyond individual scholarly authority" by invoking the words and ideas of Alexander's *Pedagogies of Crossing* via footnotes (2018, ix). *M Archive* is also written "after and with" a long list of Black women writers—including Audre Lorde, Toni Cade Bambara, Barbara Christian, the Combahee River Collective—and "after and with" "clean water," "a multitude of small and large present apocalypses," "the end of the world as we know it," and "the ways we have been knowing the world" (xi). All of these humans, events, and elements contribute to *M Archive.* With these many ways of knowing, "everything could have been different," and "the possibilities would have expanded exponentially" (6).[4]

Among Gumbs's many epistemological and ontological experiments, she identifies, in her newest work, *Undrowned* (2020), as a marine mammal in training. She chooses marine life as an aspiration, in part, because of its association with the Middle Passage, with the hundreds of thousands of African dead in the ocean. But Gumbs also wants to envision a world in which those Africans survived: "I am talking about the middle passage and everyone who drowned and everyone who continued breathing. But I am troubling the distinction between the two" (2020, 1). The term "undrowned" questions the deaths of those who died and focuses instead on those who learned to breathe underwater "in unbreathable circumstances" (2). The term also unsettles the status of survival for the undrowned whose lives were defined by slavery.

Marine biologists have always known that their subjects and allies in knowledge-making have a sophisticated system of communication and community. But the knowledge of whales is incommensurable with human epistemology. The whale "wouldn't tell us anything until we stopped knowing. stopped thinking that we knew…" (2020, 189). We humans won't understand whales, or even hear them, until we let go of our own ideas of thought and language.

> we thought we had let go of everything. we thought we were meditating right then in the midst of our thoughts. and she could hear it. every piece of held-onto knowledge clanging in our brains. every muscle memory of do right. so she waited through our impatience. our protest. our feigned indifference. she waited. and when it was over and all there was left was the sound of us not knowing nothing, then she began.
>
> (Gumbs 2018, 189)

The double negative implies that "not knowing nothing" is itself a deep kind of knowing. Or perhaps it is a silence so radical as to shush the thinking mind.

Gumbs's writings are oriented towards the impossible (or the not yet possible), which inspires me to imagine our worlds and our knowledges differently. Building from the conventional academic skills I've been taught, such as close attention to language and epistemology, I want to chart new ways of doing our work, ways that make the world a more just and inclusive place. This is a lofty goal, but I no longer see value in work that does not attempt to take us in that direction. I still don't know what this will look like, and I'm counting on everyone to help. According to Gumbs, we are "marooned to the task of trying to put things back together in a different relationship than they had ever had before. the task of believing there was some possible relation that wouldn't mean detonation" (2018, 194). This passage suggests that there could be other kinds of ontology, epistemology, and even elemental physics that might help humans put the world back together with less violence. The community, according to Gumbs, was "learning to move as if the world was hot and melting. which it was" (95). I add: still is, and still working to learn new movements.

Conclusion

These many decolonial experiments describe a "vocation not bound by the meanings and norms that constitute one's ground" but, rather, an "openness to uncertainty" (Lugones 2003, 26). Such speculations—including experiments in shapeshifting, offering new interpretations of precolonial cosmologies, and writing extensively in science fiction genres—are a frequent mode in decolonial writing, where imagining new kinds of futures is more important than truth. These radically decolonial speculations imagine, even demand, ways of doing knowledge differently, seeking out muck and play and library squirrels. Following Lugones, I would rather "elicit" and "entice" (37) readers to find new possibilities throughout this chapter rather than "presupposing" anything (20). How else do we honor the complexity of all that goes into the archive of life?

It is impossible to filter out the of distraction caused by squirrels dancing outside the Anzaldúa archive. Their material and social squirrelness remains opaque to me as they also resist the anthropomorphizing projections put upon them by the humans they encounter in their public horseplay. Squirrels form part of my experience reading the *Anzaldúa Papers,* as much as the buzzing overhead lights and the out-of-order papers. They gambol between benches, garbage cans, and grackles, following a logic I cannot access. They get very close, and I study them. They are still running around, running around, free from the brisk air conditioning and formal apparatuses of the archive. Sudden leaps of scruffy fur. I follow them.

Notes

1 Nick Cave and his partner Bob Faust operate a laboratory for emerging artists called Facility, which I read as a reference to ease and accessibility. Cave's work is intensely involved with social justice work in the communities that surround Facility, and in Chicago at large. A new reason to love his work showed up in my *Chicago Tribune* on

November 6, 2022: Cave and Faust have been collaborating with the owners of Eli's Cheesecake Company to make works of art out of mini cheesecakes as a part of their commitment to community uplift in Chicago. These are the sweetest morsels of activism.

2 Thank you, John Hawkins.

3 Scholars like Zoe Todd critique posthumanism as a trendy "Euro-Western academic narrative" that essentially reinvents a wheel, developing ideas that resonate with existing indigenous cosmologies without invoking indigenous writers (2016, 7–9). What Todd asks for is self-reflexivity about our citation practices and the academic structures that perpetually center "Western" traditions and marginalize others (18). Anzaldúa's other-than-Humanist epistemology is drawn from feminist and queer re-appropriations of pre-colonial Mesoamerica as well as a deep sense of connection between humans and the other-than-human environment. This approach is political and romantic more than it is empirical.

4 The ways in which Alexander's work guides *M Archive* are multiple and never transparent. Gumbs's title mimics the name and initials M. Alexander, making Alexander, herself, a form of archive. Each chapter ends with an endnote referring to a sentence or a phrase from Alexander's work. Sometimes this sentence or phrase appears verbatim in Gumbs's writing, but more often it does not, and we are left either to guess or simply to not know the relation between the words of Gumbs and Alexander. The opening "Note" that ostensibly explains the book's relation to the ancestors ends with "M is for . . . " followed by discordant lists of m-terms: mayhem and make-believe alongside mass incarceration might make a maybe, or maybe even a miracle.

Bibliography

Alexander, M. Jacqui. 2005. *Pedagogies of Crossing*. Durham: Duke University Press.

Anzaldúa, Gloria. n.d. "Com-positioning." *Gloria Evangelina Anzaldúa Papers, 1942–2004*. Benson Latin American Collection, The University of Texas at Austin.

———. 1987. *Borderlands/La Frontera: The New Mestiza*. San Francisco: Aunt Lute.

———. 2000. *Interviews/Entrevistas*, edited by AnaLouise Keating. New York: Routledge.

———. 2002. "Now Let Us Shift… the Path of Conocimiento… Inner Work, Public Acts." In *This Bridge We Call Home: Radical Visions for Transformation*, edited by Gloria Anzaldúa and AnaLouise Keating, 540–78. New York: Routledge.

———. 2005. "Let Us Be the Healing of the Wound." In *One Wound for Another/Una herida por otra: Testmonios de Latin@s in the U.S. through Cyberspace (11 de septiembre de 2001 – 11 de marzo de 2002)*, edited by Claire Joysmith and Clara Lomas, 92–103. Mexico City: Universidad Autónoma de México.

———. 2015. *Light in the Dark/Luz en lo Oscuro*, edited by AnaLouise Keating. Durham: Duke University Press.

Beckwith, Naomi. 2022a. "Beyond an Eye-Level Existence: An Interview with Nick Cave." In *Forothermore*, edited by Naomi Beckwith and Nick Cave, 50–56. Chicago: Museum of Contemporary Art.

———. 2022b. "Practice to Professorship." In *Forothermore*, edited by Naomi Beckwith and Nick Cave, 262–63. Chicago: Museum of Contemporary Art.

Bost, Suzanne. 2015. "Messy Archives and Materials that Matter: Making Knowledge with the Gloria E. Anzaldúa Papers." *PMLA* 130, no. 3 (May): 615–30.

———. 2022. "*The Gloria E. Anzaldúa Papers* and Radical Research Practices: A Collage." *ASAP Journal* 6, no. 3 (January): 653–78.

Cave, Nick. 2022. *Forothermore*, edited by Naomi Beckwith. Chicago: Museum of Contemporary Art.
Chen, Mel Y. 2012. *Animacies: Biopolitics, Racial Mattering, and Queer Affect*. Durham: Duke University Press.
Garber, Marjorie. 1992. *Vested Interests: Cross-Dressing and Cultural Anxiety*. New York: Routledge Press.
Gómez-Barris, Macarena. 2017. *The Extractive Zone: Social Ecologies and Decolonial Perspectives*. Durham: Duke University Press.
Gumbs, Alexis Pauline. 2018. *M Archive*. Durham: Duke University Press.
———. 2020. *Undrowned*. Chico: AK Press.
Harriday, Signe, 2019. Dir. *M Archive* Twin Cities Recap. Accessed 31 December, 2022. https://vimeo.com/325758189?embedded=true&source=video_title&owner=1580195
Huerta, Monica. 2021. *Magical Habits*. Durham: Duke University Press.
Keating, AnaLouise. 2015. "Re-envisioning Coyolxauhqui, Decolonizing Reality: Anzaldúa's Twenty-First Centry Imperative." In *Light in the Dark/Luz en lo Oscuro*, edited by AnaLouise Keating, ix–xxxvii. Durham: Duke University Press.
Lugones, María. 2003. *Pilgrimages/Peregrinajes*. Lanham: Rowman and Littlefield.
Mignolo, Walter. 2009. "Epistemic Disobedience, Independent Thought and Decolonial Freedom." *Theory, Culture & Society* 26, no. 7–8 (December): 159–91.
Muñoz, José Esteban. 2020. *The Sense of Brown*. Durham: Duke University Press.
Pérez, Laura E. 2018. *Eros Ideologies: Writings on Art, Spirituality, and the Decolonial*. Durham: Duke University Press.
Todd, Zoe. 2016. "An Indigenous Feminist's Take on the Ontological Turn: 'Ontology' Is Just Another Word for Colonialism." *Journal of Historical Sociology* 29, no. 1 (March): 4–22.
Zaytoun, Kelli D. 2022. *Shapeshifting Subjects: Gloria Anzaldúa's Naguala and Border Arte*. Champaign: University of Illinois Press.

Index

Note: Endnotes are indicated by the page number followed by 'n' and the endnote number e.g., 20n1 refers to endnote 1 on page 20.

FSC
www.fsc.org
MIX
Papier | Fördert gute Waldnutzung
FSC® C083411

Kolibri 360 GmbH
Ferdinand-Jühlke-Straße 7
99095 Erfurt, Deutschland
produktsicherheit@kolibri360.de